Crafting Science

Crafting Science

A Sociohistory of the Quest for the Genetics of Cancer

Joan H. Fujimura

Harvard University Press
Cambridge, Massachusetts
London, England
1996

Printed in the United States of America

Library of Congress Cataloging-in-Publication Data

Fujimura, Joan H.
Crafting science : a sociohistory of the quest for the genetics of cancer / Joan H. Fujimura.
p. cm.
Includes bibliographical references and index.
ISBN 0-674-17553-0 (cloth : alk. paper)
1. Proto-oncogenes—Research—History. 2. Cancer—Genetic aspects—Research—History. 3. Science—Social aspects. 4. Culture—Study and teaching. 5. Medical anthropology.
I. Title.
RC268.415.F84 1996
616.99'4042—dc20
96-32222

For Kunio and Asae Fujimura, with love and respect;
and in memory of Anselm L. Strauss

Contents

Acknowledgments *ix*

1 Introduction: Creating a New Realm *1*

2 Tools of the Trade: A Brief History of Standardized Experimental Systems in Cancer Research, 1920–1978 *23*

3 Molecular Genetic Technologies: The New Tools of the Trade *68*

4 Crafting Theory *116*

5 Distributing Authority and Transforming Biology *137*

6 Problems and Work Practices: Improvising on the Shop Floor *155*

7 The Articulation of Doable Problems in Cancer Research *184*

8 Conclusion: Crafting Oncogenes *205*

Appendix: Social and Cultural Studies of Science *237*

Notes *245*

References *281*

Index *313*

Acknowledgments

The generosity and support of a collection of inspirational scholars and friends made this book possible. I am especially grateful to Troy Duster, Anselm Strauss, and Howie Becker, who provided invaluable training, criticism, constant support, and ideas throughout the project. Lucien LeCam gave me an early introduction to the world of cancer research. Steve Martin and Satyrajit Nandi taught me tumor biology. Special appreciation goes to several people who read and commented extensively on the latest version of the manuscript, including Alberto Cambrosio, Mike Fortun, Scott Gilbert, Mimi Ito, Eda Kranakis, Bruno Latour, and Michael Lynch. I am also grateful to Laureen Asato, Nan Chico, Adele Clarke, Kathie Gregory, Patricia St. Lawrence, Leigh Star, and Rachel Volberg, who read and commented on earlier versions of the manuscript. Dick Burian, Michel Callon, Danny Chou, Nicolas Dodier, James Griesemer, Donna Haraway, Peter Keating, Karin Knorr-Cetina, Carolyn Laub, John Law, Andy Pickering, Paul Rabinow, Arie Rip, Mary Steedly, Sharon Traweek, Bob Yamashita, and Sjerp Zeldenrust each read and advised me on one or several pieces of writing that eventually turned into book chapters. I also thank several anonymous reviewers for their comments on the final manuscript. Many others gave me valuable feedback when I presented my work at conferences and colloquia. I cannot name them all, but I sincerely thank each of them for their contributions to the research and writing of this book.

I also want to thank the scientists who talked to me about their work and who gave me a place in their laboratories and classrooms. Without their generosity and candidness, this book would not exist. Laboratory members are identified by pseudonyms as I have promised them confidentiality. Those whose real names are used are public figures who have agreed to my including their names in the text. I extend special thanks to Scott Gilbert and Robert Weinberg, who read early versions of the manuscript, and an anonymous oncogene researcher, who read the final manuscript for Harvard University Press. They provided valuable commentary, but the final narrative is entirely my responsibility.

I am grateful to Lindsay Waters, Alison Kent, and Elizabeth Hurwit at Harvard University Press, who have made this book better than it would have been without their contributions.

My research and writing have been supported by the sociology departments of the University of California, Berkeley, and Harvard University; the anthropology department and program in history and philosophy of science at Stanford; the American Sociological Association; the Henry R. Luce Foundation; and the University of California Humanities Research Institute.

Kathryn Doksum, Margrete Doksum, Teresa Doksum, Per Gjerde, Shigeko Okamoto, Kathy Uno, Sandra Uyeunten, and Linda Young lived through various versions of this book and inspired me to keep working. And most precious of all is Kjell Doksum, whose steady support has been the mainstay of my work.

Crafting Science

1 Introduction

Creating a New Realm

Harold Varmus, I and our numerous colleagues have been privileged to assist as a despised idea became the ruler over a new realm. The notion that genetic changes are important in the genesis of cancer has met strenuous resistance over the years. But now that notion has gained ascendancy.

—J. Michael Bishop, "Retroviruses and Oncogenes II"

In the late 1970s, researchers published articles linking genes in viruses that caused cancer in laboratory animals to normal genes in humans. Researchers proposed that these normal human genes, called "proto-oncogenes," could be triggered to cause cancer. By 1986 the proto-oncogene hypothesis was a "fact" asserted in basic biological textbooks, and in 1989 its principal proponents, J. Michael Bishop and Harold T. Varmus, won the Nobel Prize for their oncogene work.[1] Within a few years, thousands of researchers and laboratories and hundreds of millions of dollars became devoted to proto-oncogene research. "Cancer" had changed from a set of heterogeneous diseases marked by the common property of uncontrolled cell growth to a disease of human genes.

This book is a sociohistory of proto-oncogene research in the United States with an emphasis on the collective production of scientific facts. It recounts the explosive growth of a new field of research and the creation of a novel class of genes. I examine this case of scientific innovation in terms of the daily practices in laboratories and in the context of contemporary biological science.

I do not address the question of whether the proto-oncogene theory is a true or false explanation of the cause of human cancers. Instead, I consider the activities, tools, contexts, and processes through which scientific representations are produced and accepted. This book explores the production of accepted knowledge.

My account of proto-oncogene research addresses certain technical aspects of oncogene research, but it differs from the accounts of oncogene scientists in several respects.[2] Their histories present experimental evidence to demonstrate that artificially altered proto-oncogenes transform normal cells into cancer cells. They also present evidence of altered proto-oncogenes retrieved from human tumor cells. While I give a history of this research employing selected accounts of technical achievements leading up to the proto-oncogene theory, I weave into my story an analysis of the representational, organizational, and rhetorical work done by researchers, students, sponsors, and audiences to create the "world" of proto-oncogene research.

I also consider some opposing voices. A few vocal dissenters have argued against the belief that normal genes are the roots of cancer. Proto-oncogene theorists call these scientists "heretics" and dismiss their criticisms as "sour grapes." I do not take sides or attempt to evaluate the truth of this claim. Instead, I use these alternative views as reminders that consensus, or at least common practice, is an achievement not to be taken for granted in scientific as well as in sociocultural and religious knowledge. These contrary views challenge, and thus make evident, the range of perspectives that otherwise would not be made public.

However, a critical role for proto-oncogenes in causing cancer has become fact for most biologists and many clinicians today. This development has not made available cures or therapies involving oncogenes. Although proto-oncogenes have moved into clinical practice to some degree in the form of diagnostic tools, cancer researchers caution that the path from their research to treatments for actual cancers is presently unclear.

Nevertheless, scientists, students, university administrations and departments, funding agencies, biological research supply and biotechnology companies, and Congress joined in the research effort on proto-oncogenes and their oncogenic processes. Participants from many different lines of work have come to practice a common approach in their studies of cancer. Especially in the United States the molecular biological approach, with proto-oncogene research at its center, gained an increasing proportion of cancer research support. Scientists began to use the term "oncogene bandwagon." Appropriating their term, I refer to this set of multiple cascading commitments to proto-oncogenes, and more generally to molecular genetic cancer research, as a *bandwagon.* Large numbers of people, laboratories, organizations, and resources became committed to one approach to a problem.

By 1984 the oncogene bandwagon was a distinct and entrenched phenomenon. As we will see, scientists acted on the basis of its existence, and many researchers joined the bandwagon because it was just that. This snowball effect exemplifies how oncogene research had become a phenomenon sustained by its own infrastructure and momentum. Thus, the enthusiasm for a particular research problem, and for a particular technical approach, is a social phenomenon.

This book is also a study of change and continuity in science. I portray conceptual change in science as embedded in individual and collective changes in the organization of scientific work. More specifically, I describe a *process* within which theoretical or conceptual shifts are inseparable from both the local and broad-scale organization of work and the technical infrastructures of science and society. Through this process, cancer was represented as a disease of the cell nucleus and specified sequences of DNA.

In their previous research, scientists had conceptualized cancer in different, if still genetic, terms. These earlier studies developed tools that represented "genes" in experimental systems such as inbred mice and cell lines. Inbred mice from a particular line were considered to have identical genes. Nevertheless, these genes were imagined rather than materialized. Researchers could not point to a particular material entity and say, "This is the gene that created the cancer."

In the late 1970s and early 1980s, a few tumor virologists and molecular biologists announced that they had found a proto-oncogene that caused a cell to become cancerous. Their proto-oncogene, a material form of specific segments of DNA, was a novel representation of cancer, this time in terms of "normal" cellular genes. The multitude of previous representations of chemical carcinogenesis, radiation carcinogenesis, tumor progression, and metastasis then became (re)presented in terms of this new unit of analysis. I shall describe how this new unit of analysis, incorporating both theory and material, was developed in and from previous experimental systems. Moreover, this new representation formally linked the study of cancer to the study of normal development and evolutionary theory, as well as to other fields of research. Thus, researchers crafted and recrafted the proto-oncogene theory as well as links to problems of many different scientific worlds. Bridges were built between existing lines of research, which extended and transformed them into new lines of research.

Cancer Research as Collective Work

Like other scientific fields, biomedical cancer research is a collective and interactive process. It is conducted by people interacting with one another and with other elements through time. Many of these activities take place within institutional frameworks.[3] Much biomedical research on cancer is organized along the lines of different traditions and disciplines. Shelves of books on cancer research, scores of journals and articles on cancer, and Index Medicus entries on cancer in the MedLine computerized data base indicate the expanse of contemporary biomedical cancer research fields and representations.[4] Histories of cancer research also describe a proliferation of lines of research, theories, and methods for understanding cancer causation, progression, epidemiology, and therapy. The problem of cancer is distributed among different worlds of practice, each with its own agenda, concerns, responsibilities, and conventions.[5]

Clinicians frame their problems in terms of individual cases, individual patients, and standard operating procedures: given present knowledge, how do we best treat the person? Medical researchers in the fields of radiology, epidemiology, oncology, endocrinology, neurology, and pathology work with both patients and theoretical abstractions that they construct using many cases distributed through time and space. How many patients respond to this treatment in which way? What can we say about initiation and progression of the disease when examining a number of patients over time? Basic researchers in the fields of genetics, virology, cell biology, organismal biology, molecular biology, immunology, and neuroscience work with theoretical abstractions and material models. How can we duplicate the cancer process in mice or cultured cells in order to use it as a tool for studying the disease? What are the origins of cancer? Among medical and basic researchers, the questions are broken down further. What is the role of the endocrine system in causing, promoting, or retarding the initiation or growth of the disease? What is the role of chemicals, of radiation, of viruses? What are the molecular mechanisms for the initiation and progression of the disease at the levels of genes and cells? Epidemiologists track the diseases as they appear in their different manifestations (breast, liver, colon, lung, brain, cervix, prostate) across families, "racial and ethnic" groups, occupations, countries, and regions. On the other end of the scale, pathologists examine cells in cultures taken from tumor tissues. These are just a few of the participants involved in constructing knowledge about cancer.

My ethnographic investigations concerned the interactions between participants in different worlds of cancer research. Scientists in different lines of research and treatment worked with different units of analysis, representations of data, research materials, scales of time and space, and audiences. However, many scientists had long presupposed the existence of some central "factor" linking multiple representations and types of cancers. They assumed that this central "common denominator" would eventually come to light, despite failed attempts to find it.[6]

In this context of specialized searches for the elusive "magic bullet" cure, oncogenes captured everyone's attention and imagination. Oncogene research quickly gained widespread acceptance and support across the different worlds in the early 1980s.

How did a conceptual model like the proto-oncogene theory become accepted so quickly across disparate situations, across many laboratories, and even across fields of research? I do not take consensus to be the norm in cutting-edge science. One lesson we have learned from ethnographic studies of science is that scientific practice is much more diverse and locally contingent than it was once assumed to be. Given this diversity and contingency, how did researchers from such diverse situations come to share many common practices and even agreements on proto-oncogenes?

Standardizing across Worlds

The phenomenal growth of proto-oncogene research occurred simultaneously with the production of a *package* of theory and methods. This new package consisted of the proto-oncogene theory of cancer along with standardized recombinant DNA and other molecular genetic technologies for realizing, materializing, testing, exploring, and adjusting the theory. Although these technologies included inbred mouse colonies, cell lines, and viral colonies created earlier in the century, the crafting of the proto-oncogene theory occurred in conjunction with the development of new molecular biological technologies in the early 1970s. This particular combination of theory and methods was then disseminated, adapted, and adopted across different research and clinical settings.

The proto-oncogene theory was initially constructed as an abstract notion, using a new unit of analysis, the proto-oncogene, for studying and conceptualizing cancer. The theory's conjunction of general and specific aspects was used by researchers in many extant lines of research to interpret the theory to fit their separate concerns. Scientists liberally translated the abstract notion to change and add to their laboratories and problem structures without generating major conflicts with preexisting frameworks, yet they simultaneously transmuted their previous notions and technologies.

The general aspect of the package was embodied in concepts and objects—such as genes, cancer, cancer genes, viral genes, cells, tumors, normal growth and development, and evolution—that often had different meanings and uses and were represented by different practices in diverse lines of work on cancer. These concepts and objects linked proto-oncogene research to work in evolutionary biology, population genetics, medical genetics, tumor virology, molecular biology, cell biology, developmental biology, and carcinogenesis. They were interpreted and treated differently to allow for both variability in practices and specificity within work sites. In other words, researchers in several fields of biology manipulated these concepts and objects to draw on one another's work to support, extend, and maintain the integrity of their lines of research. For instance, studies of genes and tumors were transformed by molecular biologists into studies of DNA sequences in tumor cells in vitro *and* by hormonal carcinogenicists into studies of tumor breast tissue samples taken from laboratory mice. The same genes could be studied by molecular biologists interested in cancer and by *Drosophila* geneticists interested in genes involved in normal growth and differentiation. One could say that proto-oncogenes have become a lingua franca for biologists!

These genetic studies employed recombinant DNA and other molecular biology technologies. At the time that Bishop and Varmus and their colleagues were framing the proto-oncogene concept, molecular biologists were creating new molecular genetic technologies, especially recombinant DNA techniques. The idea that cells could be transformed from normal states to cancerous states by genetic switches involving DNA sequences might have remained an idea—might never have been transformed into practice—without the development of molecular biological technologies. Bishop (1989:15) stated that "the contemporary image of the cancer cell . . . was forged from the vantage point of molecular genetics and with the tools of that discipline." According to scientists, recombinant DNA and other molecular genetic techniques provided a means by which they could enter the human cell nucleus and intervene and manipulate the nucleic acids of genes to study their activities in ways never before possible. By the mid-1980s, the researchers I interviewed had come to see these technologies, not as objects of research, but as tools for asking and answering new questions and for stimulating a flurry of work at the molecular level on questions previously approached at the cellular and organismal levels.

"Technical" standardization proved even more important for understanding the extraordinary growth of this line of research.[7] It provided a means through which techniques could be transported between labs. Even previously "low-tech" laboratories began to incorporate high-tech equipment and skills. An example of technical standardization was the dissemination of particular research materials such as molecular probes for certain oncogene sequences. After their experimental creation in research laboratories, the (re)production and distribution of probes were farmed out to nonprofit biological materials distributors (such as the American Type Culture Collection) and biological supply companies. By providing the same probes to many different laboratories, these distributors contributed to the standardization of oncogene research practices and the reproduction of practices, results, protocols, and representations across laboratories. Chapters 4 and 5 discuss the *reproductive* and *regulatory* role of recombinant DNA technologies and other molecular genetic techniques in oncogene research.

Proto-oncogene research was "standardized" even further by the introduction of experimental protocols that associated proto-oncogene theory with molecular genetic technologies. Because the proto-oncogene theory was framed in "molecular genetic language," new molecular genetic methods were required to probe questions originally framed in the "language game" of other methods.[8] Researchers untrained in molecular biology or retrovirology began to incorporate the combination of the abstract, general proto-oncogene theory and the specific, standardized technologies into their research enterprises. By reconstructing oncogene research in these new sites, a variety of research laboratories locally *concretized* the abstract proto-oncogene theory in different practices to construct new problems and new laboratory artifacts. Suddenly, novel entities such as *myc, abl, fos,* and other (onco)genes were named and defined, and novel organisms were created through gene transfer techniques. By 1989, Du Pont Corporation was advertising in science journals its transgenic OncoMouse™ (see Figure 1.1), a "technology" that physically incorporated a specific oncogene in the laboratory animal itself.

> The OncoMouse™/*ras* transgenic animal is the first *in vivo* model to contain an activated oncogene. Each OncoMouse carries the *ras* oncogene in all germ and somatic cells. This transgenic model, available commercially for the first time, predictably undergoes carcinogenesis. OncoMouse reliably develops neoplasms within months . . . and offers you a shorter path to new answers about cancer. Available to researchers only from Du Pont, where better things for better living come to life.

While the new proto-oncogene conceptual framework provided a *metaphoric* and *discursive* alliance between different lines of research, the *material* linkages were established through standardized technologies like molecular probes and the OncoMouse™. The commercial strings attached to these practical linkages contributed to the development of oncogene research. Researchers framed their problems and questions in terms of proto-oncogenes and their laboratory practices in terms of standard research protocols, recombinant DNA and other molecular genetic technologies (molecular hybridization, molecular cloning, nucleotide sequencing, and gene transfer), instruments (nucleotide sequencers, computer software and data bases), and materials (molecular probes, reagents, long-passaged cell lines,

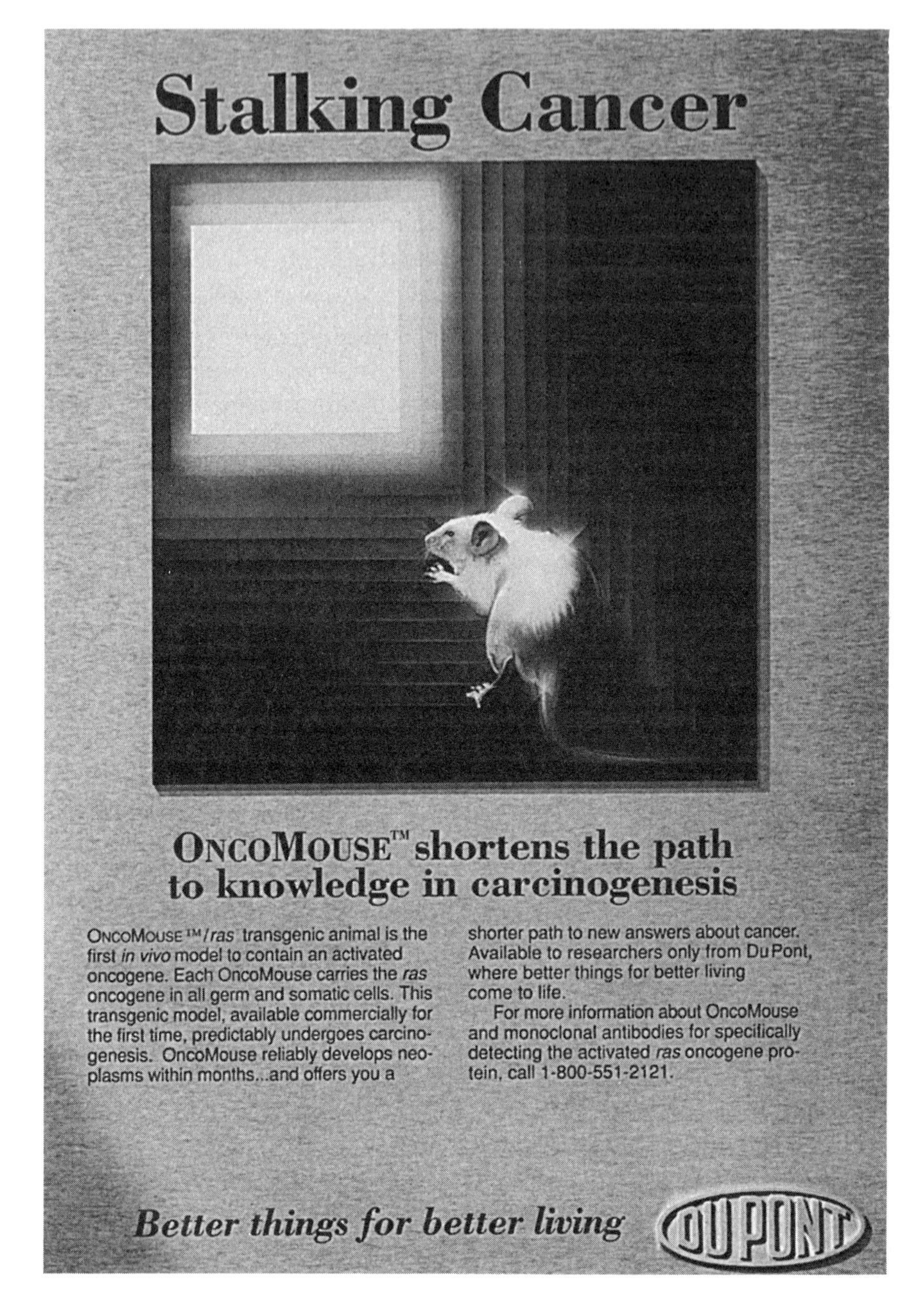

Figure 1.1 Du Pont Corporation's advertisement for a transgenic OncoMouse™, a "technology" that physically incorporated a specific oncogene in the laboratory animal itself, was published in science journals during 1989–1990. It is an example of the normal production of standardized pathological animals.

and biologically engineered laboratory animals). This conceptual, material, and practical package began to appear in many different biological laboratories. Cancer research laboratories and the practices within these laboratories had been reconstructed. Thus, the standardized tools themselves became agents for the standardization of the laboratories in a process of *co-construction.*

Granting agencies and Congress were early proto-oncogene supporters. National Cancer Institute administrators held proto-oncogenes up as the justification of their past investments in viral cancer research, a mission whose legitimacy and productivity had been questioned. They also used news of proto-oncogenes to lobby Congress for increased appropriations. Proto-oncogene research provided a new line of applied products and research tools to biotechnology businesses. University and research institute administrators used proto-oncogenes to reorganize "old fashioned" cancer research institutes into "hot" molecular biology institutes. In other words, many different constituencies jumped on the oncogene bandwagon in order to continue and augment their previous lines of work while still introducing novelty. In their view, a productive balance between novelty and standardization had been created. The transformation of these audiences and sponsors into oncogene producers, consumers, and suppliers shaped the problems pursued and the representations.

I argue, therefore, that technical standardization contributed to the production of the theoretical robustness of oncogenes. This was not a technologically determined process, because technical standardization was made possible by collective agreements about concepts like genes, proteins, and causation. These collective agreements were shaped by previous conflicts and negotiations in the histories of molecular and cell biology, enzymology, and population genetics.

The package of proto-oncogene theory and molecular biological technologies was used in the (re)construction of problems, practices, laboratories, and infrastructures. Not only were these elements of the package linked, each also changed the next. For example, Bishop's effort to "educate" physicians about the new research on oncogenes can be considered a "political" act that contributed to the acceptance of oncogenes as facts in the medical arena. But it was only one of the many different kinds of actions that contributed to the "hardening" of oncogenes, and it cannot be separated from them.

Transmuting Local Work Organizations and Practices

Transforming local practices via the oncogene package was not a simple, smooth process. It required major changes in commitment in daily and long-term practices in different situations. Changes in commitments made to particular research theories, materials, instruments, techniques, and skills in a laboratory, museum, or in the field are not trivial matters. Although small changes occur continuously—for example, machines may break down, reagents may change, new technicians and students may be hired, funding may be lost and gained—the particular constellation of skills, tools, and approaches in each work site is not so easily or frequently changed. Major changes in commitments mean transforming work organizations and "retooling" workplaces and practitioners (compare Kuhn 1970, "Postscript"). These transformations require significant time, effort, and financial resources, as I examine in this book.

I examine how laboratories and problems are transformed. I present stories of problem-solving in molecular biology and cancer research laboratories and follow the changes in the problems pursued and the solutions constructed. I analyze the relationship between local work conditions and a particular problem's structure, that is, the co-construction of problems, solutions, and situations. I describe various ways in which problems and laboratories are transmuted, including changing targeted audiences or clients and incorporating resources such as new technologies. My examples present scientific problems and solutions as the outcomes of maneuvering through the exigencies of the daily work and the longer-term organization of research designs. I give examples of researchers dealing with contingent events in their work and explore researchers' attempts to make sense of messy situations. It is often in the spaces provided by messes, contingent events, and interruptions to ongoing work that major changes in work commitments and organizations occur. Changes within laboratories are made in part by scientists with an eye to changes in practices elsewhere and in part by events and activities not under scientists' control. I provide views from the shop floor of the meshing and integrating of theory, problem, techniques, solutions, markets, and audiences.

Articulating Doable Problems

Many scientists, laboratories, and institutes changed their research agendas because, ultimately, what the scientists I studied called "doable" problems are what matters in contemporary biological science. In the 1980s many scientists and those who listened to them viewed proto-oncogene research as doable. The construction of a doable problem is the *process* of solving a problem from beginning to end. Thus, the adjective "doable" is a post hoc label attached to particular problems. Doable problems are sociotechnical achievements. The oncogene problem histories presented in this book were the outcomes of particular confluences of events and conditions, some under the researchers' control, others not.

I discuss scientists' efforts to craft a problem in terms of articulating different work activities. Articulation is the amorphous and ambiguous work of planning, organizing, monitoring, evaluating, adjusting, coordinating, and integrating activities usually considered "administrative" rather than "scientific." This work is often invisible and taken for granted until it is not done or not done well. My example of this problem-solving process draws on observations made in a biotechnology company laboratory. This case demonstrates that the world of problem-solving does not presume a separation between inside and outside, between content and context. The researchers I observed were dealing with stockholders, the National Cancer Institute, their professional and disciplinary colleagues, their different versions of the economy, the job market, proteins, and chemical reagents as part of their daily and long-term problem-solving efforts. My example shows that the world is in the laboratory, and the laboratory is in the world.

The work involved in dealing with elements from protein loops to biotechnology company stockholders is full of contingencies and uncertainty. However, the rhetoric about proto-oncogene research espoused by its protagonists omitted or softened the uncertainty and failures. They portrayed oncogene research protocols as clear-cut, fundable, and productive.

Once oncogene protocols were adopted and re-created in many different laboratories, oncogene research *became* more clear-cut, fundable, and productive. That is, the protocols acted as standardizing agents. Oncogene researchers used particular probes to identify important sequences for study. Researchers "manipulated" a particular strand of DNA through a relatively straightforward series of steps with reference to the "gene" and to its proposed role in cancer causation. They also shared a framework, a theory within which to interpret the outcomes of such manipulations as novel findings. Thus, oncogene problems became doable because standard protocols re-created in each oncogene laboratory further solidified a network of laboratories, funding agencies, materials and their suppliers, journals, and scientists working in and around oncogenes. Proto-oncogene research problems became doable through the transformation of the world inside and outside the laboratory. By 1983 proto-oncogenes and many different research worlds had been co-constructed.

Science as Collective Work and Practice

My study examines scientific change and continuity in the activities of different people and organizations collectively involved in creating and stabilizing scientific representations. I consider efforts to define and redefine the situation on the part of actors on the scene, and conflicts and negotiations between different definitions and perspectives through which stabilization sometimes occurs. I thus take an "ecological" approach to understanding science as collective action from the viewpoints of all the participants and worlds involved. The ecological approach prevailed at the University of Chicago during the first half of the twentieth century especially in the pragmatist school in philosophy,[9] the symbolic interactionist school in sociology, and the sociology of work tradition.[10] It has more recently been used to study science.[11]

I treat science as a kind of work or practice, albeit a kind that differs from, say, legal practice. The ecological approach I use begins with the premise that scientific knowledge and technologies are the result of the collective work of diverse communities. Everett Hughes (1971), for example, studied how members of different immigrant groups cooperated in the workplace while speaking different languages and holding different conventions and understandings of the world. He focused on the conflict, struggle, and negotiations over which set of conventions should govern how members of these different groups worked together. Thomas and Znaniecki (1918) similarly studied how members of Polish immigrant associations and neighborhoods attempted to install their "definitions of the situation," their views of how things should be. Like other collective efforts, the outcomes of scientific efforts are constructed through processes of negotiation, articulation, translation, debating, and sometimes even coercion through "administrative persuasion" by members of different social worlds (Becker 1974, 1982, 1986; Shibutani 1955, 1962; Strauss 1978a,b).[12] For instance, interest, and therefore cooperation, in each other's enterprises is not a given. Actors attempt to interest others in their concerns, objectives, and ways of doing things. For this to happen, social worlds or communities develop shared languages, ways of communicating with each other. In science, these shared languages often take the form of shared tools and theories.

The notion of "social worlds" ascribes meanings, commitments, and perspectives in knowledge-making both to practices and to the people practicing them. In this respect, it shares much with semiotics in viewing signs and symbols as representing the intertwined and inseparable aspects of local positions, milieux practices, and perspectives. For example, molecular biology is a (changing) set of practices around which a discipline, departments, professional associations, textbooks, and individual careers are organized and reorganized. It is simultaneously a "space" where people work. Researchers bring ideas and perspectives from their many other lives into the laboratory, just as laboratory practices help to produce the ideas and perspectives of the people practicing there. I do not mean "molecular biologist" to refer to an essence, an all-encompassing identity of a person who practices molecular biology from nine to five or even from nine to midnight. Social worlds theory includes the activities and participants involved in particular activities.

In my approach, scientific facts, theories, routines, rules, standards, and conventions are achieved orders; they are constructed of the organizations of scientific work commitments and perspectives at particular times and in particular spaces. Every instance, kind, and scale of order is achieved through complex daily interactions—conflicts, agreements, abdications, force, negotiations, manipulation, persuasion, education—among many actors working under specific local conditions. Routines are orders to be explained and not assumed. Constructing scientific facts, theories, artifacts, and procedures are temporally located collective processes.[13]

A Different Kind of Relativism

This view of science differs from that of scientific realists in that it assumes there is no unique way of representing nature or society. It incorporates assumptions from the philosophical school of pragmatism. In their arguments with philosophers of the 1930s about the nature of truth, pragmatists claimed that truth and knowledge are contingent and multiple and that all representations are incomplete and self-referential. These philosophers rejected the subject-object or knowing-known dichotomy and argued that all representations are constructed in specific situations and therefore are endowed with the viewpoints and experiences of the representers.[14] According to pragmatists, truth and knowledge are relative matters, malleable and dynamic. If people define a theory as real, they act as if it were real. As Thomas and Thomas (1928) noted, "If people define situations as real, they are real in their consequences." Thus, truth claims are real(ized) in that they have real consequences for the way people act together.

In the 1940s and 1950s, several sociologists constructed a new sociology based on pragmatist assumptions. Employing especially George Herbert Mead's work, Herbert Blumer (1969) called this sociology "symbolic interactionism." Blumer argued that humans construct meanings and realities through their interactions with themselves, with other human beings, and with nonhuman objects in concrete situations. Each of these "objects" gains its meaning and reality only through interpretation and interaction. Blumer (1969:89) suggested that "students of human society will have to face the question of whether their preoccupation with categories of structure and organization can be squared with the interpretative process by means of which human beings, individually and collectively, act in human society." Drawing from Mead, Blumer (1969:72) saw "human society . . . as a diversified social process in which people [are] engaged in forming joint actions to deal with situations confronting them." The job of sociologists, in Blumer's view, is to study these specific situations to understand the meanings and actions people make together. Everett Hughes (1971:552) similarly used pragmatist assumptions in his sociology to remind us that, under different conditions, "it could have been otherwise"; that is, our social realities could have been different. According to Hughes, sociologists should study how things become the way they are.

In my study of science I assume that any particular truth, no matter how established and privileged, is the outcome of continuous actions and interactions. Established truth in any particular situation is a resolution of negotiations or conflicts among several or many different perspectives (Mead 1938). This resolution sometimes takes the form of a hierarchy, which we can examine to see whose perspective "won" and for how long a period of time. In the case of my research, for example, certain molecular biological approaches for explaining and representing cancer have been "winning" since the early 1980s. In other cases, resolutions take the form of compromises or pastiches. In some cases, there might be no resolution; instead different approaches coexist (usually in conflict) with different amounts and sets of resources. Resolutions can be short-lived or long-term, but they are rarely, if ever, permanent. Even consensus requires maintenance.

Scientific Change and Continuity as Situated Action

The fluidity and activity of which I speak do not make scientific facts and theories ephemeral constructions. Facts and theories are situated actions that can be studied by examining the processes, the conditions, and the consequences of scientific practice in particular spaces and times. The situations I examine are framed in terms of the collective organization of scientific work.

In the symbolic interactionist tradition, there are no a priori "conditions" for action and no a priori sets of constant variables. Instead a situation (including what many sociologists might call its "context") is a contingent and interactional achievement. In each scientific work situation, nothing is predetermined. Yet, neither are situations disconnected instances. Commitments made to particular ways of doing things have consequences beyond the particular situation. In science, commitments are made to particular ways of working that are organized into disciplines (such as molecular biology and evolutionary biology), institutions (universities and biotechnology companies), methodological rules and standards (field research and experimental procedures, accepted measurements and procedures), and scientific representations (the proto-oncogene theory). These ways of working or conventions of action are both constituents and consequences of action. People create new conventions and representations in part by working with other conventions and representations to meet the demands of new situations. They often creatively reinterpret or modify earlier representations and routines to accomplish their new constructions.

Scientific Actors, Activities, and Representations

Some of the elements that constitute scientific work situations in the case of oncogene research include the actors who do the work or who support and care about the work (scientists, technicians, students, clerical staff, computer programmers and support staff), the organization of the work, material tools (materials, instruments, techniques, software, infrastructure), representational tools (models, theories), the workplaces and institutions (laboratories, universities, biotechnology companies, museums, institutes), skills, training of scientists and staff, sponsors (funding agencies, venture capitalists, Congress), regulatory agencies (Recombinant DNA Advisory Committee, Congress), and audiences. In my research I also include less corporal elements such as national policies guiding cancer research (for example, the National Cancer Act of 1971), social and cultural views of disease in general and of cancer in particular, as well as participants indirectly related to the organization of cancer research, including lobbyists and their causes (Bud 1978, Strickland 1972), medical insurance lobbies, ethicists, private industrial investment in biotechnology, and so on.

Although these elements have corporal and corporate existences, I choose to study them as they relate to activities. This approach makes no division between micro-activities and macro-processes. It blurs the distinction between "inside" and "outside" the local site of knowledge production. Thus, researcher (technician, scientist, student) work activities may include writing grant proposals, writing journal articles, injecting radioisotopes into cell cultures, injecting chickens with viral oncogenes,[15] contacting other laboratories to obtain particular kinds of cells, buying laboratory equipment, discussing the results of experiments with other laboratory staff members or collaborators, lobbying Congress for increased appropriations or special programs, lobbying the Food and Drug Administration (FDA) for permission to market genetically altered substances or organisms, lobbying and persuading the Recombinant DNA Advisory Committee (RAC), Congress, or the "public" for permission to introduce new organisms into the environment, and so on.

Action is not limited to individuals. The faculty of a medical school as a group can decide and act to reorganize an institute within the school

(Becker 1982). The National Cancer Institute (NCI) as an organization can decide to concentrate its funds in particular research directions and then fund accordingly. A presidential administration, upon pressure from lobbyists, can decide to launch a large-scale program to "combat" cancer (Strickland 1972). The same can be said for all the other formal and informal organizations involved in science-making: scientific professional societies, scientific journals, science granting agencies, academic career ladders and promotions, scientific advisory panels, scientific "expertise," scientific disciplines, scientific activist groups, and more.[16] These institutions both contribute to and result from scientific efforts.

Ecologies of Action

My approach to the study of science and technology builds on symbolic interactionism, pragmatist philosophy, and the multidisciplinary work in social, cultural, and gender studies of science (see the Appendix). I rely on several notions that help us understand the processes of constructing, stabilizing, and changing knowledge. These notions address how tools, practices, and theories are co-constructed, incorporated, and refashioned in a continual process of negotiation. I am advocating a sociology of associations that focuses on "ecologies of action" (Fujimura 1995a).

Specific concepts highlighted in this book include "traces of continuity," "distributed authority," "co-construction," "theory-methods package," "articulation work," and "doability." The notion of traces of continuity conveys the sense of how I follow traces of tools, practices, and theories that continue through time and place without assuming the determinism usually associated with tools and standards and without asserting the kind of incommensurability and discontinuity implied by Kuhn's paradigm shifts. For instance, so-called cancer viruses moved from playing the role of possible cancer-causing agents to the role of tools for exploring genetic theories of cancer. The problem became the tool of resolution. The substantive explanations for carcinogenesis *became* the next set of technologies for studying carcinogenesis. In this case, in the experimental process tumor viruses were transformed from the message (their role as possible causes of tumor growth) to medium (their role as tools in studying the initiation and growth of tumors). These viruses began to be viewed as laboratory artifacts. That is, although they were viewed as *not* representative of what actually happens in cancer growth, they were valued as tools for designing experiments and theories about how tumors are initiated and grow.

Another concept, the theory-methods package, shows how tools, practices, and theories circulate through and across worlds of practice (not only scientific worlds) and both change and are changed by their circulation. Both practices and social worlds are co-constructed in the process. Constructing a theory-methods package is a process by which interests are mutually satisfied and worlds are rebuilt.

These notions provide a path between the actor-network theory's general framework and the focus of other researchers in science studies on the contingent details of scientific work. They provide ways of discussing historical continuity and institutional and organizational contexts without invoking Machiavellian actors. Instead, I talk about power and authority as distributed among different actors, objects, and social worlds. The notion of distributed authority allows me to talk about power, not in a deterministic fashion, but in a way that still recognizes its play in science. I also present an account of how power, apparently vested in a package of theory and methods, is in the distribution of authority to various actors, objects, institutional loci, and social worlds.

Methodologically, this book spans several disciplinary approaches to present a story that cannot be written within any single discipline or through any single set of methods. I have tried to achieve a creative balance among several methods and styles of social analysis. These include cultural ethnography, ethnomethodological close description of events, historical development of a line of research, and sociological analysis. The production of knowledge is a multidimensional process whose traces can be followed through multiple cultures, institutions, actors, objects, and practices. I disrupt the divisions in sociology, history, philosophy, and anthropology of science, technology, and medicine (and knowledge in general) to move across disciplines, across worlds, across knowledges, across methodologies.

The interesting question is how and why some findings become facts. How are the facts and artifacts created? In answering this question, I point to the many events, resources, and actors that needed to come together for oncogene research to become what it did. No single person or persons, institution, event, or even process can be positioned as the creator. My story tells how oncogene research became what it did through the confluence of many events and conditions. In this confluence, we can see Bishop and Varmus's proto-oncogenes being enrolled in others' research just as Bishop and Varmus were enrolling others in their proto-oncogene research. Oncogene research resulted from the detailed activities and articulation of many actors through the exigencies of technical and organizational work. The cancer research world created Bishop and Varmus, just as much as Bishop and Varmus re-created cancer research.

My story includes some linear history of the development of the content of cancer research and oncogene science. But I also go beyond this to show how oncogenes became a significant fact and line of research—and what science is like—at the lab bench, in the office of the lab director, in the NCI director's office, in the technical facilities, on the molecular biologist's job market, in the biotechnology industry.

I present detailed descriptions of instruments, molecular biological procedures, experiments, problems, and crises in the research processes and practices. I describe the creation and subsequent development of various tools used in oncogene research. I am interested in the process of technology construction and experimentation as well as how these technologies articulated with other events.

I also stress that the oncogene package caught on so quickly in part because a large number of scientists thought that they might learn something by adopting it. Oncogene researchers were excited about the possibilities of understanding cancer at the molecular level. This knowledge was one of the major attractions of the oncogene package. The package promised technologies for producing novel findings and perhaps the keys to tumorigenesis. I view the success of this line of research as equivalent to a perception of the achievement of a productive balance between novelty and standardization.

Consequences of Actions and Representations

My use of the terms "construction," "representations," and "definitions of the situation" should be clarified. Scientific representations are constructed by scientists. They can take the form of theories, diagrams, drawings, photographs, autoradiographs, or models. Scientific representations are created and judged by sets of rules and procedures. Each scientific discipline produces rules for determining whether representations should be treated as valid entities. The community of scientists claims the license and mandate to construct and enforce such rules and procedures. Thus, scientific work includes the methodological development of rules by which "truth" is constructed and confirmed. Both representations and the rules by which they are judged are produced through scientific work. I discuss the co-production of these rules for creation and confirmation in Chapter 7.

Scientific representations are not ethereal or inconsequential because they are human (and nonhuman) constructions. Representations are real in terms of their consequences. To treat a representation as an autonomous entity is to participate in certain kinds of action or conduct. For example, molecular biology pundits like Renato Dulbecco (1986), James Watson (1990), Walter Gilbert (1991), and others have used cancer and specifically proto-oncogenes as evidence for the advances made by molecular technologies to promote and justify the Human Genome Initiative, a $3 billion project to map and sequence the human genome (as well as the genomes of other model organisms). Thus, participation in an agreement means commitment to certain kinds of conduct or action. These actions then have consequences that can further cement the "reality" of their definition.[17]

Representations can and often do become reified and stripped of the context of their production, their activities (Ricouer 1970, 1984; compare Rosaldo 1989). Histories, sociologies, and anthropologies of science have discussed how scientific theories and facts are constructed in local situations and yet stripped of any hint of their local environments when they are reified as "knowledge."[18] This reification also applies to diagrams, models, drawings, maps, and so forth.[19] Once ideas or theories become resituated, they can be used like technologies and instruments in the production of further representations. The objectified representations can become additional elements in ongoing scientific work.

Again, however, the relationship between a representation and its interpretation or use, and therefore its consequences, is not fixed. Representations change depending on who is using and interpreting them, when, and for which purposes (Duster 1981, 1990;[20] Fleck 1979; Knorr-Cetina 1981; Rabinow 1986). Indeed, the same person may interpret the idea of the gene in one way on one day and in another way on another day. As such, the study of representations should incorporate their diverse interpretations, renderings, and uses. This diversity highlights the active role of audiences. The consequences of representations—that is, their realities—are the products of cooperation, conflict, negotiation, and sometimes power struggles between audiences and producers, between readers and authors. To understand science as a way of producing realities, we also have to understand this interaction.

Scientific representations are separate from the ongoing work activity of science only by abstraction and reification. Therefore, it is incongruous to attempt to study science by first separating the cognitive aspects from their material bases. To understand scientific work and knowledge in a more comprehensive way, I have chosen to study the ideas produced and stabilized by work processes in and across the situations.

"Materials and Methods"

Given my theoretical approach, the methodological imperative is to study the activities, contexts, and processes through which knowledge is constructed, stabilized, and changed. This book is the result of several sets of research methods. It is part laboratory ethnography, part historical reconstruction of events using interviews and documents, and part sociological analysis. I observed the activities of participants and the conditions (material and social) under which these activities occur. I examined what scientists have reported about their activities in informal conversations in the lab and in interviews with me. I also inspected their formal reports of their activities (articles, graphs, statements made at conferences, textbooks, information passed on in classrooms, instruments) with the understanding that these products are "snapshot" representations of their work activities constructed for particular occasions and audiences.

I studied two laboratories in depth as a resident ethnographer. Both laboratories were located in the San Francisco Bay Area in California. One, to which I give the pseudonym the "Xavier lab," was relatively new; it had come into existence as a separate working group only a year and a half before I arrived. It was also a newcomer on the scene of oncogene research and worked on oncogene-antibody experiments. The laboratory director, "Xavier," had been trained as a protein chemist. The laboratory was part of the research division of a private biotechnology research and development company focusing on human health care products. When I began my fieldwork in their lab, the researchers were conducting experiments primarily aimed at characterizing the structures and functions of antibodies that seemed to counteract the transforming (cancer-forming) activity of activated oncogenes.

The "Yuzen lab" was located in an academic biology department in one of the universities in the Bay Area. Lab Director "Yuzen" had been trained as a virologist and molecular biologist. The laboratory had worked on viral oncogenes for over ten years and was well respected for its past research. When I began my fieldwork there, the laboratory was working on comparative studies of the activities of viral oncogenes and proto-oncogenes.

Both laboratories were relatively small, if one does not consider collaborations and, for the Xavier lab, work with other divisions in the company. Yuzen maintained a staff of between five and seven graduate and postdoctoral students. Including support staff (technicians, dishwashers, secretary, and an undergraduate student worker), graduate and postdoctoral students, and the director, the Yuzen lab had a total of ten to twelve staff members. Xavier's laboratory was composed of six members, two with Ph.D.'s in protein chemistry and four technicians with bachelor's degrees in biology. The lab members had a range of years of experience working in different laboratories and skills gained in different kinds of biological specialties.

I watched experiments as they were run and sat in on informal and formal lab meetings where lab members discussed the difficulties and results of experiments. I took part in technical procedures, although only occasionally, to get a hands-on sense of the work. I followed experiments as they were run and rerun through analysis, write-up, and presentation of results. I took courses in tumor biology to experience the classroom training of new biologists, in addition to interacting with them around laboratory work.

I interviewed participants about their different perspectives on an event or activity. I stood in the laboratory watching interactions between people and objects in the laboratory and people and things in other worlds. I sat with the technicians as they discussed their conflicts with the lab director and vice versa. At other times, I listened as technicians and the laboratory director talked with each other about their conflicts with other technicians.[21] I observed the lab's entire crew in conflict, competition, and collaboration with other laboratories within the same line of research. I talked with scientists in parallel or competing lines of research. I interviewed oncogene researchers in various parts of the country and administrators of a national funding agency.

I read the literature (academic and popular) on oncogenes, on other lines of cancer research, on cancer treatment, on the history of cancer research, on molecular biology, and on research techniques and materials. Documents included journal articles, books, biological materials and instruments catalogs, laboratory manuals, and information on organizations sponsoring cancer research. I also read newspaper accounts of the research to assess the kinds of information reaching the "general publics."

In more formal terms, the sources and methods included formal and informal interviews, observation in an academic oncogene research laboratory and a private biological research laboratory, participant observation in tumor biology courses, and documentary analyses. I attended colloquiums, public lectures, and a cancer research conference. Respondents included cancer research scientists, technicians, students, postdoctoral affiliates, administrators of cancer research institutes and funding organizations, and management of a commercial biotechnology company. I did most of the interviews and much of the fieldwork from 1983 through 1986 with additional follow-up interviews, new interviews, some hands-on experimental practice, and further documentary analysis after 1986. I conducted all of my observational work in the United States. Although my documents were not limited to literature from the United States, most were generated there.

Using these materials and drawing from the pragmatist-interactionist school and the social studies of science, I have constructed an ecologies-of-action framework in which knowledge and power, lines of scientific research and tools, are co-constructed. This study links the local production of scientific knowledge with changes and continuities in lines of research and disciplines over time. It connects daily work in laboratories, classrooms, biotechnology companies, and funding agencies with the growth of a new line of research and with changes and continuities in disciplinary concepts and practices.

Many different ecologies of action together created a bandwagon of oncogene research. This bandwagon was knit together by standardized experimental systems, packages, and other such crafted tools—the instruments of scientific expertise and power. Through their use of these "power" tools, scientists "constrained" research practices and defined, described, and constrained representations of nature and reality. However, the same tool that constrains representations of nature can simultaneously be a flexible dynamic construction with different faces in other research and clinical arenas. Standardized experimental systems and packages are dynamic interfaces translating interests and enabling shared practices and representations to be passed between different worlds. They are both subjects and objects of complex relationships among multiple participants. By examining the construction, maintenance, change, movement, and use of these tools in different situations and across situations, one can understand not only how particular representations come to be held as sacred but also the complexities of their creation. An understanding of this creation and sacralization can provide tools for creating other possible representations, other ways of knowing and practicing.

Tools of the Trade

A Brief History of Standardized Experimental Systems in Cancer Research, 1920–1978

The medium is the message.
—Marshall McLuhan

Constructing Genealogies for Proto-Oncogene Research

According to medical researchers, pathologists, and physicians, cancer is the name given to over two hundred different diseases, all of which have one property in common: uncontrolled cell growth.[1] Some authorities state that there are at least 270 different types of tumors. Both of these truisms about diseases and tumors are based on the sets of nomenclatures, categories, and the rules for classifying cases and tissues into those named categories. Classifications and categories, tissues and diseases all have changed throughout the history of cancer research. This gives us some idea of the difficulties involved in defining cancer, in determining causes, and in producing treatments. A formal and more general definition of cancer is the state in which the normal growth-controlling mechanisms are permanently impaired thus preventing growth equilibrium. Cells normally reach a point at which they stop growing. When growth continues beyond this point, the extra cells form a tumorous growth. If this tissue growth is irreversible and if its continued growth is uncoordinated with the normal tissue around it, then it is called cancer or neoplasia. However, some tumors do not fit even this general a definition. As just one example, neuroblastoma is a kind of cancer that can stop growing.[2]

During the past century scientists have studied cancer from many different perspectives using many different technologies (Bud 1978, Cairns 1978, Rather 1978, Studer and Chubin 1980). Research has focused on causes and origins, treatments, and epidemiologies—the rates of occurrence of different kinds of cancer in different communities in different regions of the world. Although there has been some success in treating a few leukemias, efforts to treat the solid tumors that make up most human cancers have made little progress. Thus, the challenge to solve the cancer problem is still alive. Hundreds of millions of dollars continue to be spent each year on cancer research by the National Institutes of Health (including the National Cancer Institute), the American Cancer Society, and other private foundations.

Rather than try to summarize all the research on cancer, which would require several volumes, I want to present a general picture of the basic research that preceded and in some ways is associated with current oncogene research, the topic of this book. To speak of the transformation of cancer biology, as I did in the last chapter, is to imply the existence of a pretransformed cancer biology. In this chapter, I present a brief narrative history of the state of cancer research from 1920 to 1978 to provide a frame of reference for the next part of my tale.[3]

This chapter also presents a theoretical perspective on the complex, intertwined relationships among scientific problems, practices, material technologies, and other historical circumstances. Scientists created tools for studying their problems of interest, but the development of the tools also influenced the questions eventually asked and answered (in the form of concepts and theories developed) in their research.

I call this multiple-directional set of influences and intercontingencies among problems, problem structures, material practices and technologies, concepts and theories, and historical circumstances "co-construction," a concept that permeates the history of cancer research problems.

This complex set of interactions was not limited by the boundaries of disciplines or research fields.[4] Just as researchers from many different lines of work and biological disciplines traded in ideas, they similarly appropriated the experimental research tools constructed for specific purposes and creatively adopted and adapted them to their own purposes. These tools included genetic (and eugenic)[5] ideas of the 1920s, inbred mouse colonies, chemicals and antibiotics developed during World War II, tissue cultures and their technical procedures, and viruses. In the narrative text and in the final discussion of this chapter, I emphasize and provide a detailed discussion of some of these tools as *standardized experimental systems* and as privileged means of producing and authenticating research results.

Prewar Cancer Research: Classical Genetics and Mouse Colonies

Basic cancer research before World War II focused on tumor etiology and transmission. Basic genetic research on cancer was aimed at understanding the causes and trajectories of tumor growth. Prewar theories of cancer included the somatic mutation theory (that cancer was due to abnormal chromosomes in cells other than germ cells; Boveri 1929), the infectious virus theory (Rous 1911a,b; Bittner 1936), and the glycolysis theory, or Warburg Hypothesis, whereby cancer results when assaults on the organism cause a normal cell to develop "an anaerobic metabolism as a means for survival after injury to its respiratory system" (Braun 1969:25, Warburg 1930).[6]

Experimental geneticists began their studies of cancer in the early 1900s on the "tails" of transplantation biology, an active field in the nineteenth and early twentieth centuries (on transplantation biology, see Cairns 1978, Löwy 1993, Shimkin 1977, Willis 1960). Researchers physically transplanted (grafted) tumors from one animal to another to investigate whether or not tumors from one animal could produce a tumor in another animal. Although most of these experiments failed, there were a few successes with a canine sarcoma in dogs, a chicken leukemia, and a chicken sarcoma. (Contemporary tumor immunologists explain these failures, "tumor graft rejection" in current terminology, as the result of histocompatibility antigens that provoke an immune response in the receiving animal to protect itself from foreign materials.)

Transplantation biology of the early 1900s led to many technoscientific developments in tumor biology that continued to be used in the 1960s and 1970s, several of which are still used today.[7] First, researchers observed that tumors change through time and can become more virulent or, in their terms, undergo progression. Second, they established large volumes of tumors that were later used in biochemical studies. Third, they constructed experimental animal strains. Rodents became a popular tool for transplantation research in the twentieth century.

Inbred strains of mice were perhaps the most important tool developed during this time for subsequent cancer and noncancer biological research. In the process of transplanting tumors, tumor biologists noticed that rat and mouse tumors could be successfully transplanted into rats and mice of the same strains but could not be successfully transplanted to animals from other strains. Neither did human tumors survive in animals. Between the 1920s and 1940s, some researchers took note of these observations and began to develop colonies of inbred strains of mice (later expanded to rats and guinea pigs) as their primary research tools for studying, among other things, the genetics of cancer.[8] They were interested in genetic factors involved in tumor etiology, heritability and transmission, incidence, and the development of specific types of cancer. Kenneth DeOme, a cancer researcher and geneticist trained in that period, attested to this trend.[9] "Genetics was the big thing to study in the 1920s and 1930s. People believed that genetics was the controlling factor in everything. No proof, but everybody believed it. In cancer research, the question was: Is genetics the controlling factor in cancer? . . . The question was whether or not the proclivity to have tumors was a genetically controlled one."[10]

To investigate the heritability of tumors, researchers created inbred strains of experimental animals. The aim of inbreeding was to produce a group of animals with "as near as possible, the same genetic composition" (called "homozygous"). These identical twinlike animals with theoretically the same genetic constitution included syngeneic and congeneic varieties. Syngeneic strains were as identical as possible. (However, some differences still remained, since the animals were only 99.999 percent homozygous.) They were produced through matings of brothers and sisters (or parents and offspring) for at least twenty consecutive generations. All offspring of each generation could be traced to a single ancestral pair. Congeneic strains of inbred animals were strains that differed from one another in the region of one genetic locus. They were produced through at least ten successive backcrosses or intercrosses to a control strain. These mice could be used to evaluate a mutation or allelic variant against a particular inbred strain background. Again, the purpose of the inbreeding was to provide a defined and consistent genotype for analysis.

Mice and rats had two important advantages as experimental animals. First, they were small in size and thus saved on space. Second, they had relatively short reproductive cycles, so work could progress at a faster pace.

The first strains of experimental mice were those with identical genetics and the propensity to develop the same kinds and compositions of tumors. One such strain of "identical" mice produced mammary tumors, while another strain of mice with identical genetics had the propensity to *not* develop breast tumors (so-called normal strains). "This became a very powerful tool, and still is . . . Now you could answer the question 'is this genetic or is it not?' It was just a matter of making simple crosses which everybody knew how to do in order to prove it" (DeOme interview). For example, if, upon making simple crosses, researchers found the same tumors in the offspring, then they concluded that that particular kind of tumor was genetically inherited.

One of the first institutes to develop research colonies for the study of mammalian genetics was the Jackson Laboratory in Bar Harbor, Maine, founded by geneticist Clarence C. Little in 1929 (Rader 1995, Russell 1981). Little initiated his experimental crosses when he was at the University of Michigan. His interest in developing inbred strains of mice began with his study of carcinoma transplantation in the Japanese waltzing mouse (Little and Tyzer 1916) and nontumorous transplants in mouse strains (Little 1924). Little and his colleagues concluded from these studies that there were heritable factors involved in the resistance of animals to tissue, tumorous and nontumorous, transplantation. In search of a way to better investigate such heritability and cancer, Little moved to Maine and began to create inbred mouse colonies.

> Little was interested in cancer and in genetics. He thought the way to study cancer was through heredity, the way to study heredity was in the mouse, and that the way to study mice was with inbred strains. The inbred strains made possible an early understanding of transplantable tumors and the role of graft rejection . . . Not even Little could have foreseen the great accomplishments of this technique to immunogenetics, cancer inheritance, cancer viruses, carcinogen testing, and studies of chemotherapy . . . Let me add to this the great value of congeneic lines, which enable the experimenter to study the effect of a particular gene (or chromosome region) without the otherwise inevitable complication of *noise* from genetic variability at other loci. (Crow quoted in Russell 1981:319; emphasis added)

By the 1940s, Little's experimental system was the *standard.* To study cancer genetics, inbred animals were the "right tool" designed to screen out the "noise" that detracted from the genetic factors of specific interest. As one example, Kenneth DeOme was hired to build the Cancer Research Laboratory at the University of California, Berkeley, according to this standard. "This laboratory coming along in the 1940s was in that era when the necessity for inbred animals was imminently obvious, so obvious that there was no question about it. To do cancer research then, said everyone, we just can't do it without having the proper animals" (DeOme interview).

One experiment by Prehn and Main (1957), for example, transplanted a mouse tumor to other mice of the same inbred strain.[11] They examined the question of whether chemically induced tumors were immunogenic in the host. Could the animal produce an immune response to protect itself from the original tumor? The researchers induced a sarcoma using the chemical methyl cholanthrene, cut the tumor into pieces, froze some, and transplanted others into syngeneic mice. The transplanted tumors grew under the skin of these syngeneic mice and were later removed. After the removal, one of the frozen tumor pieces from the original mouse was transplanted into this second mouse. The result was that the mouse rejected the sarcoma, and the interpretation was that the animal might have developed an immune response. But it was also possible that the rejection was an allogenic reaction, that is, that the two animals were not "really" identical despite coming from the same strain. In subsequent studies, when these researchers attempted to transplant other kinds of tumors to these supposedly immune syngeneic mice, they reported that the tumors did grow in syngeneic mice. They concluded that each tumor has its own distinct antigens, which they named non–cross-reacting antigens.

In these immunological experiments researchers did not consider genetic questions. They were asking other questions: Are tumors immunogenic in their natural hosts? Are there antigens in tumor cells that differ from those in normal cells? Are there immunosurveillance mechanisms against emerging tumors? If so, how do the tumors escape from this surveillance to survive and grow? To examine these questions, researchers relied on assumptions of the genetic commonality of the animals experimented upon. This genetic commonality was embodied in inbred strains of experimental animals. Especially after World War II, many other kinds of experiments were conducted using the inbred animal strains. Experiments tested whether carcinogenesis could be induced by various agents, including chemicals, hormones, and radiation.

Developing animal colonies was costly and taxing work.[12] DeOme, whose work included developing and experimenting with a colony of inbred strains of animals, discussed use of this research tool in the days before tissue cultures.

> You [couldn't] just buy [inbred animals] from a mouse catcher . . . It took fifteen years to develop inbred strains of mice. And that doesn't even include the time it took to cross them . . . Our mouse colony right now [in 1984] costs us between $150,000 and $200,000 a year. And we have a very small one now, as compared to what we used to have. We used to use here enormous numbers of animals, because we hadn't [tissue or cell] culture to do it any other way. You had to do it with the whole animals, that's all there was to it. So we had one strain, for example, [of which] we were using well over 200,000 animals per month, from that one strain. And we had thirteen strains. This is just what *we* [at our small research institute] used. (DeOme interview)

Although these centers sold animals to other scientists, most of the animals were used for in-house research. Commercial supplies of inbred animal strains (and their professional animal breeders) were not available until after World War II. Researchers in other disciplines and at other institutions who were interested in cancer usually chose to collaborate with institutes with existing colonies rather than develop their own colonies. These included, in this fifty-plus-year span, chemists and biochemists studying the effects of various carcinogens, especially chemicals and radiation, on tumor formation in experimental animals. Inbred strains continue to be used in contemporary biological research of all types, and the Jackson Laboratory continues to be a leading center for the construction of new lines and maintenance of old lines of inbred animals as well as the major experimental animal supplier. When a 1989 fire at the Jackson Laboratory destroyed many one-of-a-kind strains of experimental animals, researchers across the country reacted with concern about their future supplies. The response to the fire demonstrated the ongoing and heavy reliance of contemporary biological researchers on these animals.

The Right Tool for the Job: The Inbred Mouse Standard

How did it become, in DeOme's words, "imminently obvious" that inbred animals were the proper tool to do cancer research, even to study cancer genetics? How does a technology become designated as the "right tool for the job"? This question appears frequently in this book, since I seek to understand how scientists come to agreements about how to study nature. From my perspective "tools," "jobs," and the "rightness" of the tools for the jobs are each situationally and interactionally constructed. That is, they are co-constructed, mutually articulated through interactions among all the elements in the situation.[13] Indeed, DeOme articulated this himself: "Most of these techniques . . . develop[ed] along with the thinking. It permits you to think of new things. It permits you to think of new techniques, too. They're interactive. I don't think you can separate these as clearly as you'd like to" (DeOme interview). This interactive process hinges on the notion of *standardization*. Standardization has two meanings here that have become confounded. Clarence Little standardized the mouse colonies through inbreeding, and these inbred lines of experimental animals came to be the standard way (the right tool) to conceive of and to study cancer genetics. But he also standardized genetics.

At the turn of the twentieth century, biology was transformed into an experimental and analytic science.[14] Growing commitments across scientific and medical disciplines to the ideals of positivist empiricism translated into ideals of quantifiable and reproducible experimentation in biology.[15] Although naturalist and ecological inquiries continued in a more taxonomic vein, many new fields of biology began to pursue analytic methods of inquiry. Experimentation and analysis became the hallmarks of "science."

During this period, the inbred animals came to represent and define a tabula rasa upon which studies of cancer could be conducted. These inbred animals were considered to have "the same" genetic, or homogeneous, constitutions because they were created through crossbreeding. Thus, researchers assumed that they could study the action of chemicals, radiation, viruses, or any other potential causes of (or cures for) cancer on these animals, as well as the heredity of tumors and resistance to tumors, through transplantation research. Experimenters believed that, given this tabula rasa, they could control for genetic "factors" by using these animals. With respect to heredity, researchers presupposed that inbreeding created homogeneous tumors in the recipient animals and their donors. They could study carefully bred specific tumors across generations of inbred animals. Finally, experimenters also assumed that their experiments could be *replicated* because the animals were taken to be the same and to have "the same" tumors, since they had "the same" genes and "the same" tumor genes. They believed that they could conduct reproducible and controlled experiments by using these "genetically identical" animals as experimental vessels, as the sites for their experimental manipulations.

Standardized Experimental Systems

Inbred experimental mouse lines "realized" the concepts of genetic inheritance and genetic invariability for biologists interested in controlling for genetic factors. In other words, the ideas of invariability, homogeneity, and standardization became real laboratory (arti)facts through the development of inbred animals (and their tumors). These experimental inbred mice (and other animals) and their tumors transformed practices in multiple sites. From this transformation of practices emerged a new technical (as versus abstract) definition of genetics and cancer. The animals and the collective work behind them defined a new technical work space for realizing the concept of genetics in cancer research. These animals represented and embodied the commitments of the cancer research community to a particular definition of genetics. They were *cyborgs,* artifacts created under artificial conditions according to a formula that combined a commitment to rational scientific principles with a particular conception of cancer as a genetically caused or transmitted disease.

The resulting animals were not simply experimental animals but instead standardized experimental systems. Scientists used these "standardized" animals to construct representations that were comparable between laboratories by reconstructing laboratory work practices that were used to produce experimental representations. These practices and representations were assumed to be homogeneous across laboratories and through time.[16]

Scientific technologies such as these inbred animal experimental systems are not neutral objects through which nature is laid bare. Instead, they are meaning-laden and meaning-generating tools, just as language and writing are meaning-laden and meaning-generating media and not simply vehicles for expressing meaning.[17] These scientific technologies are highly elaborated technosymbolic systems just as artificially intelligent computer and robotic systems are elaborated symbolic systems. Scientists "create" nature in laboratories just as they create "intelligence" in computers. They create "nature" along the lines of particular commitments and with particular constraints, just as computer scientists create computer technologies along the lines of particular commitments and with particular constraints. Thus, the boundary between science and technology is blurred. To bring this theoretical discussion back to the case of inbred mice, geneticists employed experimental technologies and protocols to create novel technoscientific objects. These objects *are* nature, at least as nature is described in scientific narratives. Little's and DeOme's inbred animals cum experimental systems *embody* the phenomena; they give materiality to objects such as "genes," "viruses," and "cancer," as well as to scientific ideals such as "genetic invariance," "reproducibility," and "facts."

One could argue that the constructed inbred mouse cum experimental system limits opportunities to make inferences and extrapolations from experimental studies to "real natural" situations. Scientists talk about this problem as one of the limits of the generalizability of their findings. In contrast, theorist and historian of science Donna Haraway (1981–82, 1991) argues that there is no real-world "nature" that we can know apart from our "technical-natural" tools. Instead, these experimental systems *reinvent* nature as they incorporate sociocultural understandings.

Whichever position one takes with respect to their relationship to arguments about nature, the inbred lines of experimental mice were clearly very different from mice living in the cane fields of Hawaii. They were the result of much hard work, ingenuity, money, and purposeful as well as serendipitous and improvised efforts by researchers. They were novel phenomena constructed by scientists in their efforts to study a controlled nature in the laboratory. Scientists recognize the differences between the "artifactual nature" they create within the laboratory and the "extra-laboratory nature." However, their theories and actual work practices are almost exclusively tied to the artifactual nature.

Experimental systems result from collective and historically situated processes involving and implicating a broad set of actors, time, and spaces.[18] My approach assumes that scientific theories, facts, and technologies are produced through the collective (sometimes cooperating, sometimes conflicting, sometimes indifferent) efforts and actions of many different scientists, groups, tools, laboratories, institutions, funding agencies, janitors, companies, social movements, international competitions, and so on. Instead of speaking of work within or outside an experimental framework, of forces acting from outside to effect or create changes in something inside an experimental framework, I speak of an experimental system as incorporating and representing particular commitments.

Standardization, then, is the outcome of collective commitments and action. The tool of inbred mice lines, the problem of cancer, and a particular conception of genetics became the standard through the enormous efforts of Little, the Jackson Laboratory, DeOme, the University of California, and a host of other laboratories and sponsors and their audiences. It is often only after technologies, concepts, and laboratories have become standardized that these technologies then become the "imminently obvious" correct tool choice. This standard in turn comes to shape future collective commitments and actions.

Technical Incarnations of the Concept of the "Gene"

The genetics defined by inbred mouse (and other animal) lines is still with us, but it has also been joined by a new "molecular" definition of genetics, as we will see in the following chapters. A new set of commitments and practices has re-created a new set of tools that technically define genes and genetics in molecular terms. Before we get to that story, I present a brief discussion of early-twentieth-century ideas about the concept of a "gene" and its various historical technical incarnations in order to provide a context for reading the contemporary work on oncogenes.

> "Gene" was not a recognized word until 1909. Even then it was a word without a clear definition, and certainly without a material reality inside the organism on which it could be hung. At best it was an abstraction invoked to make sense of the rules by which inherited traits are transmitted from one generation to another. The suggestion that Mendelian "factors," or genes, are related to the chromosomal structures inside the cell, which cytologists studied, had been brilliantly argued by an American graduate student, Walter Sutton, in 1902, and independently by the German zoologist Theodor Boveri. But at that time it was an argument without direct empirical confirmation . . . In the years before . . . 1919, this connection began to acquire enough supporting evidence to make it compelling. (Keller 1983:1–2)

Here Keller portrays the gene as an abstract expression of the transmission of characteristics from parent to offspring. But as Allen (1978), Darden (1991), Jacob (1982), Keller (1983), Kitcher (1982, 1992), and Maienschein (1992), among others, have discussed, different individuals and scientific communities defined the gene differently. François Jacob (1982:226) referred to the early-twentieth-century gene as a black box. "A product of reason, the gene seemed to be an entity with no body, no density, no substance." According to Jacob, the task of genetics was to confer a concrete content to this abstract concept, "to find the nature of the substance, the physical entity, to explain how genes act." According to Allen (1978:209), when "Wilhelm Johannsen coined the term 'gene' in 1909 . . . Johannsen's word 'gene' was originally meant to be a completely abstract concept, consciously disassociated from any of the existing theories of hereditary particles." Johannsen specifically stated, "No hypothesis concerning the nature of this 'something' shall be advanced thereby or based thereon" (quoted in Allen 1978:209–210). "Johannsen's 'gene' was to be used simply as a kind of accounting or calculation unit, but was not to be considered a material body . . . Furthermore, he emphasized that individual genes were not the determiners of specific adult characters" (Allen 1978:210). In contrast, biologists Thomas Hunt Morgan, Alfred Henry Sturtevant, Hermann J. Muller, and Calvin B. Bridges specifically chose not to use Johannsen's term "gene" in their 1915 *Mechanism of Mendelian Heredity* because, argues Allen (1978:209), "of the group's commitment to the belief that the chromosome theory provides a material basis for heredity."

Even after Morgan had begun to use the term "gene" for the chromosomal theory of heredity, he did not make clear what form that materiality took. Morgan and his colleagues saw genes as "beads on a string" in line with their practice of mapping breaks and recombinations as points on a line. But they were still vague about its physical characteristics. William Castle proposed a "rat-trap" model of the chromosome, whereby genes were arranged three-dimensionally, like a big tangle of beads on a string. Well into the 1930s Richard Goldschmidt remained critical of the Morgan model and preferred to view the entire chromosome as a single entity. More in line with his work as a cytologist, embryologist, and physiologist (rather than as a geneticist), Goldschmidt stressed gene function and process over transmission and heredity. Allen (1978:274) also discussed arguments against Morgan's model by William Bateson, William Morton Wheeler, and Alfred North Whitehead, whose views were "influenced by philosophical issues arising from the advent of quantum theory in the first two decades of the century. Whitehead saw genes as purely arbitrary and abstract units, like quanta, with only mathematical reality."

In the 1950s, with the development of new biochemical technologies, DNA became the physical entity attached to the term "gene." More recently, as research has continued, the physical properties as well as the activities of a gene have come to differ depending on whom one asks. To define the term "gene" by its physical form is a simplification. Instead, I view the concept of "gene" as linked to the kinds of experimental activities carried out by biologists, the uses to which various actors (and not only biologists) put the concept, and the particular physical parts of DNA and the supporting protein structure referred to or experimented on by participants.

"Gene," then, is a concept which has different meanings depending on its situated practice and use. The concept gains its abstractness, and therefore its capacity to take on different meanings, only through the deletion of these situated activities—the work of various biologists over time and the uses that framed their work and resulted from their work. Sociologists and historians of science have been busy restoring to such representations their situations of production.

With respect to cancer, classical genetic ideas such as the somatic mutation theory should be understood in terms of the transmission of characteristics without a physical definition of the transmitting "factors." Genetic theories of cancer that came later in the twentieth century were and are also situated within the technologies of our time (see Chapter 4). Viral oncogene research as discussed later in this chapter and cellular oncogene research as discussed in the rest of this book are situated in the practices and understandings of molecular biological technologies, where genes are understood to consist of strings of DNA or RNA (see Chapter 3).

More generally, the "nature" of genes is situated in the practices of particular times and places. As human genome researchers today are learning, the nature of genes changes as they examine the structure and function of genes with new technologies. At the moment, new genetic entities are being created and the complexity of existing genetic entities and mechanisms as envisioned by researchers is increasing at a phenomenal rate. Researchers are proposing systemic views of genetic entities and actions that cannot be limited to DNA, RNA, and proteins. The relationship between genotype and phenotype is full of ambiguity, contingencies, and complexity. The nature of genes is plastic and flexible. It is interactively constructed with the technologies of particular time periods. The histories of genetic causes of cancer discussed in this book (including the mouse mammary tumor virus and the gene *src*) must be read within the technologies that gave scientists their understandings of genes.

Shifting Tools and the Instability of Meanings: Moving from Genetics to Viruses

A common occurrence with the development and use of scientific tools is that things do not always work according to one's assumptions. Through unexpected contingencies, experiments using particular research materials and technologies produce new scientific problems as well as novel and unplanned research materials. One such instance led to a technology that plays a significant role in my narration of the proto-oncogene history. In the process of making "simple crosses" of inbred mouse strains to prove or disprove the genetic basis of cancer, geneticists reported that they had discovered a mouse mammary tumor virus that apparently caused cancer in infants as it was passed from mother to infant through suckling. Mouse mammary tumors had previously been thought to be a genetically inherited cancer. However, after the experimental crosses, only infants with mothers from the high-tumor line developed tumors. Those offspring of fathers from the high-tumor line did not develop tumors (Bittner 1936, 1942).

> To everybody's horror, it turned out that the crosses didn't behave the way they should . . . with regard to [mouse] breast cancer anyway. If you had a strain that had a high incidence of breast cancer and a strain with a low incidence, and you crossed those, it should not make a difference whether they were males or females that you put into the cross. Because both males and females carried the same genotype. But it turns out that the results indicated that the most important thing was whether the female was from the high-tumor strain. If she was, the babies would have tumors. If it was the male that came from the [high-tumor] strain, [the babies would not have tumors]. (DeOme interview)

First considered indicative of an extra chromosomal factor, researchers later came to regard Bittner's data as evidence of the presence of a virus. It is now called the "mouse mammary tumor virus," or MMTV.

Researchers argued that the mouse mammary tumor virus added evidence to the viral infection theory of cancer causation, a theory that had been experimentally pursued early in the century in the form of transplantation research. Until Bittner's experiments, however, researchers had concluded that cancer was not infectious, on the basis of experimental results in which many tumors were not transplantable in animals of different strains. Jensen had transplanted mouse mammary tumors nineteen times to find that the growth at the nineteenth transplantation was the same tumor tissue as the original (Nandi Lecture 1984). The conclusion was that no infection had occurred.

Anomalous results from some of these transplantations were used by other researchers to create a new line of study. Some pathologists noted that nontumorous animals taken from high-tumor strains had "abnormalities" in their breast tissues, which they classified as precancerous tissues, which would in time develop into tumors. Studies of tissues of these nontumorous mice from high-tumor strains became the focus of investigations, for example, by DeOme's group at Berkeley's Cancer Research Laboratory. Transplanting these preneoplastic tissues from one animal to another became the modus operandi of this laboratory for the pursuit of another problem—precancerous tissues—while the infectious theory of cancer was resurrected as a viral infection theory.

Thus, as DeOme noted, technologies and materials do not remain wedded to theoretical problems and hypotheses. Although Little's inbred strains of mice became the standard tool for studying cancer genetics, they also provided the means for creating and investigating new problems around viruses and cancer.

In summary, there are two interesting points to note about the MMTV story. First, much effort was invested in exploring the genetic causes of cancer, and this effort created a particular set of tools (inbred strains of mice) for defining genes and genetics during that period. Second, and paradoxically, technologies and materials are not wedded to theoretical problems and hypotheses. Little's inbred strains of mice led to the discovery of a transmitted mammary tumor, which at first was assumed to be genetically transmitted. After examining many different kinds of crosses over time, researchers concluded that the tumor genealogies were not consistent with their theories of genetic transmission. Bittner later proposed that the transmission occurred via a virus (the MMTV), which apparently caused cancer in infants as it was passed from mother to infant through suckling. MMTV was an impetus and justification for the 1960s search for tumor-causing viruses in the major government-funded research program on viral carcinogenesis. Viruses became the popular hypothesized cause of cancer, and genetics moved into the background. In the 1980s, virus technologies were used to explore a new cause of cancer, the proto-oncogenes in normal cells, thus bringing genes back into the foreground of oncogenesis research. MMTV-induced mammary tumors were then examined using molecular biological technologies, and researchers proposed that the MMTV gene was integrated into the cellular DNA near a gene, which they named *int-1* and which was expressed in mammary tumors but not in normal mammary glands. Other experiments also implicated *int-1* as an oncogene (Varmus 1989b). Thus, the virus that serendipitously appeared early in the century in studies of genetic cancer causation reappeared as a viral cause and then reappeared again in the oncogene story as perhaps a trigger of the *int-1* oncogene. MMTV has led researchers back to genes.

Postwar Cancer Research

World War II marked a general slowdown, if not a halt, in most cancer research. Even after the war, despite Bittner's early work on mouse mammary tumor viruses, viral research on cancer did not take off until the late 1950s and 1960s. Studer and Chubin (1980) argue that this lack of attention was due in part to the National Cancer Institute's concern about promoting ideas and public fear of cancer as an infectious disease. Whatever the reason, American scientists after the war, especially those at the newly established (1937) NCI, harnessed research data collected during the development of chemical and nuclear warfare to study the impact of chemicals and radiation in both cancer causation (carcinogenesis) and treatment.

Many of these studies of chemicals, radiation, and tumor growth were done on inbred strains of mice. But the war produced another tool that greatly aided the development of tissue culture techniques: antibiotics, especially penicillin and streptomycin. By protecting both the cells in cultures and the nutritional media used to "feed" the cells from parasites such as bacteria, antibiotics helped make tissue culture techniques a standard tool in postwar experimental cancer research.

The Biochemistry of Cancer: Cultured Cells and Cell Transformation

"The incentive behind most early attempts to culture mammalian tissues in vitro was the wish to compare the properties of normal cells and cancer cells. As a result, a huge literature has accumulated in which the specific study of cancer cells has become inextricably mixed with the general problems of tissue culture" (Cairns 1978:129). The history of postwar cancer research is framed by the cultivation of novel mammalian tissue culture methods. Almost every branch of biological science appropriated cell culture to apply its particular approaches to the study of cancer through in vitro studies in tissue culture. DeOme discussed the impact of tissue culture methods on postwar cancer research and organization of work.

> Today it's different . . . You have this [problem] broken down into areas where any good biochemist . . . can carry out perfectly reasonable biochemistry on tumor tissue and ask meaningful questions in tumor biology without having to wrestle with this whole problem [of cancer in the whole animal] at once . . . You could come over here and get . . . a culture of mouse . . . tissue of various kinds—tumor and nontumor. You could take that away with you and do something about it biochemically; something

that one person, or one person and his students, could do. (DeOme interview)

Efforts to develop tissue culture techniques had begun in 1907 with Ross Harrison's (1907) nerve tissue on microscope slides (first at Johns Hopkins University, later at Yale University).[19] The association of tissue culture with cancer research began in earnest with Alexis Carrel's work on the artificial activation of growth and cell division using embryo tissue culture. In 1965, Willmer wrote that tissue culture and cancer research were historically tightly tied together.[20] He argued that tissue culture was at that time viewed as the right tool for the job in cancer research.

> In retrospect it is interesting to speculate on the effects which this work had in orientating the history of cell biology. It immediately focused attention on the method of tissue culture as one which could be extremely suitable for the study of growth and, of course, this meant for the study of cancer . . . [Tissue culture] was clearly of tremendous importance as a means of investigating, and perhaps eventually combating, the great scourge of cancer. Tissue Culture and Cancer Research were thus early linked together, both practically by the research workers themselves and also in the public mind. (Willmer 1965:3–4)

However, Willmer concluded that the popularity of tissue culture as a tool for cancer research in the end both created wasted effort and, on the positive side, aided the development of cell biology but to the detriment of cancer research.

> In consequence, Tissue-Culture laboratories were set up here, there, and everywhere throughout the world and an immense quantity of time and money was squandered on ill-directed research by adventurers who had climbed upon the band-waggon. It also meant that much of the effort in tissue culture was expended in trying to make cells grow fast and in the unorganized and rather abnormal way that they do in some malignant growths and in which they were found to do when suitably stimulated in tissue cultures. What was happening to the cells in the original tissue of the explant was at that time of minor interest as compared with the visible growth and division of the cells that moved out into the medium or onto the supporting surfaces provided. Though much of the importance in our concepts of cellular behaviour has emerged from such studies of unorganized growth, the problem of malignancy is still with us and so are many problems of cell differentiation, organization and function, for the solution

> of which Tissue Culture could have been used at a much earlier time and which by now might have been solved had not the popular appeal directed research in the way that it did. (Willmer 1965:4–5).

From "Art" to "Science"

As we see in Table 2.1,[21] the development of tissue culture methods began in the late nineteenth century and has continually been honed throughout the twentieth. It was not until the 1950s and 1960s that scientists transformed tissue culture techniques for handling multicellular systems from an "art" for a few "golden hands" to a *standardized, simplified, and routine tool* for use by many scientists. According to microbiologist Morgan Harris (1964:123), "culture methods in microbiology have long been a cornerstone for analysis, but in multicellular systems the application of similar techniques proceeded more slowly." Harris described the "revolution" from "art" to "science"—perhaps the better term here is technology—in tissue culture methods.[22]

> Owing to . . . limitations, cell culture had for many years relatively little general impact, and was regarded more as an art than as a basic research tool in biology and medicine . . . This evaluation has changed in recent years as a result of discoveries that make clear the potential of isolated cell systems for the analysis of problems in experimental morphogenesis, cytogenetic variation, cell-virus relationships, and many other types of investigations. A rapid evolution in technology has occurred, which greatly enhanced the usefulness of cell cultures in experimental research designs. New and simpler media have been devised, which in some cases are even chemically defined . . . methods have become available for the dissociation of intact tissues or other cell groups to yield suspensions that can be manipulated easily; propagation of isolated strains or primary cultures may be carried out in three-dimensional matrices, as monolayers, or in free suspension; and techniques have emerged for quantitative growth measurement, clonal isolations from single cells, and the preservation of cell populations in the frozen state. Since 1950 the expansion of these and other methods for cell culture has become particularly conspicuous. (Harris 1964:123–124)

Willmer (1965:8) similarly identified two major postwar contributions that moved tissue culture out of its "doldrums" by producing "firstly, more standardized and uniform populations of cells (i.e. pure strains like those used by the bacteriologists), and secondly, the replacement of natural media by those of known and chemically defined composition." Both the wartime

Table 2.1 Some landmarks in the development of tissue culture

1885	**Roux** showed that embryonic chick cells could be maintained alive in a saline solution outside the animal body.
1907	**Harrison** cultivated amphibian spinal cord in a lymph clot, thereby demonstrating that axons are produced as extensions of single nerve cells.
1910	**Rous** induced a tumor by using a filtered extract of chicken tumor cells, later shown to contain an RNA virus (Rous sarcoma virus).
1913	**Carrel** showed that cells could grow for long periods in culture provided they were fed regularly under aseptic conditions.
1948	**Earle** and colleagues isolated single cells of the L cell line and showed that they formed clones of cells in tissue culture.
1952	**Gey** and colleagues established a continuous line of cells derived from human cervical carcinoma, which later became the well-known *HeLa* cell line.
1954	**Levi-Montalcini** and associates showed that nerve growth factor (NGF) stimulated the growth of axons in tissue culture.
1955	**Eagle** made the first systematic investigation of the essential nutritional requirements of cells in tissue culture and found that animal cells could propagate in a defined mixture of small molecules supplemented with a small proportion of serum proteins.
1956	**Puck** and associates selected mutants with altered growth requirements from cultures of HeLa cells.
1958	**Temin and Rubin** developed a quantitative assay for the infection of chick cells in culture by purified *Rous sarcoma virus*. In the following decade, the characteristics of this and other types of viral transformation were established by **Stoker, Dulbecco, Green,** and other virologists.
1961	**Hayflick and Moorhead** showed that human fibroblasts die after a finite number of divisions in culture.
1964	**Littlefield** introduced HAT medium for the selective growth of somatic cell hybrids. Together with the technique of cell fusion, this made somatic-cell genetics accessible.
	Kato and Takeuchi obtained a complete carrot plant from a single carrot root cell in tissue culture.

Table 2.1 *(continued)*

1965	**Ham** introduced a defined, serum-free medium able to support the clonal growth of certain mammalian cells.
	Harris and Watkins produced the first heterocaryons of mammalian cells by the virus-induced fusion of human and mouse cells.
1968	**August-Tocco and Sato** adapted a mouse nerve cell tumor (neuroblastoma) to tissue culture and isolated clones that were electrically excitable and that extended nerve processes. A number of other differentiated cell lines were isolated at about this time, including skeletal-muscle and liver cell lines.
1975	**Köhler and Milstein** produced the first monoclonal-antibody-secreting hybridoma cell lines.
1976	**Sato** and associates published the first of a series of papers showing that different cell lines require different mixtures of hormones and growth factors to grow in serum-free medium.

Source: Reprinted from Alberts et al. 1983:1164, by permission of Garland Publishing, Inc.

production of antibiotics and the growth of commercial sources of cell culture media (nutrients, antibiotics, and other agents the cells need to survive) were critical for the standardization and diffusion of cell culture methods.

Constructing the Normal

The experimental goal was to compare normal with cancerous cells of the same types, including hematopoietic cells, mesenchymal cells (such as fibroblasts, myoblasts, and chondroplasts), epithelial cells, and cell lines.[23] This research required both normal and cancerous variants of the same types of cells to be grown in vitro, which proved to be a nontrivial task. Cells did (and do) not behave according to researchers' desires.

Primary cultures, for example, were (and still are) extremely problematic experimental systems. Primary cell cultures were created by taking cells from animals and transplanting them onto nutritional media. This form of culturing grew out of Ross Harrison's explant techniques (Harrison 1907). Most primary cells in contemporary biological research came from chick and mouse embryos. However, primary cell cultures were relatively short-lived, which created various disadvantages for experimentation. Researchers soon concluded that even cancer tissue cells which grew well within organisms did not necessarily thrive in cultures.

Biologists could grow normal cells in a culture and then transform them into cancer cells that would subsequently form tumors in animals. This technical achievement in turn "transformed" the experimental frames of cancer research from a comparison of naturally occurring normal cells with naturally occurring cancer cells. After World War II, this method of transforming normal cells to tumor cells became their modus operandi, and the study of cancer became the study of cell transformation in culture. Researchers examined the processes by which a normal cell became a cancer cell and compared the physiology of normal cellular functions with the pathological physiology and structures of the transformed cells. Transformation was defined as having occurred if there was a heritable alteration in the cell's phenotype. If that phenotypic alteration was associated with the acquisition of malignancy, then a malignant transformation had occurred.

"Established cell lines" were the product of, and the process that created, cell transformation studies. Cell lines were created by directly transplanting cells from primary tissues—both normal and tumorous tissues—and replating them as they divided and multiplied. Cell lines increased the length of time for experimentation, but in practice, cell lines were limited in the number of times (about fifty) the same cells could be passaged from culture to culture. Established cell lines were even longer-lived cells cultivated from variant cells that had

> a somewhat altered style of growth and prove[d] to be exempt from aging (meaning that they [could] be cultivated indefinitely). Instead of being spread flat on the floor of the culture vessel, they [were] more rounded and [would] grow to form colonies in which the cells [were] heaped one upon another. This change, which [was] called [*spontaneous*] *transformation,* impart[ed] to the transformed colony a considerable survival advantage in the final stages of growth, and so the transformed cells eventually come to dominate the culture. These are the cells that make up what are called *permanent,* or *established,* cell lines." (Cairns 1978:130)

NIH 3T3 mouse fibroblast cells are an example of an established cell line used since the 1960s to test carcinogenesis.[24] Cancer experimentalists treated these cells as normal cells. However, some debate surrounded these "normal" experimental cells in culture. Some researchers argued that such established cell lines were predisposed to malignancy; that is, they were already partially "transformed" and therefore it was not surprising that they often formed tumors when injected into animals. These cells do not "age" (they are very long-lived), which makes them quite different from normal living cells, whose senescence (or death) is inevitable. In contrast, NIH 3T6 and NIH 3T12 cells are cell lines that were thought to be tumorigenic in animals because they grew to high densities in cell culture. NIH 3T3 cells were treated as normal established cells because only a small percentage of cells were tumorigenic in animals. With the development of better primary cell cultures and other media for experimentation in the 1980s, researchers have taken a less benign view of NIH 3T3 cells.[25] Note that "normality" was defined in context. NIH 3T3 cells were defined as normal in comparison with NIH 3T6 and NIH 3T12 cell lines.

Transformation was not a neat matter. To order their work, scientists devised parameters or properties of the transformed cell that served as markers of transformation. For example, the ability to grow in low concentrations of serum was associated with a malignant change. But researchers were quick to note that there was no universal marker for malignancy. While some transformed cells might induce tumorigenesis (cause a tumor to grow) in animals, many did not. Similarly, while fibroblast cells—which grew well in culture—might be transformed by an agent, the result was not generalizable to other types of cells. Sometimes the differences between normal cells and transformed cells was so dramatic that one might have regarded them as a qualitative change, yet transformed cells were never totally different from normal cells. The phenotypic difference was quantitative. There were also morphological differences between normal and cancer cells. Normal cells tended to neatly flatten out and cling to the bottom of the cell culture plate, while cancer cells tended to be rounded or spindle shaped and crisscross over one another in a disorganized fashion. Another marker of transformation was the ability (or lack of it) of cells to grow in different kinds of cultures. Those cells that could grow in suspension cultures tended to induce tumors in animals. Later studies in the 1970s and 1980s showed other biochemical differences between normal and transformed cells, including the observation that transformed cells do not stop growing, while normal cells usually show growth limits.[26]

Researchers were not equally successful at developing experimental systems of all cells. Most established cell lines, for example, are commonly grown from rodent cells, since chick cells and human cells are difficult to keep alive in culture. Similar differences exist by cell types. Consider the skin tumorigenesis model, which was the most well-developed experimental system for in vivo studies. Carcinogens were studied using mouse skin experimental systems because they were visible and accessible, because carcinogens could be applied directly to the skin, and because the basal layer of cells were homogeneous thus making it easier to repeat studies. In contrast, skin studies in vitro were rarer, since it was (and still is) difficult to grow and maintain epithelial cells (of the skin and lining tissues) in culture. This difficulty complicated efforts to develop a good system for experimenting with cancers in epithelial cells in culture. So most cell culture studies have been conducted using fibroblast cells, not on cells of the skin, liver, or other epithelial tissues.

Nevertheless, the materials and techniques of in vitro transformation studies radically altered cancer research. The importance of cell transformation for the construction of knowledge about cancer is signified by the attention paid to the search for universal markers of transformation in culture. In order to study the transformation of "normal" cells to "cancer" cells, researchers first attempted to construct parameters that defined, and tools that materialized, the "normal" and the "pathological" (cancerous). Every few years in cancer research's recent history, announcements of a novel universal marker or parameter of transformation created great excitement. However, soon thereafter researchers encountered exceptions to their rule and would throw up their hands in despair. By the 1980s, rather than continue the contest for finding universal parameters, researchers had begun to try to use the assorted phenotypic indicators (the ways that cells behaved in cultures) that had wreaked havoc in their search for universals to instead try to use them to learn about transformation. Instead of trying to create a homogeneous tool for studying transformation, researchers decided to accept the heterogeneity of cell properties and treat them all as information in the study of transformation.

On the one hand, although transforming properties could not neatly be categorized in terms of a few parameters, cell lines provided scientists with a way to experimentally manipulate cells that was not possible in living organisms. Moreover, whereas it was difficult to isolate transformed original cells in vivo, in vitro transformation of normal cells from cell lines to cancer cells occurred 100 percent of the time, thus eliminating efforts to isolate transformed cells. They all were transformed. Cell lines also allowed the creation of homogeneous systems to simplify analysis, in contrast to the heterogeneous systems that exist within a living organism; in some scientists' views, this shift to cultured cells as the tool of the trade provided greater possibilities for reproducing experiments and therefore for constructing more robust results.

On the other hand, cultured cells were very different from the cells in tissues in living organisms (research animals or patients). In addition to their tendency to transform, cultured cells also lost the intracellular interactions that occur between cells in the living organism. The homogeneity that made them good experimental model systems also made them "unnatural" in that natural systems are heterogeneous and differentiated. Whether they were primary cultured cells or cell lines, then, the question of the naturalness of these research materials arose, as it did for inbred mice and their associated viruses and tumors. These cells in culture were also standardized experimental systems that reinvented nature. Consider another comment by DeOme: "You could come over here [to the Cancer Research Laboratory] and get . . . a culture of mouse mammary tissue of various kinds—tumor, nontumor—. You could take that away with you and do something about it biochemically. A thing that one person or one person and his students could do. Because so much of the stuff that one had to handle originally has already been handled and systematized" (DeOme interview).

Some scientists thought that the mass cultures of cells and many strains of cells produced during the postwar period were far from perfect models of cells in vivo. Willmer expressed this view:

> Mass-cultures of cells, comparable in many ways with bacterial cultures, can now be grown in chemically defined "synthetic" media, and many strains of cells have been established, some of which are strict clones, having been derived from single cells. It is, however, unfortunately true to say that some of these strains though perhaps pure in themselves, have in fact turned out, upon immunological and other tests, to be not what they were thought to be . . . It must also be confessed that, on other grounds, also the promised goals have proved to be somewhat illusory; for example from a biochemical point of view, the cells in pure strains have turned out to be surprisingly uniform in character and very different from the cells from which they were derived in the body. (Willmer 1965:8)

Thus, the abstract theoretical statements made by scientists interpreting the results of these experimental systems are statements about a created laboratory nature. Although abstract statements and theories are constituted within and of the material contexts of their production, they rarely explicitly acknowledge all the material contexts involved. Some scientists are cognizant of these difficulties and warn against, for example, the use of NIH 3T3 cells in cancer research. Other scientists explicitly mention the material production of their results in scientific articles and notebooks, but their theoretical statements delete these contexts and often generalize beyond their specific sites of production.[27]

From Biochemistry to Molecular Genetics via Viral Transformation

The development of cell culture techniques was closely linked to the growth of virus cancer research and eventually to proto-oncogenes (see Chapter 4). Without knowing it, Willmer presaged this connection in his ode to the practical value of cell cultures especially for virus research and "the present flourishing condition of the virus-growing industry," and indirectly for other research problems.

> In other respects . . . which were probably less foreseen in the beginning, the results [of mass cell culture techniques] have been of the very greatest practical value, namely in virus research . . . These developments of massed cell culture, by one method or another, have been responsible for many new techniques for growing and titrating viruses and have thus been of inestimable value, not only to the world in general for very obvious and practical medical reasons, but also indirectly to tissue culturists working on other problems and particularly those of a biochemical nature; others have also benefited indirectly because of the facilities and apparatus which have become available as the result of this popular demand greatly easing the rather tiresome technical problems which used to face the solitary worker in the field of Tissue Culture. (Willmer 1965:8–12)

However, these new techniques for growing and titrating viruses did not follow from mass cell culture techniques in a clear fashion. Indeed, they were nontrivial scientific and technical feats. One such feat for growing and titrating viruses critical to subsequent activities in cancer research was Renato Dulbecco and Marguerite Vogt's 1954 creation of tissue culture techniques for experimental research on poliomyelitis viruses in cells in culture. This work by Dulbecco and Vogt grew out of Dulbecco's intensive efforts beginning in 1950 at the California Institute of Technology to develop a tissue culture that allowed for the quantitative measurement of viruses in cell culture. Only then, argued Dulbecco, could one gain a clear picture of how viruses act on cells, and in this case how they infect embryos in culture.[28] These quantitative assays formed an important bridge to animal tumor virology of the 1950s and 1960s, as well as to the research on oncogenic viruses in the 1970s and 1980s, as I discuss below.

University of California biologist G. Steven Martin has noted that medical virology spurred the development of cell culture techniques, which in turn spurred the development of animal virology.[29] Willmer (1965:5) similarly argued that it was poliomyelitis that drove tissue culture research methods in a more productive direction. "More recent work has, of course, been along other lines, again partly determined by fear of a disease, this time by the fear of poliomyelitis, and it has amply shown that Tissue Culture had even greater possibilities in several other directions than that of cancer research." Kevles (1992) adds the influenza, mumps, chicken pox, equine encephalitis, and Herpes Zoster viruses to the list of medically related animal viruses that sparked interest and work in developing virus-cell culture systems. However, interestingly enough, Dulbecco's work on the poliomyelitis virus-cell culture assay was subsequently used to develop methods that produced viral oncogene research qua cancer research.

There is a second way in which tissue culture work on problems of a biochemical nature became closely linked to viral research in subsequent research. According to molecular biologist John Cairns,[30] viruses were more efficacious than other research tools in transforming human and chicken cell cultures.

> The appearance of spontaneously transformed variants is very common when mouse cells are being cultivated, rather rare for hamster and rat cells, and virtually unknown for human and chicken cells. This difference between the cells of different species turns out to be crucial to the study of carcinogenesis *in vitro*. Cells that undergo spontaneous transformation can be transformed into tumorigenic cells *in vitro* fairly efficiently by tumor viruses, irradiation, and chemical carcinogens. Cells such as human and chicken cells, which hardly ever undergo spontaneous transformation *in vitro*, can be transformed by viruses but not easily by irradiation or by chemical carcinogenesis. (Cairns 1978:130–139)

This virus–chicken cell relationship came to play a significant role in later narrations of the proto-oncogene story.

DNA Tumor Viruses

"DNA tumor viruses" was the name given to a class of viruses studied in the late 1950s, 1960s, and 1970s and thought to be "culprits" in cancer causation. In 1953 Ludwick Gross isolated a mouse leukemia virus and injected it into susceptible mice, which then developed parotid gland tumors. Gross (1957, 1961) next induced leukemia in rats with a virus obtained from a mouse leukemia. In 1958 cell biologists Stewart and Eddy grew the virus in cell culture and used it to produce a whole variety of cancers, including cancers in the adrenal gland, kidney, mammary gland, liver, and thymus, in many different rodent species. They named the virus "polyoma," meaning many kinds of tumors. According to the common interpretation at the time, there must exist some common mechanisms of transformation if a single virus could cause so many different kinds of cancers.

In 1960 Vogt and Dulbecco reported their experimental transformation of hamster embryo fibroblast cells in culture with a polyoma virus. Vogt and Dulbecco asserted that the polyoma DNA was integrated into the hamster DNA. In later work, they concluded that specific genes coded for by the DNA caused the transformation, and they hypothesized that the mutation or deletion of specific genes could prevent transformation. They classified the polyoma virus as a DNA tumor virus, because its genetic material consisted of DNA. Since that time other DNA tumor viruses have been studied, including the SV40, or simian virus 40, human adenoviruses, herpes viruses, Epstein Barr virus, and the hepatitis B viruses. Vogt and Dulbecco concluded that the viral DNA must persist and be expressed[31] in the host cells for them to remain transformed. If the viral DNA was removed, the cells reverted back to normal. A second important observation was that the DNA virus was cytotoxic, that is, it killed the cells. In contrast, another class of viruses called "RNA tumor viruses" could transform the cells and replicate themselves without killing the cells. In the end, RNA tumor viruses took center stage in this history of cancer research.

RNA Tumor Viruses

In the early 1900s Peyton Rous (1911a,b; Rous and Jones 1916) developed a cell culture technique that he used to study a virus (then called a "filterable agent," later named "Rous sarcoma virus") he thought to be active in causing tumors to grow in chickens. As was the case with the mouse mammary tumor virus discussed earlier, Rous's work did not generate much enthusiasm until the 1950s. Indeed, many of the researchers I interviewed and even current biology textbooks commented on the neglect of Rous's work for many years. His work did not gain recognition until 1966, when he received a Nobel Prize in medicine for his early work on Rous sarcoma. As mentioned earlier, Studer and Chubin (1980) argue that this lack of enthusiasm was due in part to various concerns of the NCI. Another reason for the lack of attention to Rous's filterable agent might have been the concern of scientists for better experimental and measurement techniques for elucidating the filterable agent as the virus Rous believed it to be.

In 1958, inspired by Dulbecco and Vogt's 1954 plaque assay for the poliomyelitis viruses, Dulbecco's postdoctoral student Harry Rubin and graduate student Howard Temin developed a focus assay for the Rous sarcoma virus that provided a way to quantitatively measure viruses and cells in culture. Temin and Rubin (1958) used both Rous's and Dulbecco and Vogt's work to improve the measure of virus titers, that is, the measurement of the concentration of Rous sarcoma viruses and Rous sarcoma cells in tissue culture. After conducting experiments using these techniques, they argued that the Rous sarcoma virus was a defective virus that could not reproduce itself, once it had infected a cell, without the help of a "helper virus" (Rubin 1964). Temin and Rubin's work initiated a great deal of excitement about RNA tumor viruses, which in turn helped develop work on viral oncogenes.

An RNA tumor virus is now classified as a retrovirus. A retrovirus has RNA, not DNA, as its genetic material.[32] In the rewritten version of Temin's original "provirus" theory, retroviruses contain a single strand of RNA that produces a double-stranded DNA in the cytoplasm of the infected cell. This conversion is aided by an enzyme called "reverse transcriptase." The DNA then moves to the nucleus of the infected cell. If the viral DNA is incorporated into the host cell's genome, the cell will then produce the virus's RNA and proteins, thus allowing for the reproduction of the virus itself (Temin 1964a,b,c).[33]

The nomenclature of reverse transcriptase (RT) signifies its location in a historical course of events in postwar biology. During the late 1950s and 1960s, the initial publications on the structure and functions of DNA as the genetic material in cells generated much excitement in all fields of biology.[34] In 1958 Francis Crick proclaimed the "Central Dogma" that DNA makes RNA makes protein. Crick announced that the process is never reversed, and Watson et al. (1987) later reiterated this view. With that framing of molecular activities in the cell's nucleus, the discovery of an enzyme that guided action in the opposite direction led to its being viewed as an enzyme that assisted reverse transcription. Scientists came to see the production of DNA by RNA in this case as an exception to the "normal action" of DNA producing RNA.[35] The two scientists who were credited with the discovery of what was then called "RNA-dependent DNA polymerase," Howard Temin[36] and David Baltimore (then at the Massachusetts Institute of Technology, currently at Rockefeller University), were awarded the Nobel Prize for their work in 1975.[37] They had presented a process that went against the accepted knowledge, the Central Dogma. This reverse transcription process provided a fascinating view of viral infection and reproduction as well as cell transformation.[38]

These and other experiments created the field of tumor retrovirology. A genealogy of research, researchers, and research technologies produces at least one relatively linear history (and there are many others) from Dulbecco and Vogt's work on tissue culture, to Temin and Rubin's work on the focus assay, to Peter Duesberg, Peter Vogt, Hidesaburo Hanafusa, and G. Steven Martin, who were all postdoctoral students with Harry Rubin in the Department of Molecular Biology at the University of California, Berkeley.[39] Working first with Temin and Rubin's assay for Rous sarcoma virus, Duesberg, Vogt, Hanafusa, and Martin demonstrated that the virus transformed cells in culture via the activities of a gene, which they eventually named *src* (pronounced "sarc," short for sarcoma).[40]

By the early 1980s, the work of these researchers had produced the following representation of retroviruses. *Gag, pol,* and *env* were three genes coded for by the RNA of RNA tumor viruses. The *gag* gene in turn coded for the production of the major structural proteins of the virus, the *pol* gene coded for the reverse transcriptase that is necessary for the virus reproduction, and the *env* gene coded for the production of the outer protein shell of the virus. While these genes were common in most retroviruses, a subclass of RNA tumor viruses also contained an additional RNA strand that apparently transformed the cells infected by the virus into tumor cells. *Src* was the first of these genes to be intensively studied and was named an "*onc* gene," but it is more commonly referred to as a "viral oncogene." By 1983 the viral *src* oncogene came to be viewed as a transduced form of a normal cellular gene that had suffered mutations through the transduction process. This normal cellular gene was then given the name "proto-oncogene." (See Varmus 1989b and Chapter 4 of this text.)

The Rous sarcoma oncogene *src* was one of many viral oncogenes to be defined and named in the 1970s and 1980s. *Src* was shown to transform chick embryo fibroblast cells in culture. *Ras* and *myc* were oncogenes first taken from viruses infecting rats and chicken tissue, respectively. However, the mouse mammary tumor virus that transmitted tumorigenesis from mother to offspring in Bittner's earlier study was later reported to contain no viral oncogene. Instead researchers accounted for its tumorigenic abilities through the insertion of its viral DNA near the proto-oncogene *int-1*. (See Chapter 4 for more on viral and cellular oncogenes.)

By 1986 virologists had identified about twenty or thirty (depending on which list one reads) different oncogenes in a class of RNA tumor viruses that were deemed responsible for the transformation of normal cells in culture into tumor cells. By 1989 the total had risen to sixty different oncogenes, but this number included oncogenes that were not located in RNA tumor viruses. Some of these genes were also shown to produce tumor growths in lab animals.

However, it was the Rous sarcoma virus that became the first critical tool and that led the way in oncogene research. In his acceptance speech for the Nobel Prize for oncogene research, Harold Varmus lauded the Rous sarcoma virus for its role in that research and echoed the comments of many researchers I interviewed.

> Several viruses also figure in our tale, but Rous sarcoma virus again has the leading role, yet another of many tributes to the pioneering work of Peyton Rous and to the principle of delayed gratification in science. The product of his diligence in pursuing a single chicken tumor nearly eighty years ago, Rous' virus remains the only retrovirus that could have satisfied the genetic

> and biochemical criteria for the work we accomplished in the era that preceded molecular cloning. (Varmus 1989a:414)[41]

Notwithstanding the ideal properties of the *src* as a research tool, according to virologists, tumor viruses generally were an ideal experimental system for the study of genetics in cancer etiology. Given congenial conditions, viruses reproduced easily and provided volumes of material for experimentation. Researchers could get approximately 100 percent transformation with these viruses in cell culture at that time, whereas it was difficult to transform cells in culture with other viruses or with chemical carcinogens. Tumor viruses were genetically simple, that is, they had a small number of genes (three or four), and were therefore an easy model system on which to conduct cancer research.

The Message Becomes the Medium

Indeed, viruses moved from playing the role of possible cancer-causing agents to the role of tools for exploring genetic theories of cancer. These viruses began to be viewed as laboratory artifacts, not as representations of what actually happens in cancer growth, but as tools for designing experiments and theories about how tumors begin and grow.

One such transformation of message to medium was Howard Temin's protovirus hypothesis. In 1971, Temin proposed that normal cells contained endogenous viruses (RNA to DNA transcription) whose task it was to move around the cell genome and generate genetic variations through incorrect transcription or by cobbling together new genes using parts of cellular genes. According to the protovirus theory, RNA tumor viruses were "minor aberrations of a device used normally in development." In other words, the virus had a regular job to do in the organism until an incorrect transcription transformed it into something that actually harmed the organism. John Cairns (1978:118) took the argument one step further by proposing that these viruses were *laboratory artifacts*. He suggested that this "minor aberration" was actually a product of laboratory manipulations, and not a transcriptional error that occurred through "normal accident" in the organism's lifetime.

> The argument has been taken one step further with the suggestion that certain of these tumor viruses are actually laboratory artifacts. For example . . . a spontaneous tumor of chickens, the Rous sarcoma . . . at first . . . could be transmitted only between closely related chickens (i.e., it could not cross histocompatibility barriers); later, however, the tumor could be

> transferred with some difficulty by cell-free extracts, and from then on transfer became possible even between species. So it does look rather as if the cancer cell came first and subsequently, because of the intense selection pressure imposed by the histocompatibility barrier, the particular set of genes responsible for the cancer became transferred to a virus that was not subject to any such barrier and could move freely from one animal to another.

Cyborgs in a Reinvented Nature: "The Chosen Pets"

As with the laboratory-produced mice and cell lines, the question again arose about the naturalness of these tumor viruses. Are current experimental systems used in cancer research artifactual productions? Are these viruses invented? Are they cyborgs in a reinvented nature? Are scientists working with artifactual viruses living in artifactual animal strains? If not natural, how relevant is the research for human cancers? Whatever the answers to these other questions, Cairns was convinced of the value of artifactual viruses. "The possibility that certain tumor viruses are simply laboratory artifacts makes them in some respects more interesting rather than less, because it means that they may be offering us a way of isolating and studying the genes responsible for the cancerous state. That seems to be the real justification for putting so much effort into investigating the tumor viruses, not the vague hope that human cancer will turn out to be a virus disease" (Cairns 1978:118).

Artifacts or not, RNA tumor viruses became vital tools in cancer research worlds especially in the 1980s. They were remarkable experimental systems that worked interactively with other experimental systems such as inbred mice, cultured cells, and laboratories. The work on these viruses also became part of the armamentarium of research efforts on many other problems in biomedicine. One such effort is the current research on another retrovirus, human immunodeficiency virus (HIV), which has been designated as the cause of the deadly Acquired Immune Deficiency Syndrome (AIDS).[42] Harold Varmus, again in his Nobel Lecture, proclaimed the research successes achieved with these retroviruses, their "chosen pets" and "vital tools."

> The story thus far confirms David Baltimore's statement of thanksgiving: "a virologist is among the luckiest of biologists because he can see into his chosen pet, down to the details of all its molecules" [Baltimore 1976:632]. Because retroviruses, *our chosen pets,* are such remarkable agents, it has been enough to train our sights on two brief questions—how do retroviruses grow? how do retroviruses cause cancer?—to have extended our concerns outward to the cellular host, as well as to have focused them inward

> upon the viruses themselves. As a result, we have entered into some of the liveliest arenas in modern biology: the genetic basis of cancer, the transposition of DNA through RNA intermediates, the control of gene expression in eukaryotes, and the molecular evidence for evolution.
>
> At this point, the study of oncogenes and proto-oncogenes has attained a degree of maturity that allows it to be conducted with astonishing little virology. Yet retroviruses remain *vital tools* for the isolation of important new oncogenes; witness in the past few years the discoveries of the *jun* and *crk* genes. Likewise, . . . seemingly exhaustive attention has been given to the life cycle of retroviruses, yet many central features are just now coming into view. Cell surface receptors for viral attachment and entry have been recently identified and show a remarkable range of biochemical properties; the proviral integration reaction has been recapitulated in vitro with nucleoprotein complexes, allowing a description of integrative precursors and intermediates; retroviruses have been recognized as pliable genetic vectors that may one day be used clinically to correct gene deficiencies, in the manner used in nature to transport host-derived oncogenes; many unexpected aspects of viral gene expression have been discovered, including translational frameshifting during the synthesis of reverse transcriptase and complex viral regulatory genes that govern the behavior of two classes of human retroviruses; and the principles of virus assembly are emerging through physical and genetic assaults on viral structural proteins and proteases. These inherently fascinating problems have now taken on a special urgency, because we are all threatened by the world-wide dissemination of a lethal human retrovirus, the human immunodeficiency virus. Thus, retroviruses continue to challenge our intellects in ways that may help us grapple with major disease, cancer and now AIDS, while also revealing fundamental features of the lives of our cells. (Varmus 1989a:427; emphasis added)

Retroviruses became the pet organism and reliable agents for producing and solving scientific problems for Varmus, Bishop, and many other practicing biologists in the 1980s. Furthermore, the choice of retroviruses as a research tool can be used to trace the links between individual laboratories, institutional and disciplinary commitments, and theoretical viewpoints.[43]

From Carcinogenesis Experiments and Somatic Mutation to Molecular Genetics

Another historical genealogy of cancer research practice can be constructed by tracing a line from late-nineteenth-century carcinogenesis studies through early-twentieth-century experimental transplantation studies to postwar carcinogenesis studies and then to the DNA transfection studies of the late 1970s and early 1980s. This line later connects with viral oncogenesis studies to form a new theory of proto-oncogenes.

Early studies of carcinogens like coal tar were used to experimentally produce tumors in animals. In the late nineteenth century, chimney sweeps in England who crawled up and down chimneys had very high incidences of skin tumors, especially on the scrotum. Research on their tumors identified coal tar and the lack of sanitation as the culprits, and later research distilled benzpyrene as the active ingredient that caused the cancers. "I've forgotten how many tons of coal or coal tar were used to get the first gram of benz-pyrene. It was in a national project in England to find out what in coal tar produced tumors" (DeOme interview). This later research was based on tests of carcinogenesis in culture cells as well as in experimental animals. After the war, as noted earlier in this chapter, scientists made use of research data from the development of chemical warfare and the atomic bomb in studying the impact of chemicals and radiation in carcinogenesis and treatment.[44]

Cultured cells were also critical to the constitution of post–World War II chemical carcinogenesis studies. Chemicals were tested for their carcinogenic effects on cells in culture. The coal tar example was followed by efforts to test in vivo many chemicals for their tumorigenic capacities. According to DeOme, the aim was to identify a common chemical element that was tumorigenic.

> The chemists said that there's got to be something common in all these things. There's got to me some common chemical configuration . . . It can't be just any old thing. Well, it turned out that we've got a five-volume set with page after page of chemicals that have been tested for their carcinogenic abilities. Hundreds, thousands of them. There isn't any common chemical structure among these. There is such great variation, one from another, that it would be almost futile today to say that there is some particular chemical structure which is capable of . . . [overriding normal growth control mechanisms]. Now what do you do? (DeOme interview)

The next step was to test radiation using the same approach.

> We began to talk about radiation in the same way. Certain kinds of radiation under certain conditions are highly, highly carcinogenic. Other kinds are not. We had known for a long time that farm workers in the south had much higher incidences of skin cancer than farm workers in the north—even if we take the white ones only in both populations. Maybe something

> quantitative with regard to radiation [was operating]. Sunlight had to be involved somehow. But there are lots of other factors . . . And you could duplicate this in the laboratory with your inbred strains of animals. (DeOme interview)

The problem, according to cancer researchers of that time, was that there were many carcinogens, most of which produced mutations and caused cancer. And researchers did not know exactly what was mutated. They asked, what controlled whether tissue in a hand continued to form a hand or whether it became tumorous? "Now where are you? UV [ultraviolet light] will produce mutations. So will a lot of the carcinogens. Most of them will produce mutations. So you are right back where you started . . . Because you don't know what the control situation is" (DeOme interview).

The search for the controlling factors and mechanism has dominated twentieth-century cancer genetics. Efforts to explain carcinogenesis through a control mechanism followed, especially in the 1950s, the somatic mutation theory mentioned earlier in this chapter. In 1912 Theodor Boveri proposed that abnormalities of chromosomes might be a cause of cancer. Although credited in later accounts with arguing that abnormal mitosis was linked to malignancy (Richards 1978:116, Bishop 1989:328), Boveri argued in 1929 that "the essence of my theory is not the abnormal mitoses but a certain abnormal-chromatin complex, no matter how it arises. Every process which brings about this chromatin condition would lead to a malignant tumor" (quoted in Huggins 1979:2). Boveri's theory framed the study of carcinogenesis studies even into the postwar period. In contrast to hereditary cancers, which were suspected of involving the germ cells, in the framework of the somatic mutation theory most cancers were caused by assaults on somatic (body) cells that were only indirectly affected by heredity (Boveri 1929). By the postwar period, the somatic mutation theory had developed into the idea that carcinogenic agents could somehow transform a somatic cell's genes in ways that permanently damaged the cell's phenotypic expression (the cell's functional activities) into the abnormal activities seen in tumor cells (Berenblum 1952:158). Cairns summarized this research of the prewar and early postwar period.

> From the very outset, the obvious interpretation of carcinogenesis was that it represented an irreversible change in the inherited characteristics of a cell (i.e., what we would now call the *mutation of genes*, though the word *gene* was not invented until well into the 20th century). Indeed, the foundations of the science of genetics were being laid down at the same time as much

> of this work was being done, and so it was natural to link the two subjects together. (Cairns 1978:65).

However, there were problems with the somatic mutation theory. One was the diversity of chemicals that could induce tumors, another the heterogeneity of their effects on tissues by both chemicals and radiation. DeOme's concerns around these "variations" are echoed in Cairns's comments.

> There was, however, a disappointing lack of coherence in the whole field of cancer research, and no underlying rules governing the response to carcinogens seemed to exist. One species would be sensitive to one carcinogen but not another, and a given carcinogen such as 2-naphthylamine would single out one particular tissue such as the bladder for attack and apparently leave all other tissues untouched. So any straightforward theory for the mutational origin of cancers seemed untenable. (Cairns 1978:65–66)[45]

Another major enigma was the long latency period before carcinogens induced cancers. How could a mutation, which was known to be an instantaneous change, be coherent with the slow, long-term development of tumors?

Writing in 1952, Berenblum was more encouraging about the possibility of explaining cancer by the somatic mutation theory. With respect to the long latency period after exposure to the chemicals, he argued that certain substances contributed to the "initiating stage" of carcinogenesis by mutation, but that more than one step was necessary to cause cancer. On the basis of a series of experiments, Berenblum proposed that "tumor-production was not a single drawn out process, as was previously believed, but that it consisted of separate stages with independent mechanisms. Certain substances, of which croton oil is a notable example, which are not themselves carcinogenic, can nevertheless cause a tumor to develop *once the preliminary stage has been instituted by a true carcinogen.* This new concept of 'precipitating action' has important practical as well as theoretical implications" (Berenblum 1952:128). Indeed, Berenblum viewed the theory as explaining the results from early transplantation studies, radiation research, and chemical carcinogenesis experiments.

> The attractive feature of the theory is that it accounts for many peculiarities of tumor growth, which most other theories fail to explain. It accounts for the fact that a tumor, once produced, goes on being a tumor; and why its properties are so faithfully transmitted to the daughter-cells even after prolonged transplantation. It explains why a tumor arises from an isolated

focus—possibly even from a single cell, in the first instance—instead of producing a diffuse cancerization; and it also explains why so many different carcinogens are capable of inducing tumors, the type of tumor produced being determined by the tissue acted upon, rather than by the nature of the carcinogen. In this way, carcinogens would appear to be nothing more than *agents that facilitate mutations.*

If this were so one should expect all carcinogens to be able to produce mutations in germ cells, and all mutating agents to be carcinogenic to ordinary cells. That X rays could produce both these effects has been known for a long time. Only recently have scientists begun to extend the investigation to chemical substances to determine whether any parallel exists among such substances between carcinogenic and mutagenic action. While the correlation was not perfect, many carcinogens were found to be mutagenic, and *vice versa.* (Berenblum 1952:158–159)

Berenblum even viewed the mutation theory as a way to link the results of chemical and irradiation carcinogenesis studies with the 1950s research on viral carcinogenesis discussed above.

The most interesting aspect of the mutation theory of cancer is the way it can be made to link up with the virus theory of cancer. It will be recalled that viruses are, chemically speaking, related to the gene-bearing chromosomes, and that according to one theory viruses are actually altered genes. According to the mutation theory, the permanently altered gene, resulting from carcinogenic action, and the tumor virus, extractable from certain tumors, are one and the same thing. (Berenblum 1952:159)

Some have suggested that viruses are gene-like entities that arose long ago from normal genes through an abnormal transformation, or "mutation," and that they have been perpetuated in their new form within living cells . . . Such "mutated genes," when entering a normal cell, would endow it with some new property that may, or may not, be harmful to it. According to this hypothesis, viruses that affect animal tissues were originally derived form the genes of animals; those that affect plants were originally of plant origin; and the "bacteriophages" (the viruses that destroy bacteria) were originally bacterial products.

From the point of view of the cancer problem, this hypothesis is highly attractive, for it is able to account for many of the discrepancies inherent in the original "virus theory of cancer." The question of "infectivity" in relation to tumor causation has thus receded into the background, while new interest is becoming focused on the possibility of relating tumor viruses with genetic influences, on the one hand, and with the action of chemical carcinogens, on the other. (Berenblum 1952:138–139)

In the 1950s, many of these studies still did not produce results that satisfied some experimentalists. In 1959, for example, Rous lambasted the somatic mutation "hypothesis" in an article responding to a discussion at an international cancer congress.

> A hypothesis is best known by its fruits. What have been those of the somatic mutation hypothesis? It has resulted in no good thing as concerns the cancer problem, but in much that is bad. It has led within the year to an authoritative statement, in the lay press, that since cancer is certainly due to somatic mutations, the possibility of having it is "inherent"; and that this being so, the most man can ever do is to palliate such malignant tumours as may have become disseminated in the body and to avoid new carcinogens as well as the old . . . Here is fatalism to blast many a hope and effort . . . Most serious of all the results of the somatic mutation hypothesis has been its effect on research workers. It acts as a tranquilizer on those who believe in it, and this at a time when every worker should feel goaded now and again by his ignorance of what cancer is. (Quoted in Studer and Chubin 1980:108–109)

Even Berenblum had carefully considered the lack of evidence for the theory back in 1952.

> Except for the circumstantial evidence that many carcinogens can produce mutations in germ cells and many mutating agents can act as carcinogens, the "somatic mutation theory of cancer" has never really been tested. It is, therefore, not a theory but merely a hypothesis. The fact that the properties of tumors are permanent characteristics does not necessarily denote that the change in the cell is due to a mutation, for we know of biological phenomena which are definitely not mutational but which nevertheless appear to be irreversible. Finally, even if the mutation theory were to account for the "initiating stage" of carcinogenesis, there is still the important "promoting stage," responsible for the conversion of the "latent tumor cells" into real tumors during the long latent period. The nature of this process is still a mystery. (Berenblum 1952:160)

In an interesting twist in the history of cancer research, Berenblum's concept of a two-stage initiation-promotion process of carcinogenesis, which he used to support the somatic mutation hypothesis, was first proposed by

Peyton Rous, a vigorous critic of that hypothesis (Friedewald and Rous 1944). Frustrated after his chicken virus research was ignored, Rous took up chemical carcinogenesis experiments from which he suggested the idea that tumors resulted from first initiation and then promotion of the transformation of cells. He suggested that the initial applications of coal tar on mouse skin caused irreversible changes in the cell populations that could then respond to subsequent applications of transformation "promoters." Berenblum and Shubik (1947, 1949) pursued this idea in England, and Berenblum's subsequent studies of croton oil suggested that the simple application of this chemical compound, in place of the complicated wounding and tarring of tissue, could accomplish the same purposes.

During this same period (the 1950s and 1960s), publications on the structure and functions of DNA generated much excitement about the possibilities of understanding cellular activities at the level of genes for all fields of biology, including cancer research. The work on the structure of DNA inspired many researchers to search for the molecular genetic causes of cancer. Some were interested in the processes by which chemicals or hormonal changes affected DNA. Others argued that radiation caused mutations in DNA. But researchers who wanted to formulate a theory about the mechanisms or common pathways of radiation carcinogenesis did not have enough evidence to do so. In the following passage, molecular biologist A. C. Braun argued that the commonalties among cancers—that, for instance, all lymphomas were "the same" even when they were caused by different agents—pointed to a common causal mechanism by which all causal agents work.

> One of the most puzzling aspects of the tumor problem is concerned with the multiplicity of diverse physical, chemical, and biological agencies that are capable of bringing about essentially the same end result. These agencies, with the possible exception of certain of the viruses, are concerned only with the inception of tumors and do not appear to be involved in the continued abnormal proliferation of the tumor cells once the conversion of normal cells to tumor cells has been accomplished. The implication of these findings appears to be that all the diverse agencies ultimately affect a common cellular mechanism(s) which, once deranged, urges the cells to continued abnormal proliferation . . . Malignancy often expresses itself in ways which are independent of causative agents. From a clinical and pathological point of view nothing distinguishes the lymphomas produced in mice by X rays, carcinogenic chemicals, or viruses. There is, moreover, reason to believe that in all cases the disease is the same with the same heritable alteration at the cellular level. If true, this would suggest that the underlying

> heritable change found in the several diversely initiated tumors ultimately converges at the same point in the cell. Where, then, could this hypothetical point be? (Braun 1969:2)

Writing in the 1970s, James Watson kept the book open on the somatic mutation theory of cancer causation. "At present there exists no direct evidence either for or against this theory, which we might best describe as cancer due to loss of an essential gene(s) function. Even without evidence, however, it is clear that somatic mutations must occur; it would be surprising if at least some did not disrupt the normal control of cell division" (Watson 1970:598).

Even into the 1970s, however, scientists argued that difficulties regarding the somatic mutation theory remained. In their view, the theory's major flaw was the evidence that many chemicals capable of causing the development of tumors in animals and transforming cells in culture were not mutagens. That is, many chemical compounds were carcinogens but did not cause genetic mutations.[46] They caused tumors to grow, but no genetic mutation could be discerned in tumor tissue.

Since Berenblum's (1952:159) early effort to explain similar difficult data from mutation experiments by the multistep process of carcinogenesis, the somatic mutation theory and the multistage concept of cancer first introduced by Friedewald and Rous have become more firmly intertwined.[47] The multistage process represented carcinogenesis as involving "at least two qualitatively different steps: *initiation,*[48] that appears to be a rapid and irreversible event, and *promotion,* a more protracted process that is at least partially reversible. In instances where repeated doses of a single chemical induce tumours, the substance is presumably acting as both *initiator* and *promoter* (thus as a *complete carcinogen).* Some agents act as promoters but appear to have little or no intrinsic initiating activity, whereas a few others have purely initiating activity" (Farmer and Walker 1985:138).

Support for the mutation theory and multistage process came in the form of many detailed studies of diverse chemical compounds as well as from a new concept called "ultimate" carcinogens. During the 1960s and 1970s researchers continued to test the idea that some chemicals could "initiate" tumorigenesis through one mutation, while other chemicals could "promote" tumorigenesis without necessarily mutating a gene. Experimental results showed that chemicals behaved differently in different types of cells and organisms, and at different time points of application. In the late 1960s, with the concept of "ultimate carcinogens," scientists argued that derivatives of original chemical carcinogens can become metabolically transformed into highly electrophilic forms (containing an electron-deficient center) that allow them to bond with similar sites in the nucleus of eukaryotic cells to mutate genes in a multistep process, thus allowing all carcinogens to eventually become mutagens (Farmer and Walker 1985, Miller 1970, Miller and Miller 1969, Miller et al. 1958). In addition, researchers extracted what they considered to be the active chemical agent (phorbol esters) in croton oil that created mutations when applied to the skin of experimental mice. Researchers also announced that they could re-create the two-step carcinogenesis process in other animal systems in vivo and in other cells in vitro.

Cairns expressed his faith in a mutational theory despite only circumstantial evidence.

> In the last few years more and more evidence has been accumulating to suggest that initiation is nothing other than mutation. Though admittedly circumstantial, it is very persuasive. Thus, the most powerful initiators prove to be the substances that are the best at binding to DNA. . . and causing mutations in various test systems. . . they produce changes in the DNA either directly, or indirectly after they have undergone certain chemical modifications that occur when they are metabolized in the body. . . It seems reasonable to assume that the first step in carcinogenesis would be the production of mutations, because this conforms to the one thing we do know about the cancer cell—namely, that in building up an expanding population of abnormal cells it must be passing on its cancerous characteristics to its descendants, and so these characteristics must in a sense be inheritable from one cell generation to the next. (Cairns 1978:93–94)

In the early 1970s biochemist Bruce Ames at the University of California, Berkeley, devised one of the most influential test systems for determining the mutagenic effects of carcinogens. What came to be known as the "Ames test" was used to determine whether chemicals (and other substances) caused mutations in mutant strains of *Salmonella* bacteria. According to many researchers, the results of the Ames test demonstrated that all (then designated) carcinogens were mutagens (Ames et al. 1973, McCann et al. 1975).[49] The test has been called an "ingenious method for detecting mutagenic chemicals using bacteria as sensitive indicators of damage to DNA" (Huggins 1979:6). Driven by desires on the part of scientists and federal regulatory agencies for simple, inexpensive, and measurable methods, the Ames method became a popular tool for studying carcinogenicity.

> The beauty of a test like this is that it costs about one-thousandth as much as a test using mice, and takes a day or two instead of at least a year. Roughly 90% of known carcinogens prove to be mutagenic to bacteria

> when activating enzymes are present—about as good a result can be obtained with any one animal species—and more than 90% of substances thought not to be carcinogenic are negative in the test. Because such microbiological assays are so inexpensive, it has now become practical for the major chemical manufacturers to investigate the potential mutagenicity of all the new compounds they wish to market, before making any major investment to produce them. Anything that comes out positive can then be withheld and tested in animals. (Cairns 1978:104)

However, after tests of many different chemicals for their mutagenic and carcinogenic capabilities, the conclusions were still ambiguous and highly debated. The Ames test has been the subject of many subsequent debates in the journal *Science*. In diplomatic terms, Huggins conceded that "the Ames procedure is an informative test of mutagenicity. It is a valuable procedure in cancer research but bacterial DNA lacks specificity for a categorical identification of a chemical carcinogen. All carcinogens cause mutations; not all mutagenic agents cause cancer" (Huggins 1979:7).

Outside the Walls of the Nuclei

Nevertheless, none of these studies could represent transformation at the molecular level. By the late 1970s, according to Cairns, the picture of tumors and cancer causation was more complex.

> A few underlying principles have emerged. For example, so many carcinogens have proved to be powerful mutagens that it seems almost certain that most forms of cancer are due, at least in part, to changes in DNA. But little or nothing is known about the functions of the genes that must be mutated, and there are enough oddities about the whole process of carcinogenesis to suggest that it is not a matter simply of piling one mutation upon another until the requisite number of genes are inactivated but probably involves something else as well. (Cairns 1978:90)

Although scientists had concluded that the heritable point might be the genes, they understood neither which genes were involved nor how they caused the development of tumors. Even accepting the initiation-promotion thesis, initiation and promotion each could require many stages. Furthermore, given the diversity in tissue responses to causal agents, Cairns concluded that if genes are involved in cancer causation, a variety of genes rather than a specific set of genes would cause cancer. He did not expect to find a common genetic pathway. (Bishop [1982] later proposed a common genetic pathway for all cancers. See Chapter 4.)

> The exact response to whole-body irradiation with penetrating X-rays is almost as variable from one species to another as the response to chemical carcinogenesis, even though it must be producing lesions in the DNA of every cell. So we are left to conclude that the routes by which cancer cells are created probably differ from one tissue to another. In other words we should not expect to find particular genes, or sets of genes, involved in all forms of carcinogenesis. (Cairns 1978:107)

Research on cancer causation by chemical compounds, hormones, and radiation in carcinogenesis, then, was primarily theoretical and biochemically descriptive in the postwar period up until the late 1970s. Epidemiological and experimental studies provided statistics associating cancers with many chemicals, with hormones, and with high levels of radiation.[50] Researchers concluded from their experimental studies that positive relationships existed between hormonal and radiation factors and cell transformation. But most experiments stopped outside the nuclear membrane.

Despite this inability to represent cancer in terms of specific causal pathways and mechanisms in the 1970s, Cairns still hoped that future research on cancer would benefit from molecular genetic technologies. "Despite our great ignorance about the basic biology of cancer, the study of carcinogens has recently been made much easier by the application of knowledge and techniques derived from bacterial molecular genetics" (Cairns 1978:90).

As it happened, another set of experiments that became significant in the development of proto-oncogene research accomplished what Cairns had hoped for. In the late 1970s and early 1980s, human DNA transfection methods were developed by Robert Weinberg's laboratory at the Whitehead Institute at the Massachusetts Institute of Technology, Michael Wigler's laboratory at the Cold Spring Harbor Laboratory, and Mariano Barbacid's laboratory at the National Cancer Institute in Washington, D.C.[51] These methods involved chemically transforming cells in culture into cancer cells, then taking DNA from malignant cells and transferring them to "normal" NIH 3T3 cells. The point was to demonstrate that the DNA in the original normal cells had been mutated into cancer genes and that these cancer genes alone, without other cellular components, could transform other normal cells into cancer cells. These DNA transfection techniques and experiments were praised for demonstrating the specific transforming capabilities of specific strands of DNA and moved the field of molecular genetic, in addition to viral, cancer research into the limelight. By the late 1970s and early 1980s, research lines from genetics and virology had together created tools with which researchers generated molecular genetic experiments on cancer and an entirely new field.

The Medium Is the Message

I have presented this brief history of several active lines of research into the genetics of cancer in the United States from the 1920s through the 1970s to set the stage for understanding the differences and similarities, the continuities and discontinuities, between earlier research on cancer causation and the new proto-oncogene research. As this history demonstrates, until the mid-1970s, researchers had no technologies for experimenting with theories of cancer in eukaryotic organisms at the molecular level. Until 1976 basic cancer research consisted of many theoretically and organizationally different lines, most of which were far from an empirical description of the molecular genetic mechanisms of cell transformation. The molecular biology of cancer was limited to studies of tumor viruses and viral oncogenes and the protein chemistry of various proteins detected in transformed cells.

Nevertheless, much of the cancer research of the twentieth century has focused on the idea of a "factor" existing within the cell that is critically involved in producing tumors. This factor, assumed to be residing somewhere in the cell, has been the "Holy Grail" of basic cancer research since the beginning of the century, just as the search for the magic bullet has been the Holy Grail for clinical research on cancer. As this chapter demonstrates, the search for this factor has taken different forms in conjunction with the skills and tools developed by scientists during different historical periods. Early in the century, this causal factor took on the name "chromosome." Later, it became the "gene," and research then shifted to questions of greater specificity: What exactly was the materiality of genes, and what exactly were the genetic mechanisms?

Between 1920 and 1976, several standardized experimental systems produced contenders for the Holy Grail. My genealogies in this chapter trace current research on cancer to several experimental systems. These experimental systems and the scientific concepts and the theories related to them were constructed in a highly intertwined fashion. It is impossible to understand the theories, concepts, and facts apart from the experimental systems used to bring them into being. Various genetic themes were linked with tumor transplantation studies and mouse colonies, tissue culture materials and methods, and tumor viruses. For example, cancer became experimentally defined as cell transformation in vitro with the construction of cell culture methods and materials. I first recognized this link during my course work in a tumor biology course. Why, I wondered, was one-fourth to one-third of the course devoted to cell culture methods and history? Ours was not a laboratory course. I soon realized that our instructor wanted to emphasize that biological knowledge about cancer is based on results of experiments carried out on and with cell and tissue cultures; that our knowledge about cancer is knowledge about the transformation of cells in vitro. He simultaneously introduced us to experimental methods as he contextualized what we were learning about cancer within the framework of those experimental methods. The instructor was teaching us to understand that our knowledge of natural phenomena is created through highly specified practices. Therefore, he emphasized the means—the materials, practices, ideas, and money (as one scientist names it)—through which this knowledge is produced. In other words, the medium is the message.

This chapter provides an analysis of the heterogeneous practices involved in creating scientific representations and objects and of the heterogeneous representations and objects involved in creating scientific practices. Biological representations and objects are the products of collective (and sometimes conflicting) arrangements that change over time. "Gene," for example, is a concept that has had different meanings depending on its situated practice and use. Currently, a new set of commitments and practices have created new technical definitions of genes and genetics in molecular terms. "Gene," then, is not an abstract concept when situated in the work of biologists, their research tools, the purposes and themes that framed their work, and the uses to which their work is applied. It is important to note that classical genetic ideas such as the somatic mutation theory of the early 1900s are not the same when applied to today's molecular genetics. Cancer theories have moved from the transmission of characteristics without a physical definition of the transmitting factors to detailed designations of pathways and actions of specific entities called DNA, RNA, amino acids, protein receptors, tumor suppressor genes, and much more.

The genealogies in this chapter show that technologies and their experimental systems do not necessarily stay with one theory, and vice versa. Elements of an experimental system can have more than one incarnation. In fact, histories of science are full of migration stories, of roving technologies, theories, and scientists. In this chapter, we have seen several examples of tools "migrating" to other problems and other solutions, just as ideas migrate to new problems and tools. For example, inbred mice colonies constructed for experimentation on the problem of the genetic transmission of mouse mammary tumors produced a mouse mammary tumor *virus* that apparently caused cancer in infants as it was passed from mother to infant through suckling. In other words, this inbred mouse technology, created to solve a genetic research problem, produced a new problem and a new line of research on viruses. In the 1960s this virus and other so-called tumor-causing viruses became the objects of a major research program on viral carcinogenesis. In the 1970s and 1980s, these same viruses became the technologies for exploring a new cause of cancer, the proto-oncogenes in normal cells.

Knowledges and technologies may also change as they move into new experimental systems and new problem domains. Thus, continuity of knowledges and technologies as they move between local situations cannot be assumed. Continuity is an empirical problem.

That a single tool can be used to create different theories and that a single theory can be used to create several different experimental systems mean that the nature-science-technology system is quite heterogeneous and inventive. Materials and technologies can at one time, in one place, be a tool for solving a research problem and at another time or place become the problem itself, as we saw in the case of cancer viruses. They can be objects acted upon by scientists, and agents acting upon science. The message becomes the medium becomes the message.

This chapter frames a history of oncogenic science as a history of experimental practices, material technologies, and theoretical and conceptual creations. Scientists employ experimental technologies and protocols to create novel scientific and technological objects. Inbred animals, tissue culture materials and methods, and tumor viruses as standardized experimental systems all embodied (gave materiality to) "genes" and "cancer." They are technosymbolic systems that reinvent nature.[52]

Situating new scientific products with(in) the experimental systems used in their development helps to illuminate the decisions researchers made about what to create, which possibilities to make realities, in their decisions of which experiments to pursue. There are no inevitabilities in scientific organization and practices, as there are no inevitabilities in social organizations and practices. Moreover, conscious intention seldom produces its desired outcomes. Thus, the move from classical genetics to the biochemistry of cells in culture to genetic manipulation and the move from animal studies to inbred animal studies to tissue culture to viruses to genes were not inevitable, entirely conscious, nor entirely contingent outcomes.

Scientific technologies then are highly elaborated symbolic systems, not neutral media for "knowing" nature. My argument here is that what we come to know as "nature" is accomplished through historical processes that are not to be taken for granted as the only ways in which nature could be produced. However, once nature is produced in particular forms, these are then assumed to be the only forms in which nature could exist. That is, these historically and socially produced forms become "naturalized." In nineteenth- and twentieth-century scientific research, standardized experimental analytic systems have created the "nature" we "know."[53]

I have told this story of the genetics of cancer research as several brief linear histories. However, the reader should note that I wrote these histories using materials written by the scientists and using histories recounted to me by practicing scientists. The narratives I have constructed are the lore of the field as told by the practicing scientists. They should be taken as just that. Scientists in current controversies construct and employ histories of medicine, technology, and science to support their arguments or to deconstruct opponents' arguments (Fujimura and Chou 1994). This is more than a debating strategy. Constructing history is one means by which scientists (re)construct rules for verifying facts and findings; constructing history is part of the verification process in science. This observation applies both to their stories and to mine.[54]

3 Molecular Genetic Technologies

The New Tools of the Trade

> *As a discipline, molecular biology is no longer a mere subspecialty of biology or biochemistry: it is the* new *biology.*
>
> —*Ajit Kumar,* Eukaryotic Gene Expression

By the early 1980s the molecularization of biology was well under way.

> Then a series of technical discoveries came together that have completely altered the landscape of possibility in molecular biology. What became practicable, for the first time, was the controlled manipulation of pieces of genetic material. Researchers could snip out sections of DNA from one organism and transfer them to another. One could cut genes out of one cell and splice them into another. The technical term for this activity is "recombinant DNA research" because what is involved is the controlled "recombination" of sections of DNA, the hereditary material . . . The implications of this for many areas of research were staggering. For example, one could now think about splicing new bits of DNA into tumor viruses, or taking them out with precision, to see how that modified their impact on the cells they infected. (Yoxen 1983:30–31)
>
> The recent surge of interest in recombinant DNA research is understandable considering that biologists from all disciplines, using recently developed molecular techniques, can now study with great precision the structure and regulation of specific genes . . . Current approaches to the outstanding problems in virtually all the traditional disciplines in biology are now being explored using the recombinant DNA technology. (Kumar 1984:ix)[1]

Contemporary biological research is even more markedly molecular. The subdisciplines of embryology, immunology, cellular biology, and evolutionary biology have incorporated molecular genetic techniques as experimental tools.[2] While high school students carry out standard molecular genetic procedures using recombinant DNA tool kits, research biologists employ similar tools to explore the molecular bases of their experimental systems.

As Chapter 2 demonstrated, until the mid-1970s, researchers did not have technologies for experimenting with theories of cancer in eukaryotic organisms at the molecular level, and basic cancer research consisted of many theoretically and organizationally different lines, most of which were far from an empirical description of the molecular genetic mechanisms of cell transformation. The "molecular biology of cancer" during this transitional period was focused on traditional virological and protein chemical studies of tumor viruses, viral oncogenes, and transformed cells.

Recombinant DNA and other novel molecular genetic technologies changed all that. In the early 1980s, many new and established biologists and cancer researchers incorporated the new genetic technologies into their laboratories. Many viewed the new molecular genetic technologies, especially recombinant DNA techniques, as "productive tools" that "opened up" new areas of study and generated novel results. Graduate and postdoctoral students and technicians said that recombinant DNA technologies were the latest, hottest techniques and would make them more "marketable" in their job searches. In the 1980s, post hoc analyses by scientists, science writers, and science business analysts pointed to recombinant DNA technologies as having reaped high returns in terms of new knowledge, techniques, publications, and academic career boosts, along with applied products, glimpses of possible new cancer therapies, and financial profits.[3] Whatever the reasons and rationales, by the mid-1980s contemporary basic cancer research was extensively molecular, with proto-oncogene research as its most prominent line of research.

In Chapters 3 and 4, I present an account of how and why researchers came to view molecular genetic technologies as potentially useful for studying cancer and other long-standing biological problems. In that account, I also discuss how these tools, in practice, were used to link problems and solutions in different worlds involved in constructing science and to link different kinds of scientific activities (detailed experiments, organizing laboratory work, interacting with various audiences). Different interpretations, representations, and reorganization of local practices were used to create linkages among researchers, laboratories, research institutions, and even biotechnology companies and diagnostic clinics. Cancer and cancer research worlds have been reconstructed through these technologies and the proto-oncogene theory.

These linkages were achieved *in part* through molecular genetic techniques, and more specifically recombinant DNA techniques, the topic of this chapter. Recombinant DNA technologies produced new substantive phenomena in scientific investigations, some of which have since been used to design potentially useful new therapies and "beneficial" organisms, which have in turn produced concerns about their potentially dangerous effects on the environment and human evolution. The technologies make possible genetic manipulations—for example, previously unfathomable crossbreeding experiments—that could not have been done using older techniques.

Although there are many technologies used in molecular genetic research, I focus on recombinant DNA techniques, because the molecular biologists I studied viewed them as the keys to eukaryotic cell genes and human cancer. Rather than simply accept and reproduce this view, however, I look at how and why these techniques came to be keys to research on human cancers. To follow this development, I examine the constitution of a few techniques and materials used in recombinant DNA technologies during this beginning period, since neither space nor time permits me to describe all the technologies used in their research.

In Chapter 2, I discussed Clarence Little's inbred animal colonies as the *standard* experimental system for studying cancer genetics. These animals became the "right tool" for screening out the "noise" that detracted from the genetic factors of specific interest. Kenneth DeOme, beginning his work in the 1930s, described them as the "proper animals" for the job. "The necessity for inbred animals was imminently obvious, so obvious that there was no question about it. To do cancer research, then said everyone, we just can't do it without having the proper animals" (DeOme interview 1984). But how did it become "imminently obvious" that inbred animals were the proper tool to do cancer research, even to study cancer genetics? How does a technology become designated as the "right tool for the job"?

I ask these questions again in this chapter. How did recombinant DNA technologies become the tools of the trade of biology, in this case cancer biology, in the 1970s and 1980s? How do scientists come to agreements about how to study nature? From my perspective "tools," "jobs," and the "rightness" of the tools for the jobs are each situationally and interactionally constructed.

Writing the History of Recombinant DNA Technologies

Scientists and historians of science often recount the efficacy of particular techniques for investigating particular problems in a post hoc fashion. In contrast, I examine the processes through which recombinant DNA technologies became *the* tools of the trade. I present two accounts of the development of recombinant DNA techniques. The first is a short history reconstructed from scientists' accounts and secondary sources written by historians of science and other observers. This history emphasizes the substantive capabilities of these tools—that is, what they could actually *do* to aid molecular biologists' efforts—as narrated by scientists, historians, commercial entrepreneurs, and science journalists.

The second history reinterprets this first one to examine the shaping, and commodification, of recombinant DNA technologies as standardized practices. I look at the part that human perceptions, desires, translations, and designs played in the scientific "discoveries" of "natural facts." I write my history of the design of recombinant DNA techniques and practices in the terms of collective work: "enrollment" and "black-boxing." I explore how molecular biologists transformed a set of artful and complex activities involved in manipulating DNA, RNA, and proteins into a set of routine tools and standard practices. By the 1980s, despite claims that recombinant DNA technologies were state-of-the-art methods, these technologies had been transformed into relatively routine and standard protocols. I do not mean to say that everyone in biology could perform these technical operations. But in contrast to situations where experimentation and customizing are still a significant part of daily work, recombinant DNA techniques were procedurally established and relatively routine by the early 1980s such that nonrecombinant technologists could learn to perform the procedures. To use a computer science analogy, they did not need to be hackers to use the programs. While many bells and whistles have been added since then, many of the basic operations were in place.

Rather than credit the transformation of recombinant DNA technologies into standard protocols only to technical improvements, I propose that the transformation also resulted from the gradual building of a network of laboratories, private enterprises, government funding agencies, practitioners (scientists, technicians, students), and material products. By adopting these tools, researchers and research laboratories have transformed their practices and their skills into molecular technologies. By funding and producing these tools, federal and state funding agencies and private industry have become part of a network of support for molecular genetic technologies. Together they have transformed a novel technique into accessible and productive tools that have in turn transformed the practices, problems, representations, and working tools of other biological subdisciplines to produce "the new [molecular] biology." In other words, "Over the next several years, the authenticity of *myc* would be ascertained by molecular cloning, nucleotide sequencing, and gene-transfer. To this day, the gene has never been defined by the strategies of classical genetics, and there is now no need to do so. The new biology is upon us" (Bishop 1989:10). Through the transformation of diverse research worlds, these technologies have been further standardized. Thus, recombinant DNA and other molecular genetic technologies have also reconstructed practices and representations in other worlds, which in turn augmented their own standardization.

In my terminology, a technology is standardized when it becomes collective conventional action, through either explicit, tacit, or coerced agreement. These practical, institutional, "sentimental," and financial commitments became solidified in the experimental protocols, research materials, and automated instruments of molecular genetic technologies.

It might appear to the reader that the development of recombinant DNA technologies was an inevitable and smoothly moving process, as if there were no other options or alternative paths, and as if the development and standardization of recombinant DNA technologies were necessary outcomes in the history of biology. This is *not* the conclusion of this chapter or this book. As I argue below, the process of standardizing technologies is a process of establishing a hegemony (Gramsci 1985) over other ways of knowing. As Steve Shapin and Simon Schaffer have so well demonstrated in their history of seventeenth-century British "science," the establishment of "scientific experimentation" as *the* means of verifying truths and facts about the natural world was the constitution of scientific authority over others of that time period, especially the natural philosophy of Thomas Hobbes and his colleagues. That franchise of scientific authority concurrently established members of the Royal Society as professional scientists and the only witnesses authorized to testify to the outcomes of experiments.

By a similar process standardized recombinant DNA technologies became the authorized technologies for fact-making. This enfranchisement was a form of hegemony in and of itself, but it also was embedded in the long-established hegemonies of scientific experimentation and more specifically genetic explanation for biological behaviors. Recombinant DNA and molecular genetic technologies are "standardized" only from particular perspectives, for example, from the perspective of those who employ and promote them as reliable tools for "discovering" new facts about nature and for producing new ways of intervening in nature.

Just as the experimental inbred animals and artificially intelligent computer and robotic systems are humanly created symbolic systems, so too are recombinant DNA technologies technosymbolic systems created by humans. That is, scientists "create" nature within their human contrived frameworks just as they create "intelligence" in computers and "identity" in inbred mice. These scientific and technological objects *are* nature, according to scientific prescripts.

However, there is one point that sets molecular biological and recombinant DNA technology research apart from previous biological research. Molecular genetic technologies can produce novel "natural" phenomena, including recombinant genes, proteins, and organisms. Previous generations of biologists have also constructed novel research materials as "laboratories" for housing experiments. These materials include inbred strains of animals, "nude" mice (with no thymus), tissue culture for experiments in petri dishes, and other research materials that cannot survive outside contrived laboratory conditions. Similarly, agricultural research has produced new "recombinants" or hybrids through crossbreeding strains and sometimes even species.[4] The difference here is that molecular genetics has produced a set of technologies that has changed the level and degree of the experimental production of nature. Molecular geneticists can create recombinants that could not have been produced through what are now perceived as "clumsy" crossbreeding methods and that can survive outside experimental laboratory conditions. They have radically shifted the world from one of genealogies in plant and animal breeding, where parents combine to produce offspring, to a world where DNA and RNA are reorganized, transferred from one organism to another, or deleted to produce novel "offspring." Thus, the molecular genetic production of nature is historically novel and creative of a new *ecology of action.*

Historical Accounts of the Development of Recombinant DNA Technologies

According to all accounts—scientific, historical, and sociological—recombinant DNA technologies provided methods for manipulating eukaryotic cell DNA.[5] Before the early 1970s molecular biologists had focused their research on prokaryotes (simple organisms like bacteria, viruses, and algae, whose cells have no defined nuclei). The new techniques permitted research on eukaryotes. Eukaryotic organisms have cells with DNA enclosed in structurally discrete nuclei and are often called "higher" organisms because their structure is considered to be more complex than those of prokaryotes. Since the cancers of interest to most researchers are in humans with complex eukaryotic DNA, techniques for recombining DNA pointed to a way to study human cancer in molecular terms. In the mid-1970s recombinant DNA technologies became the keys to eukaryotic cell genes and human cancer.[6]

In 1973 molecular biologists in Stanley Cohen's (Stanford University) and Herbert Boyer's (University of California, San Francisco) laboratories artificially recombined the DNA of two different species, a prokaryote and a eukaryote. More specifically, they recombined the DNA of a toad with that of bacterial *Escherichia coli* (Morrow et al. 1974). For practicing scientists, the Cohen-Boyer method meant that researchers could isolate and reproduce complex eukaryotic gene sequences, or pieces of sequences, en masse for study in a process called "molecular cloning."

Historian Susan Wright's (1986:315) account of recombinant DNA technologies attends to the goal, from the very beginning, of developing the commercial potentials of the technologies. She argues that "the impact of this new capacity to move pieces of DNA between species at will was immediately understood by molecular biologists in the dual scientific and practical terms that characterized the perceptions of its inventors. On the one hand, they saw the ability to do controlled genetic engineering experiments as a powerful means to open up new lines of inquiry into the structure and function of DNA . . . On the other hand, industrial applications were also anticipated." Industrial applications included the possibility that bacteria could be genetically reprogrammed to synthesize proteins normally made by higher organisms, commercially valuable proteins like insulin for treating diabetes and other pharmaceuticals. Scientists also expressed hopes for the future genetic engineering of plants and for gene therapy and vaccines for human diseases. Historian Horace Judson argues, contrary to Wright, that these practicing scientists primarily sought to use the techniques to study complex, previously confounding problems in biology. "The technology of recombinant DNA was invented and is today being elaborated not primarily to make money, nor even to cure disease, but to open up that profound and heretofore intractable problem of development and differentiation—to find genes in higher organisms and learn what they do" (Judson 1992:63).[7]

These post hoc renderings of intentions early on in the development of recombinant DNA technologies are impossible to confirm or refute. Whichever historical narrative we accept, it is clear that researchers' rhetoric about the potential of the new technologies in terms of both solutions to long-standing biological problems and applications was used to enroll allies among other researchers and funding sources. However, both narratives argue that, to accomplish either task, molecular biologists needed techniques besides recombination before they could easily manipulate DNA to do controlled genetic engineering experiments and certainly before it could be applied to other biological problems. These techniques included procedures for sequencing, mapping, and expressing genes.[8]

In 1977 molecular biologists announced that they had developed techniques that allowed them to insert a human gene into and among bacterial DNA, and then to grow the bacteria to produce more copies of the isolated foreign piece of DNA in order to analyze it (Itakura et al. 1977). The gene encoded a hormone protein, somatostatin, produced in the human brain.[9] By 1977 they had also developed faster and more efficient sequencing techniques, which they used to delineate the location of genes on the genome and later to map out the structure of entire genomes, or at least of smaller genomes. Finally, molecular biologists also developed new methods in their attempts to make the recombined eukaryotic genes, especially mammalian genes, express themselves in bacterial systems. Gene expression is the transformation of the genetic code into the proteins that make the cell function. In 1978 the first chimeric hybrid expressed, or produced, the protein somatostatin (Wright 1986:329). New gene transfer techniques—introducing foreign DNA sequences into cells in culture—allowed molecular biologists to study the functions of genes and gene fragments. "The most commonly used methods of gene transfer are cell fusion, DNA transfection, liposomes, and manual injection [or microinjection]" into somatic cells (Soprano 1984:4).[10] By 1978, molecular genetic technologies included gene cloning, sequencing, and transfer techniques for studying DNA structure and function as well as the location of genes on the DNA.

Molecular biologists, chemists, and virologists participated in the design, construction, and marketing of recombinant DNA technologies in the 1970s. Their targeted audiences included everyone from venture capitalists and multinational corporations to government sponsors and individuals suffering from genetically inherited diseases and cancer. Wright (1986) analyzed the advertising and promotion of recombinant DNA technologies by their producers that occurred before the successful construction of genetic recombinants and years before (if ever) the actual production of genetically engineered health care goods, food and agricultural products, oil spill–gobbling bacteria, and other commercially viable products. According to Wright, molecular biologists coined the term "genetic engineering" and were speaking of its practical applications and its ethical and social implications as early as the 1960s.

Yet, as late as 1985 most research and development biotechnology companies that concentrated on human health care still had few products on the market. Genentech, Cetus, and Biogen were the three largest "start-up" commercial biotechnology research and development companies. These companies were often called "genetic boutiques" or genetic engineering firms, as compared with the larger, established, and largely diversified pharmaceutical and chemical companies also investing in biotechnology research.

Furthermore, the few existing products were no more effective and were more expensive than those produced using pre–recombinant DNA technologies. For example, Genentech researchers published a bioengineering technique for producing a recombinant insulin in January 1979. But by 1986 their genetically engineered product, "Humulin," was still more expensive

($12 a bottle) than insulin prepared by extraction from pig and cow pancreas glands ($8 a bottle) (Gussak 1986). More important, biosynthetic human insulin had the much-heralded advantage of eliminating the side effects caused by the body's immune response to foreign injected insulin. Nevertheless, treatment with biosynthetic human insulin still caused an immune response in patients (Gussak 1986). In another instance, the campaign promoting interferon as a cure for cancer begun in 1971 by Mathilde Krim (a powerful lobbyist for the "cancer crusade"), select interferonologists, and the American Cancer Society also lent fuel to recombinant DNA technological and industrial development (Panem 1984; Yoxen 1983). However, although interferon continues to be used to treat a few cancers and researchers have conducted studies to assess its effectiveness in treating viral infections, it proved ineffective as the promised cancer cure. The designers of recombinant DNA technologies promoted their research by promising *potential* applied products that would increase the welfare of humankind, would be cheaper to produce, and would provide higher yields than the products then available. Their rhetorical strategies attracted supporters from both government and industry sponsors despite, and perhaps because of, the embryonic stage of its development.

Venture capitalists, multinational chemical and pharmaceutical companies, government sponsors, and the members of the general public had varied responses to the promotional campaign (Kenney 1986, Krimsky 1982, Wright 1986). Although both scientists and the lay public worried about the possible consequences of introducing the new technology, no one questioned the grand promises for both basic scientific and applied results.[11] Multinational corporations were the slowest to respond to the rhetoric and displayed the most caution. They made early investments simply as protective measures in case the new technologies did pay off.[12]

In their laboratories, molecular biologists swiftly employed the new techniques to examine the complexities of eukaryotic cell DNA, again for both scientific and applied goals. They constructed new theories about the DNA of "higher" organisms that extended and sometimes conflicted with earlier results from studies of prokaryotic cell DNA. Many molecular biologists also chose to pursue their research projects in small commercial biotechnology research and development start-up companies. Some opted to create commercial enterprises, while others chose to work for such companies as scientists, advisers, or consultants.

Despite my analytic distinction between basic and applied science here, it must be understood that the boundaries between these two categories in biological (and other scientific) research are fuzzy.[13] The division between basic and applied research even in the universities is blurry in contemporary biological research. In the 1980s and 1990s, universities have applied for patent rights on biological innovations and creations, many academic researchers are simultaneously biotechnological entrepreneurs and university faculty, and new contractual university-commercial liaisons have been forged in biotechnology research.[14] As early as 1981, a biological researcher at the University of California, Berkeley, reported that when approached by an investment firm to help build a biotechnology start-up company, he was the first "virgin" researcher they had found; the previous twenty had already established ties with commercial enterprises.[15] Since then, the relationships among universities, biotechnology companies, and multinational corporations have changed through the years to redefine the nature of scientific problems pursued in university biology departments.

Standard Tools and Routine Procedures

Methodologists still fiddle with recombinant DNA technologies and attempt new ways of doing the same thing or develop new ways to do new things. Many molecular biologists have built careers by successfully developing faster or novel techniques. New techniques are marketable commodities in the quest for positions, research funds, laboratories, and students. Indeed, the human genome projects in the United States, Japan, and Europe have primarily focused on the creation of new technologies to speed up and amplify the production rates of sequencing and mapping the genomes of human and "model" organisms.[16]

Standardized tools, by contrast, are relatively stabilized ways of acting that translate the messy biology of daily bench work into codes that are understandable in other laboratories and domains. These standard tools simultaneously reduce the uncertainties of action at the lab bench as they reduce the parameters by which "nature" is described. Scientists often assume that they will use a particular tool or technique in a particular context to accomplish a particular task. By the time tools are standardized (through Herculean feats), the number of unknown factors researchers must consider while conducting their daily experimental tasks and the number of "moments" of question, articulation, and discretion have been reduced. Researchers use these tools in a taken-for-granted manner in their work—that is, they regard the tools as relatively unproblematic—and concentrate on solving new problems. Their focus is on making the problematic unproblematic, rather than vice versa. In contrast to luck or serendipity and intensive trial and error, researchers rely on these tools for the *routine* production of data.

However, the performance of standardized techniques in the routine production of data is not necessarily easy. As any molecular biologist will report, even the simplest technique often does not work and requires more time than expected. Procedures can and do fail to yield expected results for many reasons—some never known. The techniques often are a set of sequential steps, each of which is required to work before the next step can be done. If one step does not work, researchers must repeat it until it does work. Cloning a gene usually takes six months, but it can take up to a year or more. Researchers must be able to recognize indicators and artifacts specific to these techniques and materials. Some genes, especially large ones, are more difficult to clone than others. Sometimes innovative adjustments are required. And that applies to the experienced hands. Novices, in contrast, learn to reproduce or re-perform standardized techniques such as "plasmid preps" (see Jordan and Lynch 1992), cloning of DNA, and sequencing, often through painstaking processes. Each student or technician of a laboratory creates her or his own way of making the procedures work. The "neat clean engine" of recombinant DNA technologies can also appear to be quite messy "on the ground," as we shall see in Chapter 7.[17] Yet, it must be noted that with no or little guidance other than published manuals, novices can teach themselves how to perform these techniques in ways that work.

Despite the successful recombination of genes of different species of organisms in August 1973 (Morrow et al. 1974), several difficulties still limited the basic techniques to the exploratory, cutting-edge research subject of methodologists until 1976. Although the rhetoric dating back to this time promised amazing capabilities of recombinant DNA technologies, technologists did not manage to produce standardized tools for "export" until sometime between 1980 and 1981, as marked by the first course entitled "Molecular Cloning and Eukaryotic Genes" held at the Cold Spring Harbor Laboratory's annual meeting in 1981.[18] Until 1980 recombinant DNA technologies were the research objects of high-tech methodologists, rather than the applied methods of biological researchers. It was not until 1980 that recombinant DNA technologies were available to nonmolecular methodologists in user-friendly terms in which enough "bugs" had been worked out.

Translating "Nature" into Routine Technology

In contrast to the views of historians presented above, my analysis of standardization has more in common with the concept of a black box of "irreversibly linked elements." Sociologists of science Michel Callon and Bruno Latour (1981:293) have proposed that the strength or durability of actors is based in the number of elements and relations among elements that they can enlist and bind together. "What makes the sovereign formidable and the contract solemn are the palace from which he speaks, the well-equipped armies that surround him, the scribes and recording equipment that serve him" (1981:284). According to Callon and Latour (1981:292–293),

> An actor . . . becomes stronger to the extent that he or she can firmly associate a large number of elements—and, of course, dissociate as speedily as possible elements enrolled by other actors. Strength thus resides in the power to break off and to bind together. More generally, strength is *inter*vention, *inter*ruption, *inter*pretation and *inter*est . . . All this is still "the war of all against all." Who will *win in the end?* The one who is able to stabilize a particular state of power relations by associating the largest number of irreversibly linked elements.

Enlisted elements can include bodies, materials, discourses, techniques, sentiments, laws, and organizations. For Callon and Latour, actors can be microbes, electrons, researchers, government ministries, companies, and more. Within a strong network, nonhuman entities can also "act" to change the world.[19]

Recombinant DNA technologies similarly transformed scientific work organization and biological practices through their linking and placement of elements. Recombinant DNA technologies include sets of clearly delineated techniques, carefully prepared and highly specific materials, and instruments that automatically implement a set of tasks. Each technique is organized as a set of tasks packaged into a series of preformatted steps or protocols. Some tasks are packaged more tightly than others; the series of steps are implemented automatically by instruments, or are already implemented in the form of "purified" and purposefully constructed materials.

In the late 1970s and 1980s, biologists used many different kinds of materials in recombinant DNA research. Some of the more obvious materials included restriction enzymes, plasmids, DNA probes, herring sperm DNA, reverse transcriptase, long-passaged cell lines, antibiotics, many kinds of chemical reagents (ethidium bromide, ammonium acetate, and cesium chloride to name a few), agarose and polyacrylamide gels, and more.

Restriction enzymes are a basic tool in recombinant DNA technologies. Enzymes are proteins that catalyze biological reactions. Researchers reported finding at least one restriction enzyme in each bacterial species. According to molecular biologists, each restriction enzyme recognizes and degrades DNA from foreign organisms, and thus "restricts" the entry of, and invasion by, foreign DNA. Scientists theorized that restriction enzymes evolved to preserve the bacterium's genetic integrity or purity. Whatever its role in a bacterial system, each enzyme degrades (or cleaves) the DNA of foreign organisms (without degrading its own, or "self," DNA) at highly specific points on the DNA molecular chain.[20]

Molecular biologists enlisted restriction enzymes in their problem-solving efforts for their DNA-cutting property. They used the enzymes to chop long strands of DNA into smaller fragments that can then be more easily studied. The human genome, for example, is very long and complex. It contains an estimated 3 billion nucleotide base pairs.[21] Even the viral genome is long; for example, the DNA of the SV40 monkey tumor virus consists of 5,243 nucleotide base pairs. Cutting the genome into smaller pieces allows scientists to characterize and manipulate DNA sequences more easily. More important, in recombinant DNA experiments, restriction enzymes are used to cut DNA from different organisms for subsequent recombination and cloning.

My simplified explanation masks the complexity of this material technology. To enroll the restriction enzymes in their efforts, molecular biologists sought them out and then transformed each restriction enzyme into a new entity. They spent at least ten years seeking out enzymes that cut DNA at specific points. Using lessons from enzymology, a field of research with its own distinctive history of at least one hundred and fifty years, they reconstructed restriction enzymes to suit their purposes. "Failures" in these efforts tell us more about how this construction process worked. For instance, molecular biologists worked with several enzymes that they eventually decided were less useful for their purposes because the enzymes cut DNA at random, rather than specific, points.[22] Here failure was defined in terms of the scientists' specific purposes (and not the bacteria). According to Maxine Singer, head of the Biochemistry Laboratory at the National Cancer Institute in Bethesda and an important figure in the history of recombinant DNA research, "by itself the DNA revolution was insufficient to permit detailed and designed manipulation of genetic systems . . . The enzyme revolution is as central to the successful manipulation of biological systems as the DNA revolution . . . Chemistry itself has not yet provided techniques [as does enzymology] for the precise manipulation of macromolecules" (quoted in Cherfas 1982:74). Jeremy Cherfas (1982:74), in his overview of genetic engineering, emphasizes that "the key is the ability to go beyond simply interpreting and understanding what one sees and to interfere purposefully. Without enzymes, none of that interference would be possible."[23]

Cutting and recombining DNA derived from the goal of delineating the exact nucleotide sequences of a gene, of an entire chromosome, and of an entire genome. This goal in turn was framed by decades of cumulative experimental and theoretical work on the nature (defined by biologists as the structure and function) of genes, the hereditary material.[24] The views of

DNA as the constituent of genes, of the organization of DNA, and of molecular mechanisms of heredity contributed to the creation of recombinant DNA technologies, and specifically my example of the appropriation of restriction enzymes.

Molecular biologists, then, have appropriated bacterial restriction enzymes for their particular purposes—for manipulating DNA. They use the enzymes and their activities of cleaving and annealing, severing and rejoining, which are believed to be a tool for the bacteria's genetic survival, as a tool for the recombinant DNA technologist. This enrollment was achieved through the cumulative and collective efforts of many researchers in several lines of research over many years. Common assumptions about theoretical representations of phenomena and shared conventions of representation and use were collectively negotiated and produced. The enzymes were reconstructed within scientists' contexts of representation and contexts of use, that is, through the activities and goals of restriction-mapping, gene cloning, and sequencing. Each restriction enzyme, then, comes to contemporary users as a black box embodying the cumulative work of many scientists on the enzymes' activities. Each enzyme is ready-made for immediate use, for specific purposes and under specified conditions.[25] Maxine Singer best describes this reconstruction. She states that molecular genetic engineers manipulate restriction enzymes in routine, taken-for-granted fashions until they are forced to make them themselves.

> [Biologists] will not usually concern themselves with the remarkable nature of the enzymes used to manipulate and construct precise DNA molecules. Most often those enzymes are perceived only as tools and are confronted in practice as a rather disorganized array of odd shaped tiny tubes inside a freezer. The sight inside the freezer carries no reminder that the fruits of a revolution are at hand. Nor will the investigator be reminded by the means he uses to acquire the enzymes. Some of them are now as easy to come by all over the world as a bottle of Coca-Cola—only more expensive. Others may be obtained by persuasive begging and borrowing, and, in a tight squeeze, even stealing. But when all else fails, and it is necessary actually to prepare one of the enzymes, the impact of the revolution may be sensed. Delight and amazement inevitably accompany the emergence of a clean, exquisitely specific reagent from the equally inevitably messy beginnings. Dismay coupled with awe of successful predecessors accompany the frequent frustrating failures. (Cherfas 1982:64)[26]

Having restriction enzymes in hand was not enough. Molecular biologists had then to isolate, purify, and characterize them in order to use them for scientific purposes. Characterization involves describing the exact sequences where each restriction enzyme cuts DNA. Each of these processes entailed (and some still remain) time-consuming, detailed, taxing work.

Restriction Enzymes in Cloning Procedures

Once reconstructed, isolated, purified, and characterized, restriction enzymes became a basic unit of the technologies for separating and recombining DNA of different origins. For example, restriction enzymes were crucial for isolating and cloning genes. Cloning is a recombinant DNA procedure by which multiple copies of a single gene or a fragment of DNA are produced. In the early 1980s, cloning through the recombination of the gene of one organism with DNA from another organism was a novel process. Researchers selected a genome or a larger segment of DNA in which they suspected a gene or DNA fragment of interest was located. They "cut" the genome or DNA fragment into smaller fragments with restriction enzymes and then "stuck" these pieces of DNA into plasmids. If an entire genome was cut up and stuck into plasmids, the result was what is called a "gene bank," a "gene library," or a "genomic library," depending on local terminology. The process of cutting up an entire genome or a large part of a genome was called "shotgunning" or "blasting" in molecular biological terminology. Shotgunning on the DNA of any organism produced the gene library.

Sticking pieces of DNA into plasmids led to use of the word "recombination." According to descriptions of recombinant DNA cloning techniques, DNA segments were randomly combined with a transmission agent called a "DNA vector." DNA vectors are self-replicating DNA molecules, usually bacterial plasmids, which are also cut by restriction enzymes.[27] The name "plasmid" was given to a circular, double-stranded piece of DNA that replicates outside the chromosome in cells. The DNA fragments were inserted between the cut ends of the circular plasmid, and then an enzyme, DNA ligase, was used to "seal" together the ends of fragments to the ends of the plasmid. Researchers next infected plasmid-free *Escherichia coli* bacteria with the newly created hybrids, or recombinant plasmids, to "grow up" multiple copies of these recombinants. As the bacteria multiplied at a high rate of reproduction, they made multiple copies of the recombinant plasmids, and multiple clones of the fragments of DNA, in a process called "amplification." Researchers then extracted the recombinant plasmid from the bacteria.[28]

A gene library is an unordered collection of DNA clones from a single organism. Researchers "screen," or search, the library to identify their gene or DNA fragment of interest, for example, the interferon-coding gene. Screening could take different forms depending on the information available. One could use a partial sequence of the gene of interest if it were available. Alternatively, one could use the Southern blotting technique, developed by Edward Southern (1975), which employed gel electrophoresis and nitrocellulose filters to separate bands of DNA. With the development of DNA synthesis techniques, one could use amino acid sequence information or messenger RNA to construct a DNA probe, called a "complementary DNA" (or cDNA) clone. (A probe is a single-stranded DNA or RNA molecule of a known sequence, usually radioactively labeled in the laboratories I studied. However, probes may also be immunologically labeled. The labeling allows scientists to follow the probe as it detects a gene or DNA fragment with a complementary set of nucleotide bases. See Chapter 4 for more on probes.) Through these processes, researchers could then select and isolate their gene (or fragment of a gene) of interest, put in plasmid (as discussed above), and make multiple copies (clones).[29]

Using this procedure, researchers could also cut up a gene to see which parts of the gene are active in a particular process of interest. It was (and is) also used to map the genome of an organism, the topic of incessant interest (both positively and negatively) in contemporary science and science politics. Since these enzymes "allowed" the cutting of DNA at many points and the rejoining of segments in any order, scientists used cloning techniques to map (locate and identify) genes on the genome, in case the location of a gene of interest was unknown. Researchers could use shotgunning on the DNA of any organism and have used it on human DNA. They try to locate the gene in the resulting genomic library. Molecular biologists can thus map all the genes in an organism's genome (given enough resources).[30]

According to molecular biologist Tom Maniatis of Harvard University, none of the subsequent human genetics research (including neurobiology, immunology, and cancer research) would have happened without this original work on prokaryotes and restriction enzymes.[31]

> Nathans and Ham Smith recogniz[ed] the value of restriction enzymes. Cohen . . . was deeply engrossed in the basic biology of plasmids, and it was a very complicated business, with these things rearranging and transposons [a mobile piece of DNA] jumping all over the place. But he had accumulated enough basic knowledge about how these things functioned—how they replicate, how they transpose, all this stuff—to really begin to think about how they could be used as a tool. And Boyer purifying EcoRI [a restriction enzyme] and figuring out that sticky ends could come together, and that they could be used to join pieces of DNA. There is just this whole wealth

> of basic information that derived from studying prokaryotes that made all this [human genetics] research possible. (Fortun's interview with Maniatis, 1992)

Restriction Enzymes and Cloning Procedures as "Cookbook" Technologies

From entities involved in bacterial replication, restriction enzymes became standard tools in cloning strands of DNA to be used in molecular biological experiments on all aspects of living organisms. Without cloning manuals, however, use of restriction enzymes and cloning procedures might have remained limited to a circle of a few laboratories.

In 1982 the Cold Spring Harbor Laboratory—a leading institution in molecular biology then directed by James Watson, who, jointly with Francis Crick, defined the structure of DNA in 1953—published *Molecular Cloning: A Laboratory Manual,* compiled and edited by Tom Maniatis, E. F. Fritsch, and Joseph Sambrook. The manual began as a collection of laboratory protocols used during the 1980 Cold Spring Harbor course on molecular cloning and eukaryotic genes. In recounting the history of their manual, Maniatis emphasized its unintended consequences.

> I think it was a combination of timing—that it was written at really the perfect time—and the fact that it was put together in a way that was maximally informative for [novices] . . . At the time, we didn't think it would have [such] an impact, in part because we sort of viewed the people who would use it as the sophisticated labs who do recombinant DNA. It would just be something that they would buy for their students, so they wouldn't have to give them a xerox copy of their protocol every time they came into the lab. So we thought that was its niche—just to help out existing labs . . . This manual actually was written in a way that allowed people from virtually every field to come in—with no understanding of lambda, or no understanding of bacteria—to read the manual and do it.
>
> But that was only a minor part of it. The major part of it was really introducing novices to the technology. And [now] I talk to people who say they took a trip through China, and every lab they go to in China has the cloning manual on their desk. In fact, there has developed a market for counterfeit manuals that were produced in Hong Kong. (Fortun's interview with Maniatis, 1992)

By 1984, biologists were referring to Cold Spring Harbor's cloning manual as the "Bible" or the "cookbook." At least four copies of the manual, a two-inch thick volume, floated around an oncogene research laboratory in which seven researchers used the techniques on a daily basis. I consistently found copies of the manual on benches in the lab, rather than on bookshelves.[32] The researchers asked one another which "recipe" was best for accomplishing a particular goal and often supplemented the techniques with local embellishments and adjustments, just as most experienced cooks adjust cookbook recipes. Nevertheless, the book's protocols provided the basics and the bases upon which to embellish. Today, many of its protocols are standard procedures in laboratories across the world. The manual sold 60,000 copies in its first edition and about the same number in its second edition—a phenomenal record for a specialized technical manual. Maniatis has explained its popularity in terms of its clarity and completeness, which in turn resulted from the researchers' original desire to reach novice student audiences. The consequence was that anyone with some biological and chemical background, even without specific molecular biological training, could follow the recipes and solve technical problems in their reproduction of the "dishes."

> If you had just tacked things together, they would have just started following the recipe. But since they didn't have the conceptual background, if anything went wrong, that was it. Who would they go to if they were in China, or even other places, where no one in the building does that kind of work? Whereas, the manual was written in such a way that if they read it carefully and understood it, they could actually troubleshoot . . . They could fiddle around and do things. And also, we tried to, as much as we could, coach them on troubleshooting: if it doesn't work like this, then you might try this and that . . . It was sufficiently straightforward that it did work . . . What makes the manual more than anything is the fact that it works . . . And . . . that's due to the technology, not to the manual. It's just the fact that these are methods that almost anybody can make work.
>
> I think that the manual . . . had a disproportionate impact on the use of this [cloning] technology in all these different areas . . . and I'm as surprised as anyone else. It's not that I felt we did anything brilliant, or that we could foresee what was going to happen. It just sort of happened. But, in retrospect, I think that it really did make the technology palatable and usable, so that people who had interesting problems weren't afraid to jump in and do it. (Fortun's interview with Maniatis, 1992)

Cold Spring Harbor also published other manuals to aid molecular genetic research, including *Advanced Bacterial Genetics: A Manual for Genetic Engineering; Experiments in Molecular Genetics; Experiments with Gene*

Fusions; and *Methods in Yeast Genetics.* By 1984 many other manuals and textbooks were available in biology libraries. These included *Techniques in Molecular Biology,* edited by J. M. Walker and W. Gaastra and published in 1983, and *A Practical Guide to Molecular Cloning,* written by Bernard Perbal and published in 1984. A 1984 *Nature* review commented on the Walker and Gaastra volume:

> The main readers will be final-year undergraduates (especially project students) and postgraduates embarking on research. Established research workers will find most of the book familiar, but will value its presence in the laboratory when employing a new technique or when suffering one of those irritating failures of a technique which has been working for years . . . This publication intercalates quite nicely between *Molecular Cloning: A Laboratory Manual* . . . which contains an enormous amount of practical detail, and the *Genetic Engineering* series, published by Academic Press, which contains none. (Patient 1984:763)

The proliferation of manuals presenting carefully worked out protocols for cloning and other recombinant DNA procedures has assisted the dissemination of the technologies beyond a circle of a leading edge laboratories.

Worlds Create Technologies, Technologies Transform Worlds

Private and Public Commitments to Recombinant DNA Technologies

More than their abilities to cut and anneal strands of DNA and more than their purposeful reconstruction by biologists made restriction enzymes (and cloning techniques) standard tools in molecular biology, and now in biology generally. What makes them a standard tool is their reproduction in volume, their ready availability, their position as a basic tool in the recombinant DNA kit, and the commitments made to recombinant DNA technologies by researchers, private industry, government funding agencies, and universities.

When restriction enzymes were first reshaped into molecular tools, molecular biologists isolated, purified, and characterized their own restriction enzymes. However, given that the market for restriction enzymes was a very small one when compared to markets for ready-made clothes or even compact disk players, it was amazingly soon thereafter that commercial biological supply houses began to produce them on larger scales for purchase by research and industrial laboratories. Amersham, Bethesda Research Laboratories, and Boehringer Mannheim invested in the new technology. New England Biolabs was established by scientists for the production of restriction enzymes and has since expanded its range of products. Tom Maniatis tells the story of the scientists and work that led to the development of New England Biolabs. Interestingly, but not surprisingly, this tale includes the names of the (male) scientists involved in the process, while the (female) technician doing the production and verification work remains nameless.[33]

> [Rich Roberts][34] is the one upon which [New England] Biolabs was really founded. He was their original advisor and consultant. And actually Sambrook played a very key role in this . . . Sambrook was one of the first—right at the time that Nathans and Smith were using restriction enzymes to map SV40—Sambrook was using . . . restriction enzyme sites to map recombination in adenovirus. So they had different mutants, and . . . he and Terry Grodzicker were mapping them by using restriction enzyme sites. As he got into that, he realized more and more the need of having more enzymes, because it would allow finer mapping of recombinants and so on. He's actually the one [who] encouraged Rich Roberts to do a serious systematic search for new enzymes.
>
> So, for a long time, Rich's lab was entirely directed towards identifying new activities, and they set up this relationship with Biolabs. And the amazing thing about that whole enterprise was that Rich had this one woman in his lab, a technician, who was middle-aged, and absolutely driven to do nothing but this. It was an amazing phenomenon . . . She would come in, she would work on weekends, she would work at night, and all she did was keep opening new bacteria and looking for enzymes, and when she found them, she would purify them enough to show that they were real. So this one woman was so driven to do this, and she probably is responsible for discovering a very significant fraction of existing enzymes. (Fortun's interview with Maniatis, 1992)

The name of the technician is Phyllis Myers.

As noted earlier, Susan Wright (1986), in her history of recombinant DNA technologies, examined the responses of venture capitalists, multinational chemical and pharmaceutical companies, and government sponsors to the promotional campaigns of technologists. The technologies' potential for producing profitable disease treatments, genetically engineered food and agricultural products, and other commercial products was embraced by these varied audiences. It was clear that by the early 1980s support for recombinant DNA technologies from both private and public sources was high and growing. Private industrial and governmental support contributed to the production, maintenance, and marketing and distribution of technical manuals, high-quality biological materials, instruments, and computerized data bases and software, thus making recombinant DNA technologies available to researchers across the country and the world. Venture capitalists, private entrepreneurs, government agencies, university administrations, and research institutes made commitments to recombinant DNA technologies and their attendant research problems. (I discuss the reasons for their commitments in Chapters 4, 5, and 6.) Universities and nonprofit organizations established "technology centers" where products and information could be collected or manufactured and then distributed. By 1980 securing funds for recombinant DNA research, even for equipment and instrument expenditures, was relatively easy.

The National Institutes of Health (NIH), which funds most biological research in the United States, increased its support of research involving recombinant DNA technologies from 1974 to 1987 in terms of dollar amounts and numbers of sponsored projects, as shown in Table 3.1. These figures were drawn from NIH's CRISP data base system using the code "genetic manipulation," which means any research employing or exploring recombinant DNA technologies. Although the NIH provided the largest dollar amount of support for recombinant DNA research, researchers received other federal funds including support from the National Science Foundation and the U.S. Departments of Agriculture, Defense, and Energy (Office of Technology Assessment 1984). By the mid-1980s, funding agencies were awarding more dollars to systematics laboratories investing in molecular genetic technologies than to systematics laboratories interested in more traditional and less expensive studies.

Venture capitalists and private entrepreneurs established small specialty firms that produced, for example, biochemical reagents and software for manipulating coded information on DNA. Medium-sized and large firms invested in developing instrumentation for preparing and analyzing recombinant DNA research data. Although private industry aimed its in-house research primarily at developing applied products, it also supported fundamental research in universities through research contracts, research partnerships, and other contractual arrangements. Large corporations invested a total of about $250 million into biotechnological research contracts with universities and research institutes in 1981 and 1982 (see Table 3.2). Table 3.3 presents a more detailed list of contracts between industry and academia for recombinant DNA and related research from 1974 to 1984.[35]

Many of these support firms have achieved significant economies of scale that allow them to market at lower prices and provide additional benefits to researcher clients. Bartels and Jonston (1984:43) have made a similar point concerning the disciplinary field of molecular biology. "Without apparently making any substantial new concessions, the disciplinary field was able to

Table 3.1 NIH support for genetic manipulation research

	Amounts awarded in thousands of dollars											
	1974	*1975*	*1976*	*1977*	*1978*	*1979*	*1980*	*1981*	*1982*	*1983*	*1984*	*1985*
Primary		1,412	1,699	2,405	11,440	10,514	10,845	13,952	16,136	30,848	56,572	82,700
Secondary	82	360	25,910	33,146	47,947	87,962	108,011	126,009	190,288	219,664	221,824	250,648
Tertiary		58	520	582	2,102	5,635	12,125	17,848	20,600	30,319	42,007	66,247
Total	82	1,830	28,129	36,133	61,489	10,411	130,981	157,809	210,904	280,831	320,403	399,595

	Numbers of projects											
	1974	*1975*	*1976*	*1977*	*1978*	*1979*	*1980*	*1981*	*1982*	*1983*	*1984*	*1985*
Primary		25	31	34	59	72	88	109	126	183	329	487
Secondary	1	63	203	333	467	741	877	1,102	1,367	1,516	1,545	1,664
Tertiary		2	6	6	27	66	127	169	200	273	343	502
Total	1	90	240	373	553	879	1,092	1,380	1,693	1,972	2,217	2,653

Note: Totals include projects that employ recombinant DNA technologies in primary, secondary, and tertiary degrees of centrality to the research concerns.

Table 3.2 Investments in biological research in universities and research institutes by multinational corporations, 1981–82

University or research institute	*Corporation*	*Amount (millions of dollars)*	*Contract length*
California Institute of Technology	Du Pont	0.15	—
Cold Spring Harbor Laboratories	Exxon	7.5	5 years
Cornell University	Procter and Gamble	0.119	—
Harvard Medical School	Du Pont	6.0	5 years
Harvard Medical School	Du Pont	23.0	12 years
Harvard Medical School	Seagram	6.0	—
Harvard University	FMC Corporation	0.57	3 years
Johns Hopkins Medical School	American Cyanamid	2.5	5 years
Massachusetts General Hospital	Hoechst	50.00	10 years
Massachusetts Institute of Technology	Whitehead Foundation	127.5	—
Rockefeller University	Monsanto	4.0	5 years
Salk Institute	Phillips Petroleum	10.00	—
Stanford University	Upjohn	0.1	—
Stanford University	SmithKline Corporation	0.047	1.5 years
University of California, Davis	Allied Chemical Corporation	2.5	5 years
University of California, San Francisco	Lilly	0.780	5 years
University of California, San Francisco	Merck	0.226	—
University of Maryland	Du Pont	0.5	—
Washington University Medical School	Monsanto	4.0	5 years
Washington University	Mallinckrodt	3.9	3 years
Yale University	Celanese	1.1	3 years
Total		250.5	

Source: Reprinted from Wright 1986:350, by permission of The University of Chicago Press.

attract and avail itself of considerable material resources from the biotechnology industry." This benefit has also been passed on to other biologists who have since adopted recombinant DNA technologies.

In addition to publishing companies that have fostered the commodification and dissemination of recombinant DNA techniques in widely available journals and books (manuals), many other firms and organizations have supported the dissemination of molecular genetic technologies. For example, scientists can now mail-order high-quality biological materials used in recombinant DNA experiments from private companies. These materials range from standard to customized products. Thus, instead of purifying or constructing their own materials, scientists have switched to mail-ordering of products such as restriction enzymes, modifying enzymes, and vectors required for DNA cloning from New England Biolabs (NEB) in Beverly, Massachusetts, and Bethesda Research Laboratories (BRL) in Rockville, Maryland (later renamed Life Technologies and relocated to Gaithersburg, Maryland).

> We made our own enzymes [at the beginning] . . . When I was working on the lambda operator/promoter region, we did the sequencing in England, and I actually had to come back [to Cambridge, Massachusetts] one summer and make restriction enzymes, because I had set up the columns and everything to do it, and they weren't making them. And they weren't commercially available. It's sort of funny that now you just take it for granted, these things that you pull out of your refrigerator, but people were really purifying their own enzymes. (Fortun's interview with Maniatis, 1992)

In another example, a scientist recollected a technician, Maggie Pier, who paid for her trip across Europe by taking her protocol book with her and making restriction enzymes for European laboratories. This trip was in 1977–78, when restriction enzymes were still not commercially available.[36]

NEB's 1985–86 catalog also offered kits (or "packs") for cloning and sequencing DNA. The kits included all the necessary ingredients, even detailed cloning and sequencing protocols. (Today high school students use similar kits in their biology courses!) NEB advertised their cloning kit as containing enough ingredients for running more than a hundred cloning experiments and their sequencing kit as containing enough ingredients for one hundred sequencing gels. The 1986 updated catalog advertised a DNA synthesis kit as well. Thus, scientists can now order customized materials from these and other companies (see Figure 3.1). Biologists, for example, can use customized DNA sequences as probes to find other similar DNA or RNA sequences in cells under study.

Table 3.3 University-industrial agreements for rDNA or related research, 1974–84

Date	*Partners*	*Nature of agreement*	*Size of agreement*	*Conditions*	*Source*
1. Feb. 1974	Harvard Medical School†/Monsanto	Production of TAF substance	$23 million	N.A.‡	*Science*, 2/25/77 (p. 759)
2. May 1981	Mass. General Hospital/ Hoechst	R&D in rDNA; train scientists (4/yr)	$50 million (10 yrs.)	Hoechst license to use patents	*New York Times*, 5/21/81 (pp. A1, B 15)
3. June 1981	Harvard Medical School†/Du Pont	Basic genetics research	$6 million	Harvard has patent, licensing rights	*Boston Globe*, 6/21/81 (pp. 1, 29)
4. June 1981	University of S. Florida/ Southern Biotech	Basic research	$63,600	N.A.	*Science*, 6/4/81 (p. 1080)
5. July 1981	MIT†/Whitehead Foundation	Establish institute	$125 million	N.A.	*Boston Globe*, 7/8/81 (pp. 1, 20)
6. July 1981	Michigan State University/Neogen	Develop parasite i.d. kit and related work	$455,000	Michigan to hold patent	*Commercial Biotechnology*, OTA, 1984 (p. 516)
7. Oct 1981	Washington University/ Mallinckrodt	Monoclonal antibody for cancer diagnosis	$3.88 million	Mallinckrodt has exclusive licensing rights	*Journal of Communication*, 10/16/81 (p. 10a)

8. Oct. 1981	Stanford University and University of Calif./ Engenics	Create rDNA production procedures; process development; establish new businesses	$18 million	N.A.	*New York Times*, 10/16/81 (p. A18)
9. Jan. 1982	Mass. General Hospital/ Genex	Hypertension research	N.A.	N.A.	*Bioscience*, 1/82 (p. 70)
10. Feb. 1982	Yale/Celanese	Enzyme research	$1.1 million (3 yrs.)	Celanese has exclusive patent rights	*Chemical Week*, 2/24/82 (p. 31)
11. Feb. 1982	Johns Hopkins University/Hybritech	Test antibodies made at Johns Hopkins	$1 million/yr	N.A.	*Chronicle of Higher Education*, 2/24/82 (p. 10)
12. Mar. 1982	University College (London)/Sandoz	Create new research institute	$6 million	N.A.	*Nature*, 3/4/82 (p. 4)
13. Apr. 1982	Imperial College/ Imperial Biotech	Establish new company with venture capital	$6 million	N.A.	*Nature*, 4/1/82 (p. 384)
14. Apr. 1982	Washington University/ Monsanto	Monoclonal AB research	$23.5 million	N.A.	*Nature*, 4/1/82 (p. 384)

Table 3.3 *(continued)*

Date	*Partners*	*Nature of agreement*	*Size of agreement*	*Conditions*	*Source*
15. Apr. 1982	Rockefeller Institute/ Monsanto	Plant molecular biology research	$4 million	N.A.	*Science*, 4/16/82 (p. 277)
16. Apr. 1982	University College/ Endorphin	Develop and produce pancreatic endorphin	$240,000	N.A.	*Nature*, 4/15/82 (p. 595)
17. May 1982	Cold Spring Harbor/ Exxon	Train scientists at Cold Spring Harbor	$7.5 million	N.A.	*Nature*, 5/20/82 (p. 175)
18. May 1982	Stanford/Syntex/ Hewlett-Packard	Consult via medical faculty	$3 million	N.A.	*Science*, 5/28/82 (p. 961)
19. May 1982	Stanford University and University of California, Berkeley/ Engenics	Establish biotechnology research center	$2 million	N.A.	*Science*, 5/28/82 (p. 960)
20. June 1982	Washington University/ Monsanto	Monoclonal antibody research	$1.8 million	N.A.	*Nature*, 6/17/82 (p. 529)
21. Aug. 1982	Leicester University/ British consortium of five companies	Yeast research	N.A.	N.A.	*New Science*, 8/19/82 (p. 706)

22. July 1982	Johns Hopkins University/American Cyanimid	Treatment for lung and allergy disease	$2.5 million	N.A.	*Chemical & Engineering News,* 7/26/82 (p. 31)
23. July 1982	University of Minnesota/Genetics International	Biotechnology research	$5 million	Genetics International to receive patent rights	*Chronicle of Higher Education,* 7/28/82 (p. 8)
24. Aug. 1982	MIT[†]/W.R. Grace	Amino acid-enzyme R&D	$6–8.5 million	Grace gets all licensing rights to patents	*Chemical & Engineering News,* 8/9/82 (p. 5)
25. Oct. 1982	Carnegie-Mellon University/PPG Industries	Product and process development	N.A.	PPG to transfer instrumentation to Carnegie-Mellon	*Chemical & Engineering News,* 10/4/82 (p. 4)
26. Oct. 1982	Uppsala/AB Fortia	Development of industrial applications	$36 million	Use of university as industrial base permitted	*European Chemical,* 10/18/82 (p. 2)
27. Oct. 1982	Yale/Bristol-Myers	Development of anticancer drugs	$3 million	Bristol gets exclusive marketing license	*Chronicle of Higher Education,* 10/27/82 (p. 8)
28. Nov. 1982	University of Sheffield/ Plant Science	Development of plant culturing techniques	N.A.	N.A.	*Agriculture Specialties Industries,* 11/12/82 (p. 2)

Table 3.3 *(continued)*

Date	*Partners*	*Nature of agreement*	*Size of agreement*	*Conditions*	*Source*
29. Nov. 1982	University of Wisconsin/ Cetus Madison (Agricetus)	Biotechnology R&D	$3 million	University of Wisconsin to acquire minority position in Cetus	*Biotechnology News Watch*, 11/15/82 (p. 1)
30. Jan. 10, 1983	McGill University/ Allelix (Ontario)	Nitrogen-fixation studies	$2.2 million	N.A.	*Chemical Industry Report*, 1/10/83 (p. 3)
31. Sept. 19, 1983	Columbia University/ Bristol-Myers	Basic research program	$2.3 million	Patents to be in Columbia's name; Columbia to own intellectual property; Bristol-Myers has first rights to patent	*Chemical Engineering*, 9/19/83 (pp. 19–20)
32. June 1, 1984	U.C. San Francisco†/ Chiron	AIDS virus research	N.A.	UC may receive royalties on AIDS products; patent rights under negotiation	*San Francisco Chronicle*, 12/5/84 (pp. C1, C6)

Source: Reprinted from Lappe 1984: 300–304, by permission.
†Universities on record as encouraging nonexclusive licensing of patent rights (Pajaro Dunes Conference).
‡Not available.

How to get custom DNA or peptides from a telephone—*Fast!*

Call Biosearch.

Our business *is* synthetic DNA and peptide chemistry. And we've automated these chemistries to supply you with high quality, custom oligonucleotides and peptides for prices that will fit your budget and need for fast delivery.

Custom DNA—Choose from three new pricing options to suit your budget and requirements for speedy delivery. For example, save up to 25% on our regular price if you sequence your custom nucleotide. And fast? Depending on the option you choose, we can deliver a custom product within **ten days!**

Custom peptides—We guarantee confidentiality, sequence integrity, on-time delivery, and high purity at very competitive prices. Each peptide is shipped with complete analytical data for product verification. Call us with the sequence you need and we'll furnish a quote the same day.

Additional discounts are available for multiple orders. So . . . to get the best custom DNA or peptides from a telephone,
call Biosearch. Today!

800-227-2624

Inside California:
(415) 459-3907

800-227-2624

BIOSEARCH

2980 Kerner Boulevard
San Rafael, California 94901
Telex: 703077 BIOSEARCH
Circle No. 140 on Readers' Service Card

Figure 3.1 This Biosearch advertisement for custom-tailored synthetic DNA sequences was published in 1984 in scientific journals.

Some companies specialize in even more specific recombinant DNA research tools. An interesting example of risk-taking and commitment is the company Oncor, which designed and manufactured DNA probes for oncogenes and antibodies to oncogene products. In a 1984 interview with *Genetic Engineering News,* Stephen Turner said he founded Oncor with a vision of "capitaliz[ing] on the emerging diagnostic market from recent developments in cancer molecular biology . . . There is no clear-cut link between oncogenes and clinical claims, but I'm gambling that it will happen. It was a greater risk a year ago than today. Look at *Nature;* there are four to five articles per week about oncogenes" (Johnson 1984:18). Turner also discussed investments in oncogene research by Oncogen (another small biotechnology supply firm), a collaboration between Genetic Systems and Syntex, and a joint venture between Becton-Dickinson (a large, diversified biological research supply company) and the Cold Spring Harbor Laboratory. In Turner's words, "Clearly this is a hot area. My business goal is not unique . . . In cancer molecular biology, there is a need for standard reagents in highly convenient, quality-controlled assays that researchers can use to detect human and other species of oncogenes, genetic arrangements, gene amplification and gene expression" (Johnson 1984:18).

While Turner aimed his efforts at the long-term goal of supplying clinical researchers with tools for diagnosing tumors in human patients, his immediate clients were basic researchers who used his products to manipulate genes and tumor tissue in cells and lab animals. I saw a dog-eared photocopy of an advertisement of Oncor products taped to the wall abutting one graduate student's bench in an academic basic oncogene research laboratory. The student had photocopied the advertisement from an issue of *Science,* a scientific journal that focuses on many scientific fields rather than on specialty areas. Thus, in contrast to an advertisement in *Cell,* a journal aimed primarily at molecular biologists, an advertisement in *Science* reaches researchers from many different biological and other scientific research worlds.

Science has also published information on new products and materials on the market, such as DNA synthesizers. Although *Science* does not explicitly endorse the products, it provides wide exposure to many prospective clients. *BioTechniques, BioSciences, Genetic Engineering News, Nature,* and other publications also disseminate similar kinds of information and advertisements on new recombinant DNA technological products. Although Turner specifically foresaw a larger market in clinical labs, he established a small-scale venture whose market at that point included primarily basic research clients. Many entrepreneurs followed basic research in recombinant DNA technologies and established commercial production of materials. These companies supplied researchers in both academic and commercial biotechnological research and development laboratories.

In an earlier venture, Turner co-founded Bethesda Research Laboratories in January 1976 with $30,000 of his own funds and one technician for the production and marketing of restriction enzymes. By 1980, BRL employed 150 people, had sales of more than $2.5 million, and had diversified its production to include other recombinant DNA tools and products. Turner stated that his mission was to make BRL the Sears and Roebuck of molecular biology so researchers who were not "front-runners" could gain access to the materials needed to play the game:

> "We are part of the flow of information and materials. Our mission is to supply the tools and techniques of molecular biology, wherever it may lead," [Turner] says. Turner considers that by making new techniques available as soon as they are developed, BRL can help shrink the lead time and break down the "feudalistic kind of structure" which makes it hard for those outside the elite institutions to get immediate access to what the front-runners are doing. (Wade 1980:690)

Partly because of Turner's and other entrepreneurs' endeavors, even researchers who were not at California Institute of Technology, University of California, Cold Spring Harbor, Harvard University, Massachusetts Institute of Technology, and Stanford University—the institutions housing the so-called hot labs in molecular biology—were able to acquire the materials needed for recombinant DNA research.

Researchers also exchanged materials among themselves, providing a noncommercial supply of resources. In addition, nonprofit organizations started supplying recombinant DNA research materials. For example, the American Type Culture Collection (ATCC), located in Rockville, Maryland, collects and maintains material samples from laboratories for distribution on demand. Among other materials, researchers can order plasmids and DNA sequence probes (including cloned human DNA) and pay only maintenance and shipping charges. ATCC advertised its wares in cloned human DNA sequences in the July 11, 1986, issue of *Science.*

> In 1983, members of the human genetics community petitioned the National Institutes of Health to develop a reliable and efficient means for researchers to exchange cloned human DNA. At the same time, the National Laboratory Gene Library Project decided to investigate the possibility of specific libraries and the information derived from their use. To fulfill these needs, a repository of human cloned DNA segments has been established by American Type Culture Collection . . . under contract from the National Institute of Child Health and Human Development (NICHD). Drs. Victor McKusick and Mark Skolnick are serving as advisors to the repository in addition to a board of geneticists assembled by the NICHD. ATCC will collect well-characterized probes from investigators, expand and verify the probes, and store multiple samples that will be distributed to other investigators. Active solicitation and acceptance of important probes has begun. ("Repository of Human DNA Probes and Libraries," *Science* 1986:170)

The advertisement went on to list ATCC's available human clones. At that time, the organization announced a new project to construct two complete sets of chromosome-specific libraries of all twenty-four different human chromosomal types,[37] which "should be of interest to molecular biologists studying gene structure and regulation."

Instrumentation and Informatics: Reinscribing the Links

Technological entrepreneurs have packaged some procedures in recombinant DNA technology, such as DNA synthesizing and DNA sequencing, in instruments. The "gene machine," or automated DNA synthesizer, exemplified the prodigious efforts invested by molecular biologists, commercial entrepreneurs, and venture capitalists in recombinant DNA technology. A DNA synthesizer manufactured customized synthetic oligonucleotides, or short segments of DNA (or RNA). Synthetic DNA was useful to both applied genetic engineering and to basic research scientists. For example, an article in *High Technology* described a common task in molecular biological laboratories that could be accomplished by the DNA synthesizer introduced in the mid-1980s.

> Perhaps the most common use of DNA synthesis today is to make "probes" that help locate a natural gene of scientific or commercial interest—such as that for human growth hormone, insulin, or interferon—among thousands of different genes in the DNA of a cell. Since a gene is simply a specific sequence of nucleotide bases along a continuous strand of DNA, finding a particular gene is akin to searching for the proverbial needle in a haystack. Fortunately, the ability to make pieces of synthetic DNA that are complementary to a known sequence in the desired gene has greatly simplified the search. (Tucker 1984:52)

The first automated DNA synthesizer—a beautiful example of a black box—was introduced in 1981, and by 1984 seven more sophisticated machines were on the market and operating in biotechnology companies. A 1986 article reported that "the chemistry has been refined to such a degree that the synthesis of oligonucleotides up to 50 bases in length has become *routine* and oligonucleotides longer than 100 bases have been synthesized" (Smith 1986:G63).

Molecular biologist Leroy Hood's laboratory at the California Institute of Technology developed an automated DNA sequenator, a machine that analyzed the nucleotide sequence of DNA.[38] As I noted above, cloning and identifying a gene or DNA fragment does not "reveal" the nucleotides that make up that gene and their order. Hood's instrument could "read off" the exact chemical structure of a fragment of DNA (and of an entire genome) in terms of the particular sequence of nucleotides on the chain. The sequenator was based on Frederick Sanger's enzymic sequencing method (see below). Sequencing DNA by hand is time-consuming and literally backbreaking work. Besides automating the time-consuming and technically difficult (as opposed to organizationally difficult) manual method, the sequenator increased the rate of sequencing by an order of magnitude. It "reads off nucleotide [sequences] at a rate that may soon approach 8000 bases a day, which is at least tenfold higher than is currently achieved manually and at a small fraction of the current cost per base" (Lewin 1986a:233). In essence, it could perform in one day almost the same amount of work previously performed by a skilled technician in one year. James Brown, director of the molecular biophysics division of the National Science Foundation, said that "every genetic engineering company and every large university's department of biology 'will have to have at least one of the new machines' " (*San Francisco Chronicle,* June 12, 1986).

Another major developing resource in recombinant DNA research was easy access to accumulated information. Richard Roberts at the Cold Spring Harbor Laboratory collected and annually published a list of known restriction enzymes. Molecular biologists around the world help him to keep the list up to date by notifying him whenever they have a new restriction enzyme. Similar lists of other kinds of biological information have been published for wider dissemination. These include information on DNA, RNA, and protein sequences and, when available, structure and function.

These information resources generally took the form, after journal publication, of collected volumes. But they gradually changed to the more quickly assembled and accessible centralized computerized data bases holding DNA, RNA, and protein sequence information with annotations (for example, by selected host organisms and by taxonomies of organisms). The major data bases were located at the Los Alamos National Laboratory in the United States, where the Genetic Sequence Databank, or GenBank, originated; at the European Molecular Biology Laboratory (EMBL) in Heidelberg, Germany; and at the DNA Data Base of Japan (DDBJ) in Mishima.[39] Since DDBJ was small then, GenBank and EMBL shared the job of collecting and cataloging sequence information and then pooling their information.

> All three exchange sequence updates daily via the Internet network. The three databases have also agreed to a common representation of part of the genetic information. However, GenBank has taken a leadership role in electronic submissions, timeliness of the database, network distribution, and similarity searching. Because of this close collaboration, the technologies developed at each center benefit all of the others. The collection and dis-

> tribution of genetic information is truly an international cooperative effort. (Brutlag 1994:3)

GenBank had been funded by several NIH agencies (including the National Cancer Institute), the National Science Foundation, the Department of Energy, and the Department of Defense.[40] In July 1986, GenBank contained 6.5 million base pairs of sequence information. In July 1988, it contained approximately 18 million base pairs of total DNA sequence information and approximately 2 million base pairs of human DNA sequence information (Colwell 1989:124).[41] Information in both data bases was organized in standardized, computer-readable form. "Access to the data is through distribution of magnetic tapes, floppy disks, now CD's, direct computer-to-computer and computer-to-terminal transfer over telephone lines, and computational resources [such as various software packages] . . . which provide access to both sequence-data and sequence-analysis programs" (Friedland and Kedes 1985:1172). In some cases private industry provided software; in other cases data banks custom-designed the programs.[42]

The sequence data bases provided scientists with a faster and more efficient method for accessing information needed for experiments or for interpreting experiments. The following statement by Brutlag applies also to the time period covered by this history:

> By comparing a newly determined DNA sequence with the GenBank database, researchers can tell if there is any gene, even remotely related to theirs, in a matter of minutes. By sequencing even a small part of a gene and searching GenBank for similarities, they can tell right away if they are onto something novel. Thus GenBank has changed the way molecular biologists carry out their everyday work. In the past, determining the complete sequence of a gene was usually the most laborious final step of a research project. With the availability of the GenBank database, determining the sequence of a gene is now usually the very first step taken in a research project or a medical study. Scientists shape the direction of their research based on the similar genes discovered in GenBank. (Brutlag 1994:2)

Some of the kinds of analytic searches scientists could perform using the data base system included translation and location of potential protein coding regions; inter- and intrasequence homology searches; inter- and intrasequence dyad symmetry searches; analysis of codon frequency, base composition, and dinucleotide frequency; location of AT- or GC-rich regions; and mapping of restriction enzyme sites.

However, the most common use, and the critical one for my story, was the search for sequence homologies. Researchers put their DNA or RNA sequence information into the computer to search for homologous sequences—other DNA or RNA sequences similar to theirs. If they knew the structure of a protein, they could also construct the DNA sequence of the protein's gene and then search the data base for homologies.

For example, a proto-oncogene researcher described the speed and efficiency with which two previously unrelated areas of research (on proto-oncogenes and arteriosclerosis) were "found to be related" in 1984 based on similarities found through computerized sequence data bases. Molecular biologists usually use the term "homology" to refer to similarity in two or more DNA sequences. When molecular biologists "find" homologies, the finding often means that another laboratory has already sequenced their gene and entered the sequence into the data base under another name. Redundancy is high in the data bases. When similarities that are not redundancies are found, they become the basis for further research as to what the similarities mean for functional relationships.[43] This *distributed coordination* of scientific work is emphasized in the following oncogene scientist's remarks.

> In fact, nobody has to read [pages of sequence data in search of specific information] anymore, because the computer's changed the face of that aspect of science . . . The way this is usually done is to take your sequence and plug it into the computer and ask the computer to search a gene bank, a sequence bank, for relationships. So just yesterday, for example, a fellow visiting here named Mike Brown . . . described some experiments . . . in which he was looking at the receptor for low-density lipoproteins. This is a receptor which is required to clear the blood of cholesterol. People who lack this receptor develop arteriosclerosis and myocardial infarctions at an early age. He and his colleague, Chuck Goldstein, some years ago defined the receptor. They recently purified and cloned and sequenced the gene, that is, sequenced a copy of the messenger RNA of the gene. When they plugged their sequence into the computer, they got back information that the receptor was very similar to a protein that serves as a precursor for the growth factor we've been talking about, EGF [epidermal growth factor]. So there we're dealing not with identity but with similarity. We have the information that two genes that seem ostensibly unrelated are, in fact, closely related members of a gene family.[44]

By automating procedures for identifying sequence similarities, computerized data bases also allowed scientists to pass some of their tasks on to others. In many academic oncogene laboratories in the early 1980s, undergraduate students were hired to handle much of the computerized data base work. These students knew nothing about the relevant journals, authors, and articles in the research topic area, but they could still search for sequence similarities using the computer. Thus, computer technicians joined laboratory technicians, bottle-washers, and secretaries in the "nonscientist" category of laborers in the molecular biology laboratory.

In their effort to keep apace with the ever-increasing volume of information published by researchers, GenBank and EMBL directors collaboratively explored new ways to streamline and institutionalize information entries into the systems. They asked authors to provide sequence annotation details in a standardized format and to send the information on floppy disks to facilitate input. GenBank developed the Authorin program for automated sequence entry. Since response was poor, about 30 percent, the data base directors negotiated with journal editors to institutionalize a computerized format for authors' submissions of sequences to GenBank and EMBL as an integral part of the publication process. The editor of the journal *Nucleic Acids,* who sat on the board of directors of Intelligenetics, instigated the move by requiring that authors submitting papers for publication to the journal simultaneously submit their sequence data to GenBank. " 'We need to develop a feeling among researchers that the job is not completed until the annotated sequence is in the database,' says Hamm [EMBL's data base director]. In fact, authors might soon find that having their sequence on the databases will be the only way of making their work public, as journals become reluctant to occupy page upon page of their publications with virtually unreadable sequences" (Lewin 1986b:1599).

"Logical Steps," "Efficiency," "Speed," and "Standardization"

Scientists often invoked the words "speed," "efficiency," "doable," and "productive" when explaining why they worked with molecular genetic technologies. When one thinks of science as work done in postwar economies of the West, this makes perfect sense. Why should science be any different from other kinds of work? Scientists, like technocrats and business people, attribute properties of "productivity," "speed," and "efficiency" to the technologies with which they work. Nevertheless, these "properties"—standardization, speed, efficiency, and productivity—also are outcomes of the commitments made by various parties to the production and maintenance of these technologies. A technology is efficient only when it is part of a network of commitments to particular practices, that is, not until it is formalized and standardized.

Molecular biologists consider themselves to have been very successful at packaging their technologies into well-defined, even automated, procedures that eliminate much uncertainty, discretion, and some skilled labor. The instruments and materials that constituted the standard technologies encapsulated and solidified skills and tasks, while linking procedures and materials in relatively stable relationships. That is, the standard technologies reliably (re)produced the same inscriptions across situations. Separating the steps from A to Z that were encapsulated in an instrument or in established protocols required opening up the instrument and tinkering, something not commonly done after a set of protocols has been widely adopted. For example, when using restriction enzymes for cloning, researchers moved seamlessly from the material supplier's catalog to the technical manual. While much information specifying the contexts of representation and use of restriction enzymes was provided by their commercial producers and distributors, other directions, necessary information, and helpful hints on its experimental application were "formatted" in technical manuals.

In testimony to their success, many biologists have chosen to include recombinant DNA and other molecular genetic technologies in their tool kits for examining long-standing biological problems, such as the processes of normal growth and development. They have managed this initiation into the rites of DNA technologies because of the facility of the techniques. Although this facility has not made it possible for *anyone* to pick up the Cold Spring Harbor Laboratory cloning manual and begin to clone genes, those trained in undergraduate biology, chemistry, and physics courses certainly could understand and use the manual. And, as noted earlier, as of 1995 the possibilities are even broader. Indeed, one of my sociology graduate students was hired by a technical laboratory at the Harvard Medical School to clone and sequence genes. He was specifically hired for a short term of two years (no more) and for his lack of biological knowledge. The scientists heading the lab preferred workers who were not interested in biology and therefore not likely to be "distracted" by the biology itself. In other words, they wanted smart assembly line workers.

Thus, these researchers learning to apply recombinant DNA techniques to biological problems did not have to be or become experts in enzymology, chemistry, or physics to employ the tools. Biologists untrained in molecular biology learned to perform recombinant DNA techniques in a year or less. In one case, a cell biology group "picked up" the techniques purely from working with "cookbooks" and protocols and materials received from commercial supply houses. That this could happen as early as 1986 testifies to the standardization of the techniques at that time. When possible, however, scientists and students preferred to apprentice themselves to other researchers with the appropriate skills. G. Steven Martin, protein biochemist and oncogene researcher with four students (at that time) in his laboratory sent two of them to work in laboratories on the same campus to gain recombinant DNA technological skills and then return and introduce the techniques into his lab. Martin's laboratory was exploring, among other things, the role of protein phosphorylation in oncogene activity. Phosphorylation is the process by which an organic compound takes up or combines with phosphoric acid or a phosphorous-containing group. Oncogene researchers hypothesized that transformation of normal cells to cancer cells by retroviruses and mitogenesis by growth factors proceeds by a common mechanism involving the phosphorylation of cellular proteins at tyrosine (one of the twenty amino acids). Tyrosine kinase is one of the enzymes thought to be involved in protein phosphorylation. To study the protein and gene activities in combination, the lab needed to acquire new skills.

> Two of the students are doing DNA work. The way we're doing that is they're doing a collaboration [with] labs that are already using recombinant DNA techniques. One of the students is working with [Scientist 1] in microbiology, and the other student is . . . working out of [Scientist 2's] lab. They're sort of acquiring the "recipes," as it were . . . From the other cooks . . .
>
> One student is . . . working in [Scientist 1's] lab, which is around the corner. She is doing the protein work over here and . . . the DNA work around the corner. Ultimately, another student will be moving there in a few months' time. She'll be moving back here, setting things up here. That's ultimately the only way in which it's practical to make that sort of change . . . in the absence of me doing a sabbatical. Because I just don't have the time to go [to their labs]. I can't spend all day there. So there are basically two ways of making a switch like that, in practice. One is to do a sabbatical in someone else's lab, and the other is to import the technology through a student or postdoc. And I think the former . . . is the better one, but it's just not possible in this particular instance. (Martin interview)

Martin described the process of learning recombinant DNA technology as learning a whole new set of recipes.

> In the last eight to ten years, we've been doing protein biochemistry—purifying proteins, characterizing protein phosphorylation. [That is, we've been] basically working with proteins, which involves mostly classical biochemical techniques of protein separation, protein analysis. And we're

slowly getting into essentially recombinant DNA work, techniques for handling nucleic acids, which involves a whole different set of technologies.

> Although the physical manipulations are not that different, it's just a whole set of different, new techniques . . . You have to know what artifacts [to look for] . . . You have to know the properties of a whole new set of enzymes. You're still pipetting, basically. You're still running columns, you're still running gels . . . The new procedures are, in terms of manipulations, only trivially different. It's not as though I were a pathologist, used to cutting sections and now doing something completely different. Nevertheless . . . the materials are different, the procedures are different. It's just as though you were a Chinese cook and you suddenly now started cooking French [cuisine]. (Martin interview)

Another good example of a highly praised and widely disseminated molecular genetic technology is DNA sequencing. There were two sequencing techniques commonly used during the late 1970s and 1980s. One technique, employing enzymes, was developed by Frederick Sanger and A. R. Coulson (1975); another more chemically based process was developed by Allan Maxam and Walter Gilbert (1977).[45] Both sequencing methods allowed researchers to sequence in months what would have taken years with earlier techniques. The Sanger-Coulson method (with the addition of dideoxynucleotides two years after their initial work) was more successful than the Maxam-Gilbert method, in part because (according to one biologist), "anyone could do it, while one had to be a decent organic chemist to do the Maxam-Gilbert method well." Later, both methods were automated, thus making them easily available to nonexperts. Both Sanger and Gilbert won Nobel Prizes in 1977 for their techniques. Gilbert gave an example of these straightforward tasks in DNA sequencing when accepting the prize for his and Maxam's sequencing method.

> To find out how easy and accurate DNA sequencing was, I asked a student, Gregor Sutcliffe, to sequence the ampicillin resistance gene . . . of *Escherichia coli* . . . All he knew about the protein was an approximate molecular weight, and that a certain restriction cut on the [pBR322] plasmid inactivated that gene. He had no previous experience with DNA sequencing when he set out to work out the structure of DNA for his gene. After 7 months he had worked out about 1000 bases of double-stranded DNA, sequencing one strand and then sequencing the other for confirmation . . . The DNA sequencing was correct. Sutcliffe then became very enthusiastic and sequenced the rest of plasmid pBR322 during the next 6 months, to finish his thesis.[46]

Cherfas (1982:124) reemphasized the point of facility and efficiency: "[It took] thirteen months, for the entire sequence of 4362 bases on both strands, by a student with no previous experience of the method." Today automated DNA sequenators reduce uncertainty one step further than Gilbert's method. Now the operator need only know what solvents and reagents to put into the instruments. Tasks that were once considered Ph.D. thesis problems are now routinely performed by machines (W. Gilbert 1991).

According to industry manufacturers and molecular technologists such as Leroy Hood, the advantages of automated DNA synthesis and sequencing were similar to those of other automated manufacturing processes. First, automation sped up the process of DNA synthesis. Second, it relieved laboratory staff of this detailed preparatory work and freed them to undertake other tasks. Synthesizers and sequenators could work at night, when most laboratory staff members were asleep, which further sped up the laboratory's work. Third, since variables were more regimented, the technology also tended to more reliably reproduce results. Finally, it reduced the expertise needed to produce DNA.[47] The machines even automated many monitoring and evaluative tasks and thus eliminated the need for tacit knowledge and expertise, for example, detailed knowledge of indicators and possible artifacts. Analytical feedback systems alerted and automatically shut off "runs" in case of errors, thereby conserving costly reagents and preventing damage to the system. By the time scientific tasks were packaged into instruments, they had become standard tools operable by nonspecialists. According to one biotechnology industry analyst,

> Manual DNA synthesis is a true art that can only be accomplished by a technician trained in organic chemistry; yet such knowledge is rare in the biological laboratories where most gene fragments are made. In contrast, any technician can be trained to operate an automated synthesizer in a few days. Since the instrument is preprogrammed, the operator has only to fill the reagent and solvent reservoirs, enter the desired DNA sequence, and purify the end product. As a result, the rate-limiting step in biotechnological R&D becomes the biological experiments, not the preparation of the needed DNA tools. (Tucker 1984:50)

Even into the 1980s, however, protein synthesizers were quite expensive, and they remain so today. In 1986 microprocessor-based DNA synthesizers cost between $20,000 and $70,000.[48] Added to the initial capital investment costs are maintenance and materials costs. Thus, although many private industrial laboratories bought synthesizers, most academic laboratories manufactured their own DNA probes in-house or purchased custom-synthesized DNA probes from commercial biotechnological firms. Those academic laboratories with access to synthesizers shared them with other laboratories. (In some instances, the time saved combined with advancements in the instrument and future price deflations made the machines affordable even to academic laboratories.)

Given enough time, scientists can learn arcane skills developed in other subdisciplines. But time was (and remains) a rare resource in contemporary biology. The decreased time frames have in turn reconstructed time frames and problem choices in modern biology. A senior investigator described the constraints on investigators in contemporary biology. "Researchers have to convince the funding sources that their studies will produce progress, results, within a political time span. If you tell them not to expect progress in five years, they won't fund you . . . The whole structure of science is pushing for quick results."

The journal *BioTechniques* has a regular "Biofeedback" feature that asks readers to send in their "ideas, tips, hints, warnings, or pertinent questions" for improving the "speed, performance or convenience in common bioresearch laboratory procedures or manipulation," clearly indicating the emphasis on increasing efficiency and speed. The article featured in the May/June 1984 issue was entitled "A Rapid Small-Scale Procedure for Isolation of Phage [Lambda] DNA," which the authors described as "technically easy" (Benson and Taylor 1984:126). The reliance on fast and efficient technologies reduced time frames for problem-solving, which pushed researchers to continue to use fast and efficient technologies. Thus, short time frames and "fast" technologies were co-constructed.

Logical Steps and Standardization: Properties of Collective Work and Practice

Scientists did not believe that recombinant DNA technologies eliminated all experimental uncertainty. In most recombinant DNA experiments observed, however, researchers recited the litany of steps they needed to perform in order, for instance, to clone or sequence a gene. They believed in the order of technologies for handling DNA as opposed to the technologies for handling proteins. As a result, they attributed failures to inattention to detail or contaminated materials, and not to problems with the techniques. The following quotation indicates their respect for precision in molecular research.

> Q was continuing an experiment from yesterday. He was ligating two DNA pieces together [joining DNA strands together using the enzyme ligase].

> This time he took the time to do his calculations carefully: exactly how much of each DNA to mix together, the correct ratio. He had not been so successful before, because he was in too much of a hurry. This time he did two pages of calculations . . . His calculations are based on "Ligation Theory," page 286 of the Cold Spring Harbor's *Molecular Cloning: A Laboratory Manual.* (Field notes, Xavier lab)

For researchers, standardized technologies are translated into the "logical steps" of oncogene problem structures and possible results. A graduate student recited these virtues to me:

> Everyone knows that it would be worthwhile to sequence this gene. It's obvious what should come next. There are logical steps in this work. 1. Identify the protein [involved in transforming normal cells into cancer cells]. 2. Clone the gene associated with the protein. 3. Sequence the gene. For each step, there's always a logical next step to do. And you can always find money to sequence. We get enough grant money to sequence what we want to. It may not solve the cancer problem, but it will give some information. (Field notes, Xavier lab)

However, the "logic" and "obviousness" of the experimental steps are located in molecular biologists' narratives and not in, say, the structure of the gene.[49] Conventions are collectively understood and agreed-upon ways of doing things (Becker 1982). The logical sequence of steps in recombinant DNA research discussed above is logical only insofar as contemporary biologists collectively agree to represent nature in this fashion and via this set of protocols. That researchers could specify these logical steps meant that molecular biologists have organized and packaged their work into a sequence of tasks that became adopted and accepted in many laboratories as the "industry standard." Should one attempt to open a sequencing machine and reorganize or eliminate steps from A to Z, one would find it necessary to argue with many people and convince many professions, research institutes, individual scientists and technicians, biological supplies manufacturers, and so on, to change their ways of working.

Neither does a clear understanding of the logical steps of an experiment mean that the work will be done. In one case, a graduate student who wanted to sequence a gene as his next logical step could not persuade his laboratory adviser to allow him to do it. The adviser judged that the experiment could have taken one and one-half years and was technically difficult. A co-worker explained the adviser's reasons for this decision in terms of resource investments and possible payoff.

> You wouldn't do something that you know six other labs could do, because they might beat you. The credit goes to whoever gets the results first, so you don't want to risk it. You don't begin [an experiment the results of which] you could wake up and find someone published the next day. Or, worse yet, [the published results] come out in a journal just as you're finished with the experiment. That's using up your resources and time with no credit. (Field notes, Xavier lab)

In other words, just because an experiment is feasible does not mean that it will be done. Scientists judge the feasibility of a particular problem on more than clearly defined tasks or steps. In this case, the laboratory director weighed the potential payoff in terms of novel findings against the skills of his student, the lab's resources, competition from other labs, and his responsibility to adequately train his student in a reasonable time frame.

This example shows that, from the vantage point of an individual laboratory, the "bandwagon effect" is not a clear-cut provocation to work on a particular problem. Another respondent (see Chapters 4 and 7) discussed the limiting effect of the bandwagon on his laboratory. Competition in a bandwagon situation became a limitation on work in his laboratory when other laboratories "leap-frogged" his laboratory in the study of specific problems generated by his lab.

Trade-offs: Efficiency and Complexity

Just as the logic of this experimental process must be placed in the organization of scientific work, so must the narrative tropes of "efficiency," "speed," and "standardization." How fast is fast, how efficient is efficient? What becomes efficient and fast? Does this efficiency and speed in some areas of scientific work mean inefficiency, lags, and obstructions in other areas of work? Does standardization mean that flexibility and customizability are lost?

For example, established cookbooks, protocols, and instruments were not so black-boxed as to prevent some scientists from opening up the package and tinkering with the techniques. In fact, whereas most biologists limited their modifications to the selection and maintenance of favorite recipes among the several available for accomplishing particular goals, some biologists did much more. These tinkerers improved on old recipes and constructed entirely new techniques. The improvements and new techniques were often made within the context of established techniques where parts of old recipes were used to construct new ones. Producing shortcuts was also a favored technical project. However, these efforts to customize techniques for application to new problems or to produce shortcuts or new techniques were limited to a small subset of molecular biologists. Most molecular biologists were applying the standard techniques to new problems.

The technologies, then, have produced a division of labor in which some molecular biologists are skilled and committed enough to "hack"—to open up black boxes and customize the tools for their own purposes—while others must rely on "hackers" or stick to the standard technologies. This division of labor is stratified in contradictory ways. In this first case, hackers hold the upper hand. Other biologists must rely on them should they need to modify the technologies in order to better solve their problems, speed up their work, and deal with breakdowns in the technologies. However, for some biologists, taking the time to customize techniques may mean losing the race in solving another biological problem. For them, these technologies are simply tools for accomplishing other goals, solving other problems, and technical hackers are "merely" technicians in their scheme of the world where science and scientists are privileged over technology and technicians.

Hence, these narrative tropes of "efficiency," "speed," and "standardization" are double-edged. For all their use in enabling and empowering, they are simultaneously associated with limitations, loss of flexibility and customizability, and obstructions and delays in other processes. As the iconoclastic biologist Harry Rubin has argued, older, slower methods often force scientists to think through their problems more carefully. He also argues that these technologies are inefficient for dealing with particular kinds of problems, especially problems of complexity. In his view, when dealing with problems of complexity, efforts to formalize and standardize situations and tools often produce inefficiency. Efficiency, then, depends on the frame of the problem and molecular biologists' views of their technologies' efficiency.

Standardizing the Tools

Molecular biologists enrolled a large number of elements—objects, people, representations, and relations—in their recombinant DNA technologies and built strong linkages among them. The linkages formed the network, the collective commitments, within which objects such as restriction enzymes, automated DNA synthesizers, and computerized data bases became standard tools for creating representations of cancer. Without this web of commitments, restriction enzymes would have been available to just a few laboratories. With the commitments of many new parties, restriction enzymes and recombinant DNA technologies more generally have become standard tools for producing biological representations.

By 1982 novel recombinant DNA and other molecular genetic technologies were represented in cookbook manuals providing technical recipes, and by 1984 they were consolidated in instruments automating many recombinant DNA tasks. Private industrial biotechnology and biological supply companies produced and distributed materials, techniques, and instruments used in recombinant DNA research. Universities trained students who became technicians in biotechnology companies and university laboratories, as well as students who became researchers qua consumers of the technologies.

In sum, the collective activities and commitments of molecular biologists, enzymologists, venture capitalists, entrepreneurs, funding agencies, university administrations, governments and their international economic battles, enzymes, and cancer researchers (as we shall see) created, produced, and supported a set of standardized, transportable tools for manipulating eukaryotic DNA. These commitments formed an infrastructure of skills, technologies, and material production that in turn allowed recombinant DNA technologies to be used in other lines of research. Researchers began to use recombinant DNA technologies to study other long-standing and new biological problems because they were "productive." But my point is that "productivity" is also the *outcome* of this network or infrastructure, and not just its cause. While biologists endow recombinant DNA techniques and instruments with the properties of "productivity," "logic," "speed," and "efficiency," these properties of the objects and tools are defined through the commitments made to these technologies.

When researchers use these technologies today, they engage more than techniques, materials, and instruments for experimental manipulation of DNA. They also appropriate and extend the commitments of suppliers and consumers; the labor of biotechnological workers; the assumptions, concepts, and theories of the disciplines that produced these techniques and instruments; agreements among biologists about the structure and function of genes, DNA, RNA, and proteins; and the status and legitimacy of physics and chemistry, disciplines that have strongly influenced theories and approaches of molecular biology.

Molecular cloning techniques and gel electrophoresis technologies, DNA synthesizers, and DNA probes cannot be evaluated as separate, individual entities that determine the directions of biological research and representation. These technologies have producers who construct them within their frames of reference and consumers who employ them within their frames of reference. Techniques have no power in and of themselves. They require other elements and relations to function. A computer software package—for example, a word processor—would be useless without the appropriate hardware, the skills for adapting the package to the particular operating system, tasks adjusted to the way the package organizes tasks, operators, and more. It would not be marketable if it ran on only a few little-known computer systems, if it was too expensive, or if it required that tasks be set up in an arcane way to interface with the program. Similarly, the "technically best" mousetrap does not get to trap the mice in buildings across the nation. The winning mousetrap is the one that manages to secure the commitments of manufacturers and suppliers who have good advertising and provide easy availability and access. Only then will households and building maintenance offices be able to consider qualifications such as low cost, ease of use, and effectiveness at trapping mice.

Recombinant DNA and other molecular genetic technologies, then, marked a cultural geography of forces (materials, tools, skills, protocols, sentiments, commitments) that in turn reconstructed practices in many different worlds and stabilized molecular biological representations of nature.[50] As recombinant DNA technologies were constructed and adopted, they reconstructed the world of biological research in a dialectical process. This process of dialectical standardization relates to the hegemonies discussed in the introduction to this chapter. However, within my general framework, power and authority are not concentrated in the hands of a few scientists but are instead distributed among different actors, objects, and social worlds. In Chapter 5, I present an account of how power becomes vested in a package of theory and methods that *distributes* authority during a particular historical period to various institutional loci. Recombinant DNA technology contributed to this authority through its own hegemonic cum standardization process. In this process, other cancer research fields have been neglected.[51]

Although I argue that standardization has hegemonic effects, I simultaneously question the uniformity of its effects on all parties. Recall Wade's assessment that "BRL can help shrink the lead time and break down the 'feudalistic kind of structure' which makes it hard for those outside the elite institutions to get immediate access to what the front-runners are doing." This democratic, anti-elitist attitude is commendable, especially in times of tight research budgets. However, even these efforts to "reduce costs" are couched within rhetorics and institutional pushes to adopt particular molecular technologies that were and are still expensive. Researchers in university laboratories with limited budgets were and are often quite resentful of the prices and legal controls over the technologies they are told they "must" have. Even the researchers I interviewed in a then relatively successful biotechnology company resented having to pay annual license fees of $10,000 to Stanford University and the University of California for the rights to use recombinant DNA techniques. Prices are not low in molecular biology.

This "democratization" of tools via standardization has had another consequence, however underdetermined. The standardization of recombinant DNA technologies has circumscribed the "job" for the "tools."[52] As Walter Gilbert has remarked, "Developmental biology now looks first for a gene to specify a form in the embryo" (1991:99). Developmental biology has become molecular developmental biology. As embryologist Scott Gilbert told me, these molecular biologists have "pheno-extracted" the organism.[53] In earlier days, embryologists studying development would work on a daily basis with sea urchins. They had to "get their eyes used to" the embryos in order to observe any changes, or, in Evelyn Fox Keller's words for Barbara McClintock's approach to her cytogenetics research, allow time to develop "a feeling for the organism" (Keller 1983). The sea urchins were recalcitrant and did not always do what the embryologists wanted them to do. For example, Scott Gilbert's laboratory was four weeks behind schedule because their particular sea urchins turned out to be sterile. No embryos, no experiments. Molecular developmental biologists no longer have sea urchin tanks in their laboratories. They work with clones of organisms, what Scott Gilbert calls "purified nature." The messiness and randomness of the organism are gone, but so is the "object" of study. Although Scott Gilbert is right about the embryo's disappearance from the laboratory, the "object" has always been constructed throughout the history of biology. The very construction of the laboratory radically changed the means and methods for describing and explaining "nature." What we have here then is *another* construction of the object of study, the tools for its study, and—as we shall see in the next chapter—the explanatory model.

4 Crafting Theory

By the end of the 1960s the first phase of expansion of molecular biology was over. Some of the key issues, like the nature of the genetic code, had been sorted out, which made the selection of the next set of comparable problems rather difficult. What questions could be as important? The early 1970s were also a time of some anxiety in science. Research budgets were starting to fall in real terms for the first time since the war. Money was available in the field of cancer research, but all the approaches to that problem looked formidably difficult.

—*Edward Yoxen,* The Gene Business

A flurry of research activity on proto-oncogenes ensued in the 1980s as researchers claimed that proto-oncogenes provided at least partial answers to many long-standing biological problems. They proposed that proto-oncogenes were the keys to the previously locked doors in normal differentiation and development, cell proliferation, cancer (especially human cancer), and even evolution. More significantly, the oncogene theory (that normal cellular genes—the proto-oncogenes—somehow go awry and turn into cancer genes, loosely called "oncogenes") appeared to provide a way of linking all of these problems through a unified approach.[1]

The history of viral oncogenes and the history of recombinant DNA technologies are part of the genealogies told by proto-oncogene researchers when they recount the history of their field. Viral oncogenes, recombinant DNA technologies, and their various narratives are the bricks and mortar of the crafting of the proto-oncogene theory.

The proto-oncogene theory derived, to a great extent, from two lines of research that forged a common thesis about the molecular mechanisms of cancer causation. In the late 1970s, tumor virologists announced that they had found a class of genes in the normal cell that could be triggered to transform the normal cell into a cancer cell. The genealogy of viral oncogenes traces its roots to viral transformation studies, which in turn came from cell transformation studies, which owed their existence to experimental systems for culturing viruses. The proto-oncogene proposal was later supported by molecular biologists using DNA transfection experiments. DNA transfection experiments traveled a different historical route from late-nineteenth-century carcinogenesis studies through early-twentieth-century experimental transplantation studies to postwar carcinogenesis studies and then to the DNA transfection studies of the late 1970s and early 1980s.

Tumor virologists and molecular biologists announced that the proto-oncogene theory encompassed and unified many past and contemporary areas of cancer research and that further investigation using the oncogene framework would produce explanations at the molecular level for problems previously pursued in classical genetics and in chemical, radiation, hormonal, and viral lines of research on cancer. They claimed to have developed a molecular explanation for many different types (classifications) and causes of cancer. Ultimately, they asserted, their research might lead to a common cure, a "magic bullet" for cancer. They also proposed their theory could account for both normal and pathological (cancerous) growth and differentiation.[2]

However, claiming that one's theory unifies many lines of research does not mean that others will agree with the claim and rush to pursue experiments based on the theory. By widening our focus to follow both "unifiers" and the "unified," we can construct a story that also includes the perspectives and practices of both.

Between Tumor Virology and Evolutionary Biology

Tumor virologists developed the concept of normal cellular genes as causes of cancer in part by borrowing the concept of gene conservation as employed by evolutionary biologists.

In the 1970s, tumor virologists reported that they had found specific "cancer" genes, which they named "viral oncogenes," in the viruses that transformed cultured cells and caused tumors in laboratory animals. This experimental work was done using traditional methods and technologies of virology and molecular biology to investigate RNA tumor viruses.[3] As other laboratories joined in this line of research and explored other viruses, they reported discoveries of more viral oncogenes. However, after a decade of research, many investigators and administrators concluded that these viral oncogenes caused cancer only in vitro and in laboratory animals. No naturally occurring tumors in animal and human populations were credited to viral oncogenes.[4]

In 1976 J. Michael Bishop, Harold T. Varmus, and their colleagues at the University of California Medical Center at San Francisco (UCSF) announced they had found a normal cellular gene sequence in various normal cells of several avian species that was very similar in structure to the chicken viral oncogene, called *src* (Stehelin et al. 1976).[5] (Recall from Chapter 2 that the *src* viral oncogene was the outcome of work done in the late 1960s and early 1970s by researchers based at the University of California, Berkeley, on the molecular biology of viral transformation.) Bishop and Varmus's laboratory had constructed a "probe," which they named *src* RNA, and used it to "search" the chicken genome for a "match." A probe is a synthetically constructed strand of DNA called an "oligonucleotide," to which a radioactive element can be attached to serve as a marker or a way to track the probe.[6] The probe attached itself to a similar sequence in the chicken's DNA. Two years later, through the same search process, Bishop and Varmus reported that they had also discovered complementary DNA sequences related to the *src* viral oncogene in normal cells in many different vertebrate species from fish to primates, including humans (Spector, Varmus, and Bishop 1978).[7] They named this cellular gene the "cellular *src* oncogene." Bishop and Varmus and their collaborators suggested that the viral gene causing cancer in animals was transduced from normal cellular genes by the virus; that is, the virus took part of the cellular gene (probably the messenger RNA version without the introns), and made it part of its own genetic structure. Furthermore, the transduced gene sequence was not necessary for viral reproduction. A significant difference between the viral and cellular sequences was the lack of introns in the viral DNA (translated from the viral RNA), whereas the cellular *src* DNA had these intervening sequences. Based on their research and that of others, Bishop and Varmus proposed that some qualitative alteration (through point mutation, amplification, chromosomal translocation) of this normal cellular gene may play an important role as a cause of human cancer.[8] Before these experiments, decades of efforts to link viruses to human cancer had been unsuccessful.[9]

The UCSF team constructed a theory that turned an earlier hypothesis on its head. According to Bishop (1982, 1989) and Varmus (1989a), they had originally planned their experiments to test the virogene-oncogene hypothesis proposed by viral oncogene researchers (Huebner and Todaro 1969,

Todaro and Huebner 1972). In the end, Bishop and Varmus conjectured that the gene was *originally* part of the cell's normal genome, rather than a viral gene recently implanted by viruses into the normal cell (viral infection). They hypothesized that this gene became part of the cell's genome early in the organism's evolutionary history and thereafter was retained as genetic baggage. In short, they proposed that the viral oncogenes came from normal cellular genes rather than vice versa.[10]

> If cells contain genes capable of becoming oncogenes by transduction into retroviruses, perhaps the same genes might also become oncogenes within the cell, without ever encountering a virus. By means of accidental molecular piracy, retroviruses may have brought to view the genetic keyboard on which many different causes of cancer can play, a final common pathway to the neoplastic phenotype. (Bishop 1989:7)

> From these findings, we drew conclusions that seem even bolder in retrospect, knowing they are correct, than they did at the time. We said that the RSV transforming gene is indeed represented in normal cellular DNA, but not in the form proposed by the virogene-oncogene hypothesis. Instead, we argued, the cellular homolog is a normal cellular gene, which was introduced into a retroviral genome in slightly altered form during the genesis of RSV. Far from being a noxious element lying in wait for a carcinogenic signal, the progenitor of the viral oncogene appeared to have a function valued by organisms, as implied by its conservation during evolution. Since the viral *src* gene allows RSV to induce tumors, we speculated that its cellular homolog normally influenced those processes gone awry in tumorigenesis, control of cell growth or development. (Varmus 1989a:420)

In 1989, when Bishop and Varmus won the Nobel Prize for their work, newspapers summarized their thesis.[11] Dolores Kong wrote in the *Boston Globe* of October 10, " 'It has been known for a long time that cancer was in some sense a genetic disease. The importance of our finding is that one can identify explicitly the genes that play a role in cancer,' said Varmus. As a result of their contribution, more than 50 different genes involved in cancer have been identified, he said." In another article in that day's *Globe* Richard Saltus explained, "Bishop and Varmus discovered that the source of the gene [in a cancer-causing virus] had been the chickens themselves. An apparently normal gene was pilfered by the virus and was now able to cause cancer when it infected the birds."

In other words, Bishop and Varmus proposed that the gene that caused cancer was part of the cell's *normal* genetic endowment. They based their claim in part on "evolutionary logic." Since the gene was reportedly found in fish, which are evolutionarily quite ancient, they argued that the gene must have been conserved through half a billion years of evolution. Bishop and Varmus constructed a mutually creative relationship between their research and evolutionary biology. In addition to drawing on evolutionary arguments, they also injected their theories, inscriptions, and materials into the wealth of research, debates, and controversies in evolutionary biology.[12]

> Transduction by retroviruses is the only tangible means by which vertebrate genes have been mobilized and transferred from one animal to another without the intervention of an experimentalist. How does this transduction occur? What might its details tell us of the mechanisms of recombination in vertebrate organisms? What does it reflect of the potential plasticity of the eukaryotic genome? Can it transpose genetic loci other than cellular oncogenes? Has it figured in the course of evolution? How large is its role in natural as opposed to experimental carcinogenesis? These are ambitious questions, yet the means to answer most of them appear to be at hand. (Bishop 1983:347–348)[13]

Their critics simultaneously based their arguments on the theory's "evolutionary illogic." Why would a cancer gene be conserved through evolution? At first, the announcement of normal cellular *human* genes homologous to a *viral* oncogene was greeted with some skepticism.

> The first couple of years [after the discovery] were difficult. [Our findings that viral oncogenes had homologous sequences in normal cellular genes] were extended with some difficulty to a second and third gene . . . and then it was rapidly extended to all the rest [of the twenty known viral oncogenes]. We had to overcome a bias in the field. Our findings were first . . . Well, they were rationalized. It was hard for us to come to grips with the idea that a gene carried by a chicken virus that caused cancer was also in human beings. It didn't make sense. *Why would we have cancer genes as part of our evolutionary dowry?* (Bishop interview)

Nevertheless, Bishop argued that their proposal was also "evolutionarily fine."

> Our first evidence that human beings had this gene . . . although it evolutionarily looked just fine, there are a lot of biologists who don't really accept the evolutionary logic . . . So until the gene was isolated from humans and

shown to be the same as what we'd starte
At the outset, there was a lot of skeptic
found the same gene in human beings.
amused me. Everyone was perfectly happ
even mice, but it wasn't supposed to be in humans. I don't know why. But there was a lot of resistance to that. (Bishop interview)

Evolutionary logic was used here to both support and criticize their proposed explanation. On the one hand, the conservation of cancer genes did not make "evolutionary sense," while, on the other hand, the location of gene sequences similar to the viral oncogene in many different species pointed to evolutionary success of the gene sequence. The way out of this apparent inconsistency, as it turned out, was to show that the gene sequence was essential to normal growth and development.

Links to Developmental Biology

Normal growth and development are research problems that form the basis of developmental biology. This has been, and remains, an established and popular field of biological research. At the time of the initial announcements by Bishop and Varmus, they proposed that their "normal" proto-oncogene had something to do with cell division. Later, as researchers in molecular biology and biochemistry on normal growth and development began proposing the existence of growth factor genes based on research on growth factor protein, this work became associated both theoretically and concretely with Bishop and Varmus's work on oncogenes. For example, Russell Doolittle, a molecular evolutionist and chemist at the University of California at San Diego, and Mi[illegible] Cancer Research Fund in London both rep[illegible] platelet-derived growth factor (PDGF, a pro[illegible] for normal growth and development) was r[illegible] or the protein product of the *sis* oncogene [illegible] Doolittle et al. 1983, Waterfield et al. 1983). [illegible] reported finding that the epidermal grow[illegible] was identical to an oncogene *(erbB)* prote[illegible] and Bishop group.

This link between normal growth factors and proto-oncogenes provided an evolutionarily acceptable explanation for finding that potentially cancer-causing genes were conserved through time, as Bishop argued in a 1983 review article and in a 1984 interview.

> The logic of evolution would not permit the survival of solely noxious genes. Powerful selective forces must have been at work to assure the conservation of proto-oncogenes throughout the diversification of metazoan phyla. Yet we know nothing of why these genes have been conserved, only that they are expressed in a variety of tissues and at various points during growth and development, that they are likely to represent a diverse set of biochemical functions, and that they may have all originated from one or a very few founder genes. Perhaps the proteins these genes encode are components of an interdigitating network that controls the growth of individual cells during the course of differentiation. We are badly in need of genetic tools to approach these issues, tools that may be forthcoming from the discovery of proto-oncogenes in Drosophila and nematodes. (Bishop 1983:347–348)

> And it took us a while to convince people that [these genes] might have a different purpose in the normal body . . . But if something went wrong with them, they would become cancer genes as they were in the virus. (Bishop interview)

Bishop expanded the number of research problems in his laboratory from the study of one viral oncogene to studies of several viral oncogenes and their related proto-oncogenes. By 1984 some members of the laboratory were asking questions regarding the normal functions of the proto-oncogenes in developmental biology.

> My laboratory doesn't much resemble what it was ten years ago . . . The work's evolved in response to progress in the field. You get one problem solved, and you move on to something new that presents itself. A number of people in my laboratory are explicitly interested in normal growth and development. They're here because we believe that the cellular genes we study are probably involved in normal growth and development. And I wasn't studying cellular genes involved in normal growth and development fifteen years ago . . . There is a conceptual and probably mechanistic connection between cancer and development. But I'm not a developmental biologist, and I haven't read seriously in the field. There are people in my laboratory who will probably become developmental biologists as they fashion their own careers. (Bishop interview)

The links between viral and cellular oncogenes and developmental biology were concretized in a student who worked both in his laboratory and in a *Drosophila* genetics laboratory. "I have a major collaboration with another member of the biochemistry faculty here, a Drosophila geneticist, because we use genetic analysis in Drosophila to try to see what the genes we study do in development. And I'm not a geneticist, and he's not a student of oncogenes, so that's a necessary collaboration. We have joint students between us" (Bishop interview).

Mutual Translation: Molecular Biological Oncogenes and Tumor Virological Oncogenes

In 1978, soon after the announcements by tumor virologists Bishop and Varmus, a few molecular biology laboratories attempted to develop gene transfection techniques to study cell transformation.[14] Gene transfection methods at that time involved extracting samples of DNA from transformed cells and introducing them into "normal" cells to see if the gene for transformation could be transferred into the normal cells.[15] These researchers soon reported that they had found cancer genes similar to Bishop and Varmus's proto-oncogenes.[16]

In one set of experiments, molecular biologists in Weinberg's laboratory at the Whitehead Institute at the Massachusetts Institute of Technology first exposed "normal" mouse cells to DNA from mouse cells that had been transformed by chemical carcinogens.[17] The outcome, as reported by the researchers, was the transformation of the normal cells into cancer cells. They then extracted DNA from these transformed cells and used them to transform normal cells. The results were used to claim that the normal cells were transformed by the transfected DNA and that the transfected DNA therefore contained active cancer genes (Shih et al. 1979). These and other researchers then applied this gene transfection assay as well as secondary DNA transformations to human tumor cell lines and primary tumors. After cloning and isolating the genes, they claimed that the human DNA sequences obtained through transfection similarly were transforming agents (Der, Krontiris, and Cooper 1982; Parada et al. 1982; Santos et al. 1982). Weinberg (1983:127A) concluded from the experimental outcomes that "the information for being a tumor cell [was] transferred from one [mammalian] cell to another by DNA molecules."

> The successful isolation of transforming DNA in three laboratories [of Barbacid, Cooper, and Weinberg] by three different methods directly associated transforming activity with discrete segments of DNA. No longer was it necessary to speak vaguely of "transforming principles." Each process of molecular cloning had yielded a single DNA segment carrying a single gene

> with a definable structure. These cloned genes had potent biological activity . . . The transforming activity previously attributed to the tumor-cell DNA as a whole could now be assigned to a single gene. It was an oncogene: a cancer gene. (Weinberg 1983:130)

During this time, many molecular biologists were attempting to clone transfected cancer genes, but only those at a few laboratories had the skills and tools to do so. These few laboratories, especially those of Robert Weinberg at the Whitehead Institute at the Massachusetts Institute of Technology, Michael Wigler at the Cold Spring Harbor Laboratory, and Mariano Barbacid at the National Cancer Institute in Washington, D.C., raced to be the first to clone transfected human cancer genes. Researchers in Weinberg's laboratory eventually managed to clone Earl Jensen's bladder carcinoma gene (the EJ human bladder cancer gene), only to find that it was the same gene (called T24 human bladder cancer gene) cloned a month earlier by researchers in Wigler's laboratory (Goldfarb et al. 1982) and by Barbacid's laboratory (Der, Krontiris, and Cooper 1982). More significant, soon after cloning the EJ gene, a [illegible] aboratory conducted homology searches fo [illegible] nd found that it was similar to the *ras* (rat [illegible] t al. 1982). Weinberg was not happy about [illegible] t instead of finding a new cancer gene, We [illegible] le of the human cell transforming activity [illegible] month after Chiaho Shih secured his EJ p [illegible] to the Weinberg lab what neither Bob nor [illegible] nted to hear . . . The human cancer genes [illegible] old genes they'd been working with for years. They were the genes that Weinberg had tried so desperately to put behind him" (Angier 1988:110).[18]

As one of a handful of research groups with the skills and technology for transfecting mammalian cell DNA, Weinberg's lab had few competitors doing such cutting-edge work on human cancer genes. The homology with viral oncogene *ras* meant that retrovirologists could use the well-honed tools to study the EJ human cancer gene. (As a subtext, this homology meant that Weinberg's laboratory had found not a new human cancer gene but another member of Bishop and Varmus's set of proto-oncogenes. Thus if a Nobel Prize was awarded for this work, it would go to Bishop and Varmus and would not be shared by Weinberg.)

> The Weinberg scientists were justified in their despair. They had recently convinced themselves that they were breaking new ground in cancer biology and they thought they had to share their field with only a scattering of

> investigators. In light of Luis's discovery, however, the Weinberg lab knew that every retrovirologist in the United States and abroad would converge on human oncogenes. Whenever scientists sniff the possibility that their work relates to human cancer rather than just to rat or chicken cancer, they pounce. Human cancer is where the grants are, and cancer research gets noticed. Earl Jensen's bladder carcinoma gene—that is, the *ras* gene—was about to become the most fashionable gene in molecular biology. A scientist wouldn't even have to be a passable transfector to study the oncogene; he or she could approach it through the experimental graces of the Harvey sarcoma virus. (Angier 1988:115)

After these initial reactions passed, the true race was on.

> Once the transfection artisans had accepted the threat of stepped-up competition, they realized that the homology between their genes and viral genes was, in fact, a scientific bonanza. Now they didn't have to bother with messy biochemical means to find the protein encoded by the oncogene:[19] Ed Scolnick had alr ten years
> that he'd worked on the *ras* *s* protein
> . . . He and others knew th searchers
> even knew where in the cell nnection
> between EJ and *ras* was of ushed the
> field ahead by several years ever, that
> Scolnick didn't know about *as* gene,"
> said Weinberg. "Now *there* 6)

After subsequent research, Weinberg's laboratory reported that a single point mutation had caused this single normal *ras* gene to become a cancer-causing gene (Tabin et al. 1982).[20]

Weinberg was more sedate than Angier in his writing of these events. In 1982, he tentatively proposed that his transfected oncogenes were of a class with the oncogenes reported by tumor virologists Bishop and Varmus.

> A second question concerns the relation of these oncogenes to those which have been appropriated from the cellular genome by retroviruses and used to form chimeric viral-host genomes. The most well known of these genes is the avian sarcoma virus *src* gene, the paradigm of a class of more than a dozen separate cellular sequences. Do these two classes of oncogenes, those from spontaneous tumors and those affiliated with retroviruses, overlap with one another or do they represent mutually exclusive sets? Although the answer to this is not yet at hand, it will be forthcoming, since many of

> the sequence probes required to address this question are already in hand. (Weinberg 1982:136)

Weinberg (1982:135) reported that while "the study of the molecular biology of cancer has until recently been the domain of tumor virologists," it now was also the domain of molecular biologists. In 1983, he and his associates claimed to have confirmed the equivalence between these sets of oncogenes.

> Two independent lines of work, each pursuing cellular oncogenes, have converged over the last several years. Initially, the two research areas confronted problems that were ostensibly unconnected. The first focused on the mechanisms by which a variety of animal retroviruses were able to transform infected cells and induce tumors in their own host species. The other, using procedures of gene transfer, investigated the molecular mechanisms responsible for tumors of nonviral origin, such as those human tumors traceable to chemical causes. We now realize that common molecular determinants may be responsible for tumors of both classes. These determinants, the cellular oncogenes, constitute a functionally heterogeneous group of genes, members of which may cooperate with one another in order to achieve the transformation of cells. (Land, Parada, and Weinberg 1983:391)

Bishop tentatively supported the claims of Weinberg and Geoffrey Cooper.

> Weinberg and Cooper have evidently found a way of transferring active cancer genes from one cell to another. They have evidence that different cancer genes are active in different types of tumors, and so it seems likely that their approach should appreciably expand the repertory of cancer genes available for study. None of the cancer genes uncovered to date by Weinberg and Cooper is identical with any known oncogene. Yet it is clearly possible that there is only one large family of cellular oncogenes. If that is so, the study of retroviruses and the procedures developed by Weinberg and Cooper should eventually begin to draw common samples from that single pool. (Bishop 1982:92)

Indeed, the list of oncogenes as constructed by review articles expanded to include the genes studied by Weinberg and Cooper. In 1984 the list included over twenty oncogenes (including viral oncogenes) and their homologous proto-oncogenes. By 1989 the list had expanded to sixty. According to Bishop (1989:11, emphasis added), "the definition of proto-oncogene had now become more expansive, subsuming any gene with the potential for conversion to an [activated] oncogene—by the hand of nature in the cell, or *by the hand of the experimentalist* in the test tube."

To summarize, a few molecular biologists constructed an equivalence between their cancer-causing genes and the proto-oncogenes of tumor virologists. They argued that their cancer-causing genes were in the same class of cancer-causing genes reported by tumor virologists. This representation expanded the category of proto-oncogenes to include genes that had been transformed by chemicals reported to be carcinogens in volumes of previous studies on cells, on whole organisms, and especially on humans. The work in Weinberg's laboratory linked carcinogenesis studies, human cancer, and oncogenes. His was one of the first laboratories to claim to have found cellular oncogenes in human tumor cell lines. This finding simultaneously provided a new link between Bishop and Varmus's oncogene and carcinogenesis studies. As researchers embraced one another's work, the concept of a normal gene as a cancer-causing gene stabilized.

There have since been many debates among oncogene researchers about specifics, but the general proto-oncogene theory became the most popular theory of cancer causation in the 1980s. (See Figure 4.1 for an early schematic diagram of oncogene activation through the relocation of a part of a chromosome to another chromosome.)

Links with Clinical and Epidemiological Genetics

Oncogene researchers also proposed links with clinical and epidemiological genetics that they argued might lead to treatments and possibly to cures. For example, Bishop stated, "Medical geneticists may have detected the effects of cancer genes years ago, when they first identified families whose members inherit a predisposition to some particular form of cancer. Now, it appears, tumor virologists may have come on cancer genes directly in the form of cellular oncogenes" (Bishop 1982:91).

In 1982 Bishop also proposed that the proto-oncogene might lead to "a final common pathway" for all cancers that would link and explain research in chemical, viral, and radiation carcinogenesis and normal growth and development. (See Figure 4.2 for Bishop's schematic drawing of this final common pathway and Figure 4.3 for another schema published in the *New Scientist* in 1983.) "Normal cells may bear the seeds of their own destruction in the form of cancer genes. The activities of these genes may represent the final common pathway by which many carcinogens act. Cancer genes may

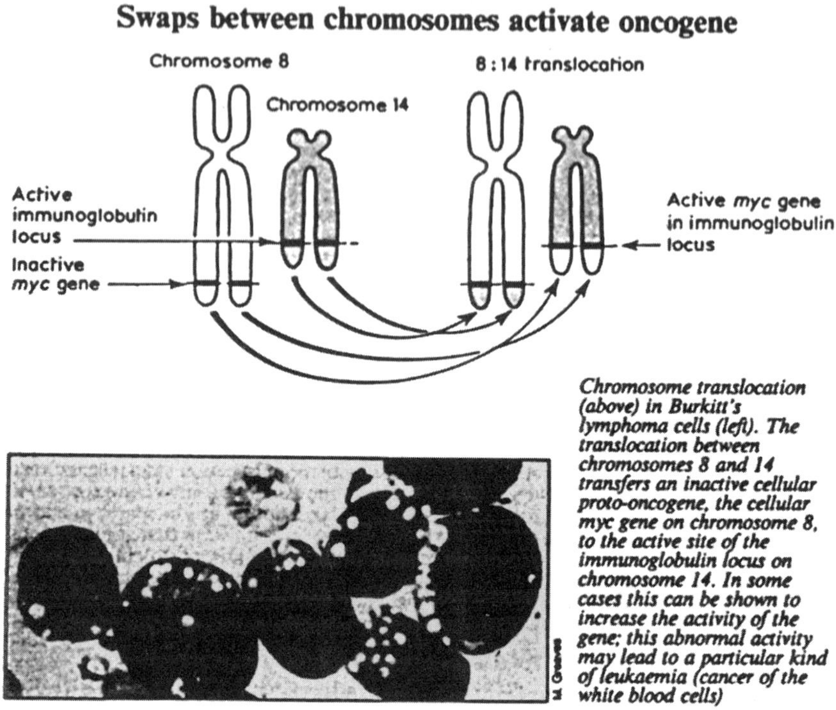

Swaps between chromosomes activate oncogene

Chromosome 8
8 : 14 translocation
Chromosome 14
Active immunoglobulin locus
Inactive *myc* gene
Active *myc* gene in immunoglobulin locus

Chromosome translocation (above) in Burkitt's lymphoma cells (left). The translocation between chromosomes 8 and 14 transfers an inactive cellular proto-oncogene, the cellular myc gene on chromosome 8, to the active site of the immunoglobulin locus on chromosome 14. In some cases this can be shown to increase the activity of the gene; this abnormal activity may lead to a particular kind of leukaemia (cancer of the white blood cells)

Figure 4.1 An early schematic diagram of oncogene activation by chromosomal translocation. From Robertson 1983:689; reprinted by permission.

be not unwanted guests but essential constituents of the cell's genetic apparatus, betraying the cell only when their structure or control is disturbed by carcinogens" (Bishop 1982:92). Viruses, he continued, were merely the tool that made visible this underlying mechanism. "At least some of these genes may have appeared in retroviruses, where they are exposed to easy identification, manipulation and characterization" (Bishop 1982:92). Bishop reiterated this view in his Nobel Lecture: "By means of accidental molecular piracy, retroviruses may have brought to view the genetic keyboard on which many different causes of cancer can play, a final common pathway to the neoplastic phenotype" (Bishop 1989:7).

Weinberg similarly speculated broadly that the proto-oncogene theory accounted for findings in many lines of cancer research. "What is most heartening is that the confluence of evidence from a number of lines of research is beginning to make sense of a disease that only five years ago seemed incomprehensible. The recent findings at the level of the gene are consistent with earlier insights into carcinogenesis based on epidemiological data and on laboratory studies of transformation" (Weinberg 1983:134).

In a volume entitled *RNA Tumor Viruses, Oncogenes, Human Cancer and AIDS: On the Frontiers of Understanding,* the editors Philip Furmanski, Jean Carol Hager, and Marvin A. Rich proclaimed that "we must turn these same tools of molecular biology and tumor virology, so valuable in dissecting and analyzing the causes of cancer, to the task of understanding other equally critical aspects of the cancer problem: progression, heterogeneity, and the metastatic process. These are absolutely crucial to our solving the clinical difficulties of cancer: detection, diagnosis and effective treatment" (1985:xx).[21]

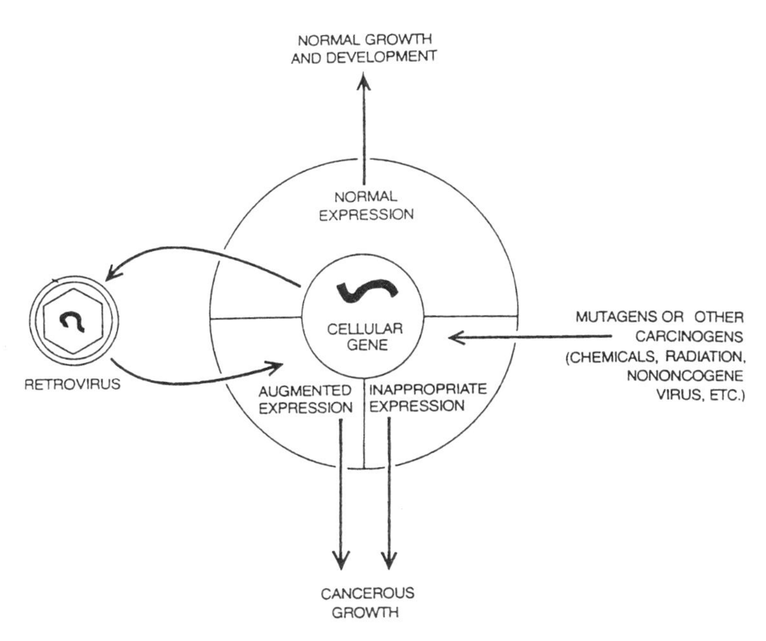

CANCER-GENE CONCEPT, supported by oncogene data and other preliminary evidence, suggests a unifying explanation for various forms of carcinogenesis. The common central element is a group of cellular genes required for normal growth and development. Transplanted into a retrovirus genome (*left*), such a gene becomes an oncogene. Cancer can also result if the cellular gene is affected by any of a wide variety of mutagens or other carcinogens (*right*).

Figure 4.2 J. Michael Bishop's schematic drawing of his vision of "the final common pathway" by which many different carcinogens act to create the neoplastic phenotype, a tumor. From Bishop 1982:91; copyright © 1982 by Scientific American, Inc.; all rights reserved.

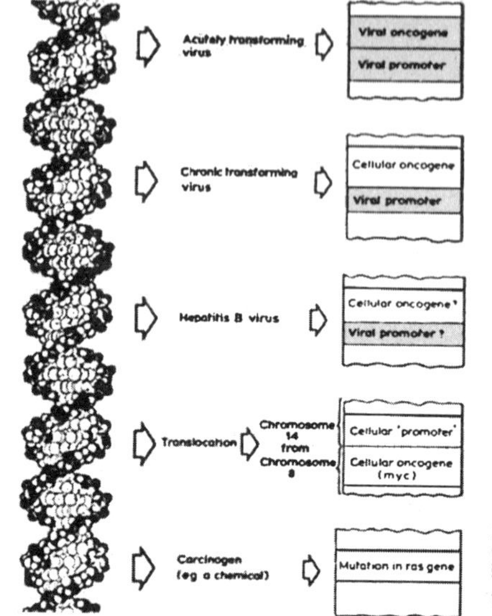

Figure 4.3 Another schema of "the final common pathway" where "the perversion of genes" is implicated in the path to cancer. From Robertson 1983:690; reprinted by permission.

Tumor Virology and the National Cancer Institute: Constructing Continuity with the Past

Proto-oncogene announcements also provided music for institutional ears. In their investigations of RNA tumor viruses, Varmus and Bishop reported finding specific genes in the viruses that transformed cultured cells and caused tumors in laboratory animals. Most of the tumor virology research was funded by the National Cancer Institute's Viral Cancer Program (VCP), which was established as a contract program in 1964 on the premise that many human cancers were virally induced. The NCI is one of the many divisions within the National Institutes of Health. The NIH administers federal health policies and maintains a phalanx of research facilities, staff, and resources in the area around Washington, D.C. The National Cancer Institute focused on the role of viruses in cancer etiology through heavy funding commitments to the VCP. Many virologists and molecular biologists were funded by the NCI through this program to study what are now known as DNA tumor viruses and retroviruses (or RNA tumor viruses).

> Frank Rauscher was head of NCI [then]. He was a virologist . . . a molecular biologist. And he said, "Cancer is the result of a virus. And if we can isolate and understand the virus, we can solve cancer. And we'll do it in two years." I had dinner with him in Chinatown over in San Francisco in—I forget what year it was now—and he was sure that we were going to solve the cancer problem in two years. It was very simple. He probably wouldn't like to remember that now. But he set out to do it, and he developed the so-called Virus Cancer Program at NCI . . . He went to Congress, and he got a lot of money for this program. He developed a team. It was kind of directed research. And the team went around the country and got all kinds of scientists in all different fields to focus on the viruses associated with human cancer. I was one of them, Bishop was one of them, Varmus was one of them.[22]

The Virus Cancer Program promoted by Rauscher had been designed to run along the lines of "scientific management" by industrial sociologist Louis Carrese and physician Carl G. Baker. However, despite years of effort, researchers were unable to convincingly link most naturally occurring tumors in animal and human populations to viruses or viral oncogenes.[23] Viral oncogenes were reported to cause cancer only in vitro and in laboratory animals. The following comment by DeOme is only one of many made in this vein. After a whole literature on chemical carcinogenesis was produced, "the virologists got into this [cancer]. My friend Stanley, who was then head of the Virus Laboratory—Nobel Laureate, would say publicly that all tumors would be found to be produced by viruses. I called him up and said, 'why do you make a statement like that?' Nevertheless, the virologists were at it and very successfully. The only animal they had trouble with was man . . . He didn't ever have virus-produced tumors" (DeOme interview).

In 1971 the National Cancer Act was initiated and signed into action by President Richard Nixon (Rettig 1977).[24] Despite the lack of evidence that viruses produced human cancers, the act continued to fund allocations to virus cancer research. This move drew heavy criticism from biological and biomedical scientists. Controversy raged in scientific circles and in the public press over the huge sums of money concentrated on the virus cancer program. In the views of many critics, viruses were considered to be an unsubstantiated and unlikely cause of human cancers. Furthermore, throwing money at a disease in what was viewed as a politically motivated "cancer crusade" was seen as the wrong way to do science.[25] A very prominent cancer researcher stated his views thus: "I had and still have mixed feelings [about the War on Cancer]. There's no doubt it brought a lot of people into this area, supplied a lot of people with money. But it probably sponsored more crappy research than almost any scientific campaign I can remember. Some of the stuff that was being done, because there was a lot of money available, was abysmal. I'm not certain that there were enough review panels at that time" (Sander interview).

Many scientists criticized the contractual dispensation of research funds on the basis of concerns about the quality of work produced through such scientific management and about inequitable distribution of funds primarily to NIH laboratories and scientists. Critics also expressed concern about the growth of the NCI at the expense of other institutes within the NIH.[26]

In response to the criticism, an ad hoc committee was constituted to review the VCP. In 1974 the committee, headed by Norton Zinder of Rockefeller University, submitted an extremely critical report to the National Cancer Advisory Board (NCAB), which oversees the work of the entire NCI. In 1980, as a further consequence of this and other in-house controversies, NCI leaders decided to break the VCP up and integrate the pieces into other NCI programs. This overhaul had cumulative negative effects on viral cancer research funds.

Given this controversial history, it is not surprising to find that several administrators and scientists in the early 1980s used the proposed role of proto-oncogenes in causing human cancer to justify past investments in viral oncology. Bishop, for example, stated: "The study of viruses far removed from human concerns has brought to light powerful tools for the study of human disease. Tumor virology has survived its failure to find abundant viral agents of human cancer. The issue now is not whether viruses cause human tumors (as perhaps they may, on occasion) but rather how much can be learned from tumor virology about the mechanisms by which human tumors arise" (1982:92).

James D. Watson, molecular biologist and then director of the Cold Spring Harbor Laboratory, similarly used viral oncogenes and human homologs to justify the earlier NCI Tumor Virus Program that funded much of the research at Cold Spring Harbor. "Given the still prevalent unfair public misconception that the NCI Tumor Virus Program was a failure, and the new strong possibility (fact?) that most if not all of viral oncogenes have their human counterparts, the time is more than ripe for NCI to point out how well the public purse has, in fact, been used" (quoted in DeVita 1984:1).

Vincent T. DeVita, Jr., then director of the National Cancer Institute, agreed with Watson. "I would concur with Dr. Watson's views. Recent discoveries of retrovirus oncogenes and their human homologs make it reasonable for one to state that few areas of research have been so fruitful. We are closer to understanding the underlying abnormality of growth that is cancer than the architects of the NCP [National Cancer Program] could have imagined in 1971." Extending his discussion from the VCP to the entire NCP and the contract system of funding, DeVita stated that

> we have often been asked if the NCP has been a success. While I acknowledge a bias, my answer is an unqualified "yes." The success of the Virus Cancer Program which prompted this essay is a good example. Since its inception, this Program has cost almost $1 billion. If asked what I would pay now for the information generated by that Program, I would say that the extraordinarily powerful new knowledge available to us as a result of this investment would make the entire budget allocated to the NCP since the passage of the Cancer Act worthwhile. There may well be practical applications of this work in the prevention, diagnosis, and treatment of cancer that constitute a significant paradigm change. The work in viral oncology has indeed yielded a trust fund of information, the dividend of which defies the imagination. (DeVita 1984:1–5)[27]

Both oncogene researchers and cancer research administrators argued, then, that the "new" oncogene research would be based on the "extraordinarily powerful new knowledge" produced by past investments. The viral cancer genes constructed from the investments of the NCI in the Viral Cancer Program during the 1960s and 1970s have in the 1980s and 1990s become human cancer genes through the proto-oncogene theory and recombinant DNA technologies. Viral cancer genes with no previous connection to human cancer have become human cancer genes. The NCI's and James Watson's earlier choices and predictions have been proven fruitful and justified, while Bishop's theory has gained credibility from DeVita's and Watson's translations. Here we see a case of mutual translation for mutual benefit.

Indeed, DeVita used the proto-oncogene theory to justify the entire National Cancer Program in 1984. DeVita told me that he used oncogenes to sell the general future program of molecular genetic research on cancer to Congress:

> Molecular genetics is a term nobody in Congress understands, really. Oncogenes they know. How do they know? I tell them. I can explain oncogenes to them much better than I can explain molecular genetics. When I point my finger at a Congressman, I say, "Mr. So-and-So, you and I both have genes in us, which we believe are the genes that are responsible for causing cancer." It gets their attention. They say, "My God! What do you mean I have genes in me . . .?" I have to explain it to them. If I tried to explain molecular genetics, they'd fall asleep on me.[28]

DeVita could explain oncogenes to the members of Congress in part because the term "gene" meant something to them, whereas they had no frame of reference for interpreting "molecular genetics" at that time. "Gene" then did symbolic work for DeVita. It was a shared concept, even if the specific definitions might have been different for DeVita and the various members of Congress. Although we cannot assume that the symbolic meaning of the term was the same for everyone, "gene" has come to signify a power to affect outcomes in phenotype.[29]

The audience listening to DeVita's discussion of oncogenes understood these genes to be the first step in the process toward tumor and death. However much their views may differ in terms of "scientific" nuance or sophistication, the statement that we all have genes that can be triggered to cause cancer can engender great fear in congressional members and their constituents because they share the basic notion of genetic causality. This fear, however, was laid to rest by the claim that a unifying pathway to all cancers may exist and that there may be ways to intervene in this pathway. Thus, DeVita found "oncogenes" useful for translating the interests of the sitting members of Congress.

National Cancer Institute administrators supported and promoted the proto-oncogene theory for several reasons. Their sponsors were Congress and the public it represented, including other scientists. The proto-oncogene theory provided them with both the justification for past research investments in the VCP and a product to present to Congress. Justifying their past was part of the effort to build their credibility in the present and for the future.

The National Cancer Institute supported and promoted the proto-oncogene theory and molecular genetic approaches to cancer in general. From 1978 the NCI increasingly awarded its basic research funds, versus clinical and educational funds, to molecular biological studies of cancer. Before 1983 the National Institutes of Health had no category for oncogenes in its computer data bank of funded projects.[30] In 1983, NIH instituted an oncogene category and listed the number of sponsored projects at fifty-four and the number of dollars disbursed to oncogene projects at $5.5 million. In 1984 Director DeVita summed up NCI's investments in molecular genetic cancer research.

> [In 1984] we had $198 million in molecular genetics . . . [That figure] includes oncogenes, but it also includes people who are walking up and down the genome, tripping on oncogenes but looking for something else. And they're going to find the regulatory elements that control the oncogenes, [which is] really the major step. Oncogenes have told us something very

important, but now what you want to find out is what regulates these genes so that you can use this information to turn them on and off. (DeVita interview)

There is other evidence of NCI's commitment to molecular genetic approaches. In 1981, while NCI leaders were in the process of reorganizing research at the Frederick Cancer Research Facility, they decided to shift the facility's emphasis. "We put three or four crackerjack oncogene scientists up there, and they're up there cranking out the data and having a fun old time," said DeVita. NCI also appointed a viral carcinogenecist who had worked with George Todaro, one of the originators of an early version of the proto-oncogene theory, to the position of associate director of the entire NCI. He oversaw the Frederick Cancer Research Facility and specifically kept track of oncogene research progress (Shapley 1983:5).

Theory as Collective Work

J. Michael Bishop, Harold T. Varmus,[31] and their colleagues, working in the late 1970s, argued that their proto-oncogenes accounted for findings in many other lines of cancer research and represented a unified pathway to cancer in humans and other "higher" organisms. However, this model per se could not account for their success. Instead, Bishop and Varmus and their colleagues crafted and recrafted the theory by constructing equivalences between previously unequivalent units of analysis—for example, between genes controlling cancer and genes controlling normal growth and development. For other lines of research such as those under the auspices of the Virus Cancer Program, they constructed continuities through time and locales while introducing novelty into the existing lines of research. The theory ultimately provided alternative ways of studying and explaining biological activities (such as carcinogenesis) at the molecular level and gave form to other scientists' efforts to recraft their existing research using this new unit of analysis.

Proto-oncogene theorists did not do all of this by themselves. Instead, they and their colleagues created working relationships between the proto-oncogene theory and projects (both past and present) in evolutionary biology and population genetics, medical genetics, tumor virology, molecular biology, cell biology, developmental biology, and carcinogenesis, as well as with past and present agendas of established institutions and funding agencies. While early oncogene theorists suggested potential connections between their theory and questions in other fields of biology, researchers in these other fields soon, on their own, extended their own lines of research by linking them to proto-oncogene research. These interactions produced a robust proto-oncogene theory and the oncogene line of research.

Traces of Continuity

Bishop and Varmus's creativity was in turning earlier virus studies on their heads. Their theory held that, contrary to the existing hypothesis that viruses had transferred their cancer-causing sequences to humans, viruses had gained their cancer-causing sequences from normal cellular DNA. The movement of genetic elements reversed directions literally and figuratively, but the same or homologous gene sequence continued through time and space. Critically, their work implicated the same (or similar) DNA sequences that previously had been studied extensively and that had gained enormous material and sentimental support from many researchers, organizations, and institutions.

The crafting of the proto-oncogene theory was situated within a network of historical and contemporary commitments to particular representations and representational practices in cancer research, such as inbred animals, cell lines, Temin's proto-virus ideas, the provirus theory, viral cancer research and collected materials, new molecular genetic methods, and prior commitments of funding agencies. I have traced the continuities in research tools, practices, and concepts that produced proto-oncogene research. With this network and on these historical pylons, by 1983 the new "unifying" proto-oncogene theory of cancer had been adopted and used as the basis of investigations in several new and established laboratories in several lines of biological and biomedical research as well as by the National Cancer Institute.

5 Distributing Authority and Transforming Biology

> *One must rather conduct an ascending analysis of power, starting . . . from its infinitesimal mechanisms, which each have their own history, their own trajectory, their own techniques and tactics, and then see how these mechanisms of power have been—and continue to be—invested, colonised, utilised, involuted, transformed, displaced, extended etc., by ever more general mechanisms and by forms of global domination.*
>
> —*Michel Foucault,* Power/Knowledge

How did oncogenes become "the rulers of a new realm"? Arguing that one's theory unifies many lines of research will not make others agree with the claim and rush to pursue experiments based on the theory. Theoretical constructions alone cannot account for the growth of oncogene research. Neither can technologies, however glorious, explain the "oncogene revolution." Instead, the set of commitments made to the proto-oncogene theory by researchers in several lines of work and by the National Cancer Institute, combined with timing and a lot of hard work, produced a bandwagon of new commitments to oncogene research.

As earlier chapters have shown, by 1982 the proto-oncogene theory had won commitments from researchers in several different lines of research. And oncogene researchers had proposed the proto-oncogene theory just at the time when recombinant DNA and other molecular genetic technologies were being standardized and attracting attention in most biological subdisciplines.

By 1982, this theory and these methods together constituted a *package* that began to attract new investigators, students, and private industrial laboratories to study oncogenes. The combination of the oncogene theory and recombinant DNA technologies provided the tools with which researchers could build new research. They could play with, extend, challenge, and reconstitute both the statements made by these early oncogene theorists and their own ongoing research enterprises.

Standardized recombinant DNA technologies, the proto-oncogene theory and its historical pylons, including various previous and extant institutional, organizational, material, theoretical, and political commitments and contexts, together constituted a package of theory and techniques that served as a dynamic interface linking different worlds across time and space. In analytic terms, the package provided a flexible set of conventions for action into many different situations. This set of conventions provided broad outlines for action and recreated research problems in laboratories across the United States.

I portray the bandwagon of oncogene research as co-constructed with the package. The development and maintenance of this research and the package of conventions gave form to and modified each other through time. As the package captured the interests of and came to be reconciled with members of other lines of research, students, funding agencies, and suppliers, the proto-oncogene theory and technologies were (re)constructed in many other laboratories.[1] Proto-oncogene research protocols and molecular genetic technologies have transformed practices in many laboratories and lines of research, consequently changing the kinds of knowledge and objects they produce.

The Oncogene Bandwagon

In 1984 the journal *Science* published an article stating, "the evidence implicating oncogenes as causes of human cancers, although still circumstantial, has been accumulating rapidly during the past few years" (Marx 1984:2). Oncogene research articles crowded the pages of general science journals like *Science* and *Nature,* in addition to journals specializing in biochemistry, molecular biology, and cancer research. More generally, modern biology had become molecular biology.

Oncogene research was a distinct and entrenched phenomenon. Researchers referred to the "oncogene bandwagon" in conversations. Scientists acted on the basis of its existence. Oncogene research had grown to the point where it was sustained by its own momentum. That is, many researchers joined the bandwagon because it *was* a bandwagon, or, as one prominent proto-oncogene researcher called it, a "goldrush."

> *Researcher:* Whenever a branch of science is presented with a large number of possible experiments, people rush in and do them, because people want

to have results. And there have been, in the last five years in particular, an incredible number of profoundly exciting experiments that have come out of work with oncogenes. It's like the Goldrush of 1849. "There's gold in them there hills!" There's still a lot to do, and there are interesting experiments for people with various proclivities.

Interviewer: In comparison with other lines of research in cancer research, are you saying—

Researcher: Are there other lines? . . . What I'm suggesting is that there are very few people who are not profoundly influenced by thinking about oncogenes. And even those who would consider themselves not primarily students of oncogenes or molecular biologists, have nevertheless been forced for one reason or another to confront the significance of oncogenes. I don't see how you can work in cancer research and do anything reputable without considering your work in relation to the general notions that have been developed by this vast army of people who are working on the problem. (Sander interview)

Actors from diverse social worlds had committed their resources to molecular biological cancer research. These commitments included: (1) very large increases in funding allocations;[2] (2) designated positions in academic departments, research institutes, and private industrial laboratories; (3) accessible training and tools, including knowledge, standardized technologies, materials, and instruments; and (4) a cadre of researchers training in molecular biological skills. In other words, by 1984 an infrastructure of skills, funding allocations, committed researchers and teachers, positions committed to molecular biologists, biological material suppliers and supplies, and even whole companies and research institutes committed to oncogene research problems had been established. This infrastructure then constrained and influenced the decisions of new investigators. It served to maintain previous commitments as well as to gain new commitments.

Particularly in cancer research, a lot of science has been done in the last couple of years with the assumption that these cellular genes are cancer genes and that everything has been checked . . . You are forced to run with the pack to some degree to compete for grants. To publish at the same rate and same speed . . . [It also affects] the direction of your work . . . If everyone says that gene is the cancer gene, then you are more likely to look at it again and say, "is it indeed the cancer gene and how does it differ from this and how is it related to that?"[3]

A senior investigator commented negatively on this "snowballing bandwagon" (thus mixing metaphors).

> I think the reason for [the problem] has to do with the way science is funded. And the way young scientists are rewarded . . . [A] youngster graduates and . . . gets a job in academia, for instance. As an assistant professor, he's given all the scut work to do in the department . . . In addition, he must apply for a grant. And to apply for a grant and get a grant . . . nowadays, you have to first show that you are competent to do the [research], in other words, some preliminary data. So he has to start his project on a shoestring. It has to be something he can do quickly, get data fast, and be able to use that data to support a grant application . . . so that he can be advanced and maintain his job . . . Therefore, he doesn't go to the fundamental problems that are very difficult . . . So he goes to the bandwagon, and takes one little piece of that and adds to that well-plowed field. That means that his science is more superficial than it should be. And that's bad for the field of science. (Oakdale interview)

However, other investigators spoke of this situation as a positive process through which solutions to problems "get nailed down." They regard others working on "fundamental problems" as flailing along without knowing how to ask, much less solve, their problems.

Molecular biological cancer research by 1984 appeared to new investigators to be the research line of choice based on the collective commitments of many other actors, including researchers, funding agencies, research institutes, and private companies. According to these committed participants in science-crafting, their decisions to pursue oncogene research were based on their goals: to construct doable problems, produce novel findings, build careers, produce marketable products, and, in the words of Everett Hughes (1971), build successful "going concerns." Recombinant DNA and other molecular biological technologies and theories of cancer framed around DNA constituted a package that researchers and organizations could use to get their work done.

Participants based their evaluations of oncogene research as useful to getting their work done on four conditions: (1) the proto-oncogene theory offered the chance to pursue research on *human* cancer; (2) the package of proto-oncogene theory and recombinant DNA technologies provided a pathway to exploring new, uncharted territory ("sexy" problems); (3) researchers could incorporate the new, "hot" standardized recombinant DNA technologies into their laboratories at relatively economical start-up costs; (4) work in some laboratories had led to "dead-ends" or "roadblocks," while work in some cancer research institutes had been criticized as "old-fashioned." The package of proto-oncogene theory and recombinant DNA technologies could be reconciled with the complex, diverse, and dynamic organizational and intellectual requirements, goals, and conditions of work of researchers in several different lines of work.

Enrolling Allies to Oncogenes via the Theory-Methods Package

Oncogene theory proponents enrolled allies not only by reporting that their results accounted for findings in many other lines of cancer research, but also by framing and posing new doable problems on oncogenes for other researchers to investigate. They posed questions that scientists could experimentally investigate using recombinant DNA and other molecular biological technologies; that laboratories were already organized and equipped with resources to handle, or could relatively smoothly import the requisite resources to do so; and that satisfied significant audiences.

The proposed problems were both specific and general. Researchers could immediately begin experimentation on specific problems, while thinking of possible ways to translate more general problems into specific experiments.

Bishop's article "Cellular Oncogenes and Retroviruses" in the 1983 *Annual Review of Biochemistry* is an excellent example of mapping proposed problems onto established laboratory organizations and available technical skills. He first summarized work in several other lines of cancer research and then presented proposals for research that linked oncogenes with these other lines of work in cancer research and in biology generally, including experimental carcinogenesis, evolutionary biology, normal growth and differentiation, medical genetics, and epidemiology.

Oncogene theorists sent their probes for oncogenes to other laboratories and to suppliers, thus facilitating oncogene research in other labs by providing standardized tools. A postdoctoral fellow in Bishop's laboratory described the increasing demands for their probes: "We've had so many requests for our probes for [two cellular oncogenes] that we had one technician working full-time on making and sending them out. So we finally turned over the stocks to the American Type Culture Collection." While Bishop may have been quite astute and enthusiastic about promoting proto-oncogene research, clearly other researchers were *interested*. Enrollment is explicitly mutual.

The probes were more than physical materials. As discussed in Chapter 3, they were "black boxes," designed with reference to specific hypotheses about their involvement in cancer causation. More significant, they embodied the specific approach to a problem of the laboratories in which they had been constructed. Exporting probes is one way to standardize the world outside. With Bishop and Varmus's probe, one is more likely to find what Bishop and Varmus reported finding and less likely to find other possibilities than if one constructed a probe of one's own. Not only does the probe select for a particular reality it also preempts other candidates. Any researcher can call or write to ATCC to order the probes at the cost of maintenance and shipping. Thus, oncogene researchers made the tools for testing and exploring their theory available and accessible to a host of other researchers.

Oncogene theorists also taught and talked about their work to students and researchers in other biological disciplines. A molecular biologist described the positive response of cell biology conference participants to an oncogene talk. According to this respondent, most of the conference participants, uninitiated in the complexities of oncogene research, were awed by the lecture and unable to evaluate the difficulties and complexities in the data.

> I was at the cell biology meeting this year in San Antonio. [Previously] cell biology meetings really [didn't] have anything to do with cancer research. Cell biologists as a rule study different parts of the cell and tissues. It's not a cancer research oriented field really at all . . . They have big talks at these meetings that are sort of "what's new" and "what's interesting." This year two out of three talks on the first night were on oncogenes and the relation between the PDGF homology with an oncogene and the multistep oncogene theory where they took two different kinds of oncogenes and they complemented each other and then caused the tumor. So those were two of the three main topics . . . I was one of the few people there who actually know about the research in terms of being able to criticize it scientifically . . . So [to] . . . the general scientific public this sounds really good and everybody is all excited and they think we've got a handle on this and we can figure it out. And I'm sitting there thinking, "what about the fact that spontaneously transformed cells that cause tumors don't score on this assay?" You've never been able to find a gene in a spontaneously transformed cell that you can detect in the 3T3 cell assay, even when you use spontaneously transformed 3T3 cells. And yet those cells cause a tumor. And what about the fact that these genes aren't seen in many tumors? They might be an artifact of tissue culture when you culture the tumors, because you usually use tumor lines instead of primary tumors. And all that kind of stuff. Very few people know that much information to sit there and criticize. So everybody's very excited. The researchers themselves who are doing the work, the people who came there to give the talks, they're just flushed with success.[4]

Oncogene researchers also spoke about their work in the popular media. In 1984 *Newsweek* correspondents acclaimed the new proto-oncogene research and even discussed its potential for producing diagnostic aids and treatments.

> Such discoveries shed important light on the fundamental processes of cancer as well as the growth and development of all forms of life. In the future, they will surely lead to better forms of diagnosis and treatment. The presence in cells of abnormal amounts of proteins caused by gene amplifications, for example, could lead to sensitive new tests for certain kinds of cancer. As for treatment, scientists envision the development of drugs designed to specifically inhibit oncogenes. These would be far better than anticancer drugs that indiscriminately kill normal cells along with cancerous ones. "We would," says Frank Rauscher of the American Cancer Society, "be using a rifle rather than a shotgun." (Clark and Witherspoon 1984:67)

Commitments of Students, New Investigators, and Established Researchers

The growing commitments of tumor virologists, molecular biologists, and the NCI to oncogene and related molecular biological research became, in turn, further provocation for students, new investigators, and even researchers established in other lines of work to frame their theses and research problems in similar terms.

Besides considering interesting intellectual questions and the problem of curing cancer, new researchers had to attend to career development contingencies in making problem choices. The immediate foreground was filled with the exigencies of their daily work lives: researching and writing Ph.D. theses, establishing and maintaining laboratories and staff, publishing and gaining tenure, writing grant proposals, attracting and training students. In this situation, constructing doable problems that produce results someone will publish is a practical and pressing concern. Thus, desirable "cancer research" became "doable" research (see Chapter 7).

Given the structure of other commitments, students and beginning researchers gained major advantages for establishing their careers and laboratories by choosing to investigate problems under the rubric of oncogene research. By 1983 these advantages included clearly articulated experiments, research funds, high credibility, short-term projects, increased job opportunities, and the promised generation of further doable problems. The "down side" of this situation was that students and postdoctoral researchers had little space for maneuvering. Following other options could have meant removing themselves from "the game."

In 1982 the combination of proto-oncogene theory and molecular biological technologies provided clear problem and experimental protocols. By the end of 1983, a graduate student in an oncogene laboratory could explain his work as a set of logical steps. Students learn these logical steps from their laboratory directors. A senior oncogene investigator described his research problem formulation as a textbook reductionist, logical approach.

> It's always seemed to me, because I don't see any other way to go about it, that you have to employ a reductionist approach. You have to say, "I now know that out of the thousands of genes in a given cell, Genes A, B, and C are the genes that are screwed up in a cancer cell. And what's wrong with them is that this base has changed that way, and this gene has moved to that chromosome, and the effect of these changes—which you can spell out with nucleotide sequences—is to make a protein which has an altered kinase activity, or an altered location in a cell, or there is an abundance of the protein, or the protein's not properly regulated by some other factor which I now call 'Y' . . ." It's the description of those changes [that] in the long run lead to, first of all, fundamental understanding of how cell growth is regulated, and secondly, to clinical insights . . . how to make diagnoses at the earliest possible time [and how] to think about the ultimate in strategies—blocking the activity of genes that are instrumental in cancer. (Sander interview).

Obviously, research is rarely so neat and tidy as in this senior scientist's post hoc descriptions. However, the clarity of the descriptions of oncogene research trajectories speaks to the "logic" created by oncogene investigators as compared with the research process in many other lines of biological work where standardized packages are not available.

New investigators raised the credibility of their work by constructing problems using cloning, sequencing, Southern blots, and other molecular biological technologies. Latour and Woolgar noted that "credibility is a part of the wider phenomenon of credit, which refers to money, authority, confidence and, also marginally, to reward." For example, "the mass spectrometer is . . . an actual piece of furniture which incorporates the majority of an earlier body of scientific activity [in physics]. The cost of disputing the generated results of this inscription device [is] enormous" (Latour and Woolgar 1986:242).

Biologists understood that few scientists would be willing to attack findings based on results from technologies borrowed from physics or chemistry (e.g., liquid scintillation spectrometers), because physics ranks high among scientific "hierarchies of credibility" (Becker 1967). Although power is negotiated, the relative power of the negotiating parties matters (Strauss 1978a). Since many molecular biological methods came from physics and chemistry, they carry with them high credibility or more "cultural capital" (Bourdieu 1977), which increases oncogene researchers' negotiating power. That is, the credibility of molecular biology, physics, and chemistry becomes embedded in their "facts."

Careers of new investigators were also linked to molecular biological technologies. Since the early 1960s molecular biology technologies had steadily been adopted in many fields of research. By the early 1980s very few fields of research and biological disciplines lacked strong molecular components. University biology departments, research institutes, commercial biotechnology companies, and pharmaceutical companies vied for the best molecular biologists in the world (Stokes 1985, Wright 1986). Expansion of genetic screening programs for inherited diseases extended the impact of molecular biology to the public, including private citizens, insurance companies, and governments (Duster 1990, Holtzman 1989, Nelkin and Tancredi 1989, Suzuki and Knudtson 1989).[5] To graduate and postdoctoral students, then, oncogene research was linked to job opportunities. A postdoctoral student training in an oncogene laboratory explicitly stated the view that his research on oncogenes, with one foot in cancer and the other in molecular biology, would definitely get him a job in either the university or in industry.

Cancer research similarly was a well-funded arena, with a $1.2 billion budget for the NCI alone in 1984 ($1.3 billion for 1987). With additional funds from other National Institutes of Health, the National Science Foundation, the American Cancer Society, and other private foundations, cancer research was a thriving enterprise. At the intersection of the molecular biology and cancer research, oncogene research provided new researchers with important resources for building careers. A graduate student who had been studying cellular immunology and shifted his thesis problem to investigate an oncogene talked about the growth of commitment to the new line of work:

> I studied cellular immunology and am now doing c-*myc* [cellular *myc* oncogene] research. My professor said, "There's the funding, go for it." So I did. I wrote a grant to fund this oncogene research project. When you write

> a grant, it also forces you to get into it, to think up innovative ways of approaching the problem. Why not go for it? It makes sense to use the funding that's there to do your work. If it turns out to be significant work, then good. If not, you can always change later.[6]

Oncogene research, like other recombinant DNA–assisted research, also provided results in relatively short time frames. Most oncogene researchers published from two to fifteen articles a year depending on their professional status, the size of their laboratories, and the number of collaborations. These short time frames added fuel to the growth of oncogene research and drew attention away from other areas like cell biology. Several respondents blamed this focus on short-term payoff on the supporters of research rather than on the researchers themselves. Five-year projects were too "long-term" and not fundable. Of course, the concept of a long-term project in part resulted from the acceptance of short-term projects in oncogene research.

Finally, new and more established investigators reported that they engaged in oncogene research because of promised intellectual payoff and new generations of downstream questions. Reports of novel oncogene findings provided glimpses of new ways to study difficult and challenging problems. For example, in 1983, *Nature* published an article in its "News and Views" section entitled "Oncogenic Intelligence: The *Ras*matazz of Cancer Genes" (Newmark 1983). The article was just one of many published between 1978 and 1985 announcing exciting new findings from oncogene research. Researchers constructed novel, intellectually exciting representations, which they then used to construct further experimental questions.

The intellectual excitement was not limited to oncogene research but extended to other molecular biological research. Almost all respondents, independent of their political views about how the new molecular biological technologies should be used, echoed this excitement. A respondent, who had been conducting protein biochemical research, explained the excitement in terms of a whole new scale of analysis opening up to scientists. This new scale was the outcome of molecular biological technologies that allowed researchers not just to observe but to actually *change nature* in the laboratory.

> You can ask certain sorts of questions which you can't really answer with just the biochemical methodology . . . Genetics essentially involves modifying what's already there, rather than simply describing what's going on. It allows you to ask much more specific questions about which components of the system are necessary to do what. Recombinant DNA technology is starting to allow one to ask those sorts of questions in animal cells, tumor

> cells . . . questions which there is as yet no other way of approaching. (Martin interview)

Established researchers found the possibilities for exploring new scales of analysis useful. Consider a senior investigator who had been studying the effect of radiation on transforming cells in culture. After much excitement about the oncogene theories of carcinogenesis, he sent his student to train in recombinant DNA techniques in a nearby laboratory in order to test two hypotheses: first, whether radiation played a role in the mutation or transposition of one or several proto-oncogenes and, second, whether radiation damage to cells made it easier for the viral oncogene to become integrated into the normal cellular genome. The graduate student gained the benefits enrolled in the proto-oncogene theory, the senior investigator imported new skills and a new line of research into his laboratory, and radiation and oncogenes became linked.

Commitments of Research Institutes

In Chapter 4, I described commitments made to oncogene research by the National Cancer Institute. By 1986 other cancer research institutes had changed their agendas by hiring researchers trained in molecular biology and establishing the proper facilities. Among them was the Memorial Sloan-Kettering Cancer Center, the country's oldest and largest research and hospital complex devoted exclusively to cancer, which shifted its research focus from immunology to molecular biology. A molecular biologist, Paul Marks, was appointed to head the organization, and he replaced the old leadership with molecular biologists. A respondent outside the organization commented on Marks's agenda: "So now you have Memorial Sloan-Kettering in lockstep going toward molecular biology." Marks himself stated: "You have the feeling now that this research is making inroads toward the control and cure of the disease . . . Most of the answers to cancer lie down on the level of the genes, in our understanding of how cells differentiate and divide . . . You have to work with people who've been trained to think like that" (quoted in Boffey 1986:27).

Commitments of Private Industry

Despite the uncertain commercial payoff from oncogene research, several large pharmaceutical companies and major research and development (R & D) companies committed funds, researchers, and laboratories to oncogene research and recombinant DNA technologies (Koenig 1985). These pharmaceuticals included Hoffman-La Roche, Inc., SmithKline Beckman Corporation, Merck and Co., and Abbot Laboratories. Investing R & D biotechnology companies included Genentech and Cetus[7] and especially smaller companies aimed specifically at oncogene products including Oncogene Science, Inc., Oncogen, and Centocor Oncogene Research Partners.

One respondent, a vice-president of research and development at a commercial biotechnology company, regarded these commitments as efforts to "get in on the ground floor." Even if a particular company is not the home of the desired new discovery that leads to a patentable diagnostic or therapeutic product, it will have established the infrastructure for early entry into the race to produce the final commercial product(s). In 1985 a research director at Hoffman-La Roche stated, "If you're interested in [oncogene] products, you can't afford not to be in the race now" (Koenig 1985:25). Entering the oncogene market was regarded as both a race and a gamble.

These commitments to oncogene research on the part of private industry paralleled the commitments of new researchers. These commercial investments provided both job opportunities and affordable research tools for new investigators. Indeed, oncogenes, academic oncogene research commitments, and private industrial commitments interactively produced one another. A more colloquial way of putting this is that the scientific elite—the researchers, laboratories, institutes, and biotechnology companies—engaged in crafting science used the package to bootstrap themselves and their product, the proto-oncogene, into stronger positions.

How Did Oncogenes Become the "Rulers of a New Realm"?

Roads Not Traveled

In symbolic interactionist studies of science, we are fond of saying, "things could have been otherwise." We take this maxim from Everett Hughes, an early interactionist. However, it is difficult to demonstrate how things could have been otherwise, since we know only what happened, and even that we know only partially if at all.

The question posed in this chapter is, why did this line of research succeed? Why did so many people, organizations, and institutions join in oncogene research, rather than, say, in ecological or environmental research on cancer? Scientists could have taken the alternative path of environmental studies of cancer once the Ames test mentioned in Chapter 2 had been developed. Beginning in the early 1970s the Ames test was used to determine the mutagenic effects of carcinogens such as chemicals. There were some controversies about the reading of the test results; although all carcinogens cause mutations, not all agents that mutate DNA cause cancer. Despite controversies, during this early period, Bruce Ames's test provoked much excitement and interest. According to some of the scientists I interviewed who worked during this period, cancer researchers could have "gone environmental" rather than molecular at this time. Instead, many chose to study the molecular genetics of oncogenes.

During this same period, Sidney Epstein was studying the political and policy-making issues surrounding environmental carcinogens and published his ideas in *The Politics of Cancer* in 1978. Epstein argued that the causes of cancer were known and preventable, and that the barriers to prevention were political, not scientific.[8] Proctor (1995) argues that the war against cancer still continues to go badly in part because the priorities and practices of cancer research are shaped by political priorities, ideological gaps, and interests and disinterests; by government and industrial support; and by professional or institutional parochialisms. He recounts a history of cancer debates to show how politics intervened in research to redirect attention away from environmental causes of cancer and, to his mind, shaped our currently dismal inability to determine the causes of cancer despite billions of dollars and many years spent on the research.

The rejection of the environmental approach in the 1970s is just one of many points where researchers, organizations, political efforts, and monies could have turned to cleaning up the environment of carcinogens rather than to the molecular genetics of cancer. Indeed, the road to proto-oncogenes is constituted of choices made every day by many participants.

In another alternative scenario, proto-oncogenes could have been the outcome of a different path of development. According to Scott Gilbert, the work on "somatic cell genetics" in the 1970s marked another turning point. This term in relation to cancer research is now used generally to refer to any research dealing with mutations in non–germ line cells that lead to tumors. However, in the late 1960s and early 1970s it was the name of a specific research program that failed. These "somatic cell geneticists" studied the mechanisms of cell fusion in great detail with the plan of fusing cells to do gene mapping and to study development. For example, in the mid-1970s Frank Ruddle at Yale and Robert Kleep at Johns Hopkins wanted to map cancer genes by fusing cancer cells with nonmalignant cells. They developed a set of "sexy," novel techniques to pursue these plans.[9] These researchers might have also created "oncogenes." In fact, they did not and instead the tumor virologists and molecular biologists portrayed in this chapter did.[10]

Other "outlier" research such as studies of mouse teratomas and teratocarcinomas and solid state carcinogenesis could have taken cancer research in another direction. Biologists could not explain these tumors in genetic terms, since they appeared to involve no changes in the DNA. Briefly, mouse teratoma and teratocarcinoma studies involved the transplantation of embryonic cells from their normal place in the body to an ectopic site.[11] This manipulation produced a teratoma and carcinoma without causing any apparent changes in the DNA. Solid state carcinogenesis involves the planting of a solid piece of plastic or metal into rodent body tissue.[12] While such introduction of a solid piece of plastic or metal induces a tumor, the introduction of a piece with holes in it or of ground up plastic does not produce a tumor. No biochemical changes or changes in the DNA were known to be involved in carcinogenesis here. In addition, studies of metastasis claimed that cells change so quickly during the process of metastasis that genetic explanations were improbable.

To explain these nongenetic cancers, some researchers have invoked the term "epigenetic" (Rubin 1985). This term has been used to refer to holistic and systemic interactions beyond the realm of explanation by reductionist science, as well as to specific mechanisms or events that result in changes in the phenotypic expression of the genetic code. However, there were very few studies of epigenetic mechanisms of oncogenesis at all and few that were accepted and treated seriously. The researchers involved in these studies did not generate doable problems (Chapter 7).

The Package of the Proto-Oncogene Theory and Molecular Genetic Technologies

Of course, the phenomenal growth of proto-oncogene research was in great part facilitated by the package of theory and methods used to generate doable problems. This new package consisted of the proto-oncogene theory of cancer and recombinant DNA and other molecular genetic technologies for testing and exploring the theory. However, by opening up this package, I tell a story of the mutual production of the package and the extended network of commitments made to oncogene research.

The proto-oncogene theory was constructed as an abstract notion, a hypothesis, using a new unit of analysis to study and conceptualize cancer. Researchers in many extant lines of research used this abstraction to interpret the theory to fit their separate concerns, all under the rubric of oncogene research. Researchers translated the general theory into concrete research problems in their laboratories without contradicting the existing basic framework.[13] Moreover, the concrete expression of the theory was framed by oncogene researchers in the terms of recombinant DNA and other molecular biology technologies, which by the early 1980s were accessible to other biology laboratories. This combination of the abstract, general proto-oncogene theory and the specific, standard technologies was used to generate novel doable problems. By locally concretizing the abstraction in different practices, researchers with ongoing enterprises (re)constructed the new idea and the new methods in new sites, thus further extending the network.

Indeed, the growth of oncogene research was both the cause and the consequence of this capacity for maintaining the integrity and continuity of the interests of the enrolled worlds while providing them with new tools for doing new work. Laboratories in many different biological subdisciplines and medical specialties viewed the theory-methods package as a means for constructing new doable problems and an opportunity to augment or replace their old, well-known routines with "sexy" new recombinant DNA techniques. At the same time, the proto-oncogene theory did not challenge the theories to which the researchers had made previous commitments. Indeed, the new research provided them with ways of triangulating evidence using new methods and a new unit of analysis to support their earlier ideas. These views of oncogene research were "realized" through the efforts of these researchers and, in turn, this realization further extended the reach of oncogene research and the complexity of the theory. The increasingly complex theory is today taken to be the best representation of cancer as well as normal growth and development at the molecular and cellular levels. Yet, it is impossible to separate this "best" representation from the tools and practices of the scientists who judge it to be so. These scientists are oncogene specialists whose work forms the basis of the complex version of the original proto-oncogene theory.

In the grants and programs arena, the director of the National Cancer Institute used this new research to justify past investments in the Virus Cancer Program, whose legitimacy and productivity had been questioned, and to lobby Congress for increased appropriations to the NCI. He presented the new research to Congress as promising hope for a possible cancer cure for their constituents.

University administrators used oncogene research to reorganize cancer research institutes now deemed "old-fashioned" into "hot" molecular biology institutes.[14] Of course, the very label "old-fashioned" resulted from the growing popularity and credibility of molecular biology. Biological supply and biotechnology R & D companies saw in this new research an opportunity to develop products and markets in the then slow biotechnology business. These companies promoted their products as the answer to the cancer problem. As Du Pont's transgenic "OncoMouse™" advertisement states: "OncoMouse reliably develops neoplasms within months . . . and offers you a shorter path to new answers about cancer" (see Chapter 1).[15] All of these different actors used proto-oncogene research to maintain and extend their lines of work and simultaneously extended the reach of proto-oncogene research and molecular genetic technologies.

The package was accessible to more than the elite laboratories. By the early 1980s, recombinant DNA technologies had been standardized into a set of protocols encompassing specific tasks, procedures, materials, and a few instruments (with more instruments created in the mid-1980s). The technologies and protocols had built-in methods for suggesting what was to be done to which material for what reason or purpose and with which outcome. Molecular biologists had constructed tools for manipulating DNA in eukaryotic organisms (including humans) that were adopted by other laboratories. Through this process, state-of-the-art tools became conventional tools used by many laboratories. These conventional tools were more accessible to researchers in other biological specialties, to new investigators, and to researchers far from the laboratories where the technologies were first created (University of California, San Francisco; Stanford University; California Institute of Technology; Massachusetts Institute of Technology; and Harvard University).[16] Another important aspect of accessibility is funding. By the early 1980s, funding agencies and commercial enterprises had committed more resources to the molecular genetic studies and were giving research grants and contracts to many laboratories proposing to incorporate the novel tools.

The standardized tools, available funds, a host of researchers and laboratories and institutes, specialized biotechnology companies, the proto-oncogene theory, and the actual oncogenes were thus co-produced. The theory-methods package provided procedures for a relatively straightforward construction of doable problems, or what Kuhn (1970) would call "normal science." Through the extension of oncogene research from a few sites to a host of laboratories, private companies, and government supporting agencies, the package has come to represent and facilitate collective work by members of different social worlds while creating oncogenes as stable facts. The package and the collective work behind it define a conceptual and technical work space that was less abstract, more structured, less ambiguous, and more concrete than that defined by the original conceptual model alone. Standardized molecular genetic technologies further defined the conceptual model, and the conceptual model framed the use of the standardized technologies. Such co-definition of the range of possible actions and practices has created a greater degree of "fact (and skill) stabilization." The package has served as a dynamic *interface* between multiple social worlds and concurrently represents the contingent articulations of oncogene research at different sites.

The package, then, has assisted the transformation of practices in multiple social worlds and the emergence of a new and at least temporarily stable definition of cancer. This molecular genetic view of cancer is maintained by the reorganization of commitments and practices of cancer research as well as other biological research worlds. The linkages thus created, in turn, have shaped subsequent commitments and work organizations made by researchers, laboratories, research institutions, biotechnology companies, and even diagnostic clinics. Twenty years ago the word or concept "proto-oncogene" did not exist. Today oncogenes are facts in undergraduate biology textbooks and the building blocks of new research programs, theories, and diagnostics. The world has been reconstructed to include proto-oncogenes.

The power and authority attributed to the theory-methods package are not concentrated in the hands of the few scientists who created it, but are instead distributed among different actors, objects, and social worlds. This "distributed authority" makes the power of the proto-oncogene "paradigm" more stable and entrenched. In most discussions of power and hegemony, we see the domination of many by a few. In this situation, we still recognize the play of power in science, but the power of the package of theory and methods is in the *distribution* of authority to various actors, objects, institutional loci, and social worlds.

(Re)Constructing Cancer

"Cancer" is a delimited field in science. In the case of oncogene research, it is what the package of proto-oncogene theory and molecular genetic technologies constructs it to be. Proto-oncogenes, molecular technologies, and cancer are co-constructed in laboratories. The "cancer" spoken of by scientists working in their laboratories is not the "cancer" suffered by relatives, friends, and strangers. When I asked one of my respondents to draw a picture of cancer for me, he asked me whether I meant the cancer that he studied or the cancer that his sister was enduring.

Recall the example in Chapter 2 of the mouse mammary tumor virus. Inbred mice colonies used to investigate the problem of the genetic transmission of mouse mammary tumors instead found that a mouse mammary tumor *virus* apparently caused cancer in infants as it was passed from mother to infant through suckling. This inbred mouse technology, created to solve a genetic research problem, produced a new problem and a new line of research on viruses. In the 1960s this virus and other so-called tumor-causing viruses became the objects of a major research program on viral carcinogenesis. In the 1970s and 1980s, these same viruses became part of the technologies for exploring a new cause of cancer, the proto-oncogenes in normal cells. The explanatory model for mouse mammary tumors moved from a genetic factor, to a virus, and then again to a gene called *int-1*.

> Found in the milk of inbred mice with a high incidence of mammary cancer over fifty years ago in Holland and at the Jackson Laboratories in Maine, MMTV was the first mammalian retrovirus to be discovered; it remains the only efficient viral agent of mammary carcinoma, and thus a model for one of the most common of human cancers . . . MMTV-induced mammary tumors are quasi-clonal growths of virus infected cells. To ask whether the tumor cells result from insertion of viral DNA near a heretofore unknown proto-oncogene, Roel Nusse examined many tumors to find one with only a single new provirus; he then cloned that provirus and its flanking cellular DNA in *E. coli.* An unfamiliar gene, which we call int-1, was nearby, and it was expressed in that tumor and several others with nearby insertions, but not in normal mammary glands.
>
> But this was not sufficient to implicate int-1 as an oncogene. First there was the circumstantial force of repetition: over three-quarters of mammary tumors in the C3H mouse strain harbor insertion mutations in the int-1 locus. Then Tony Brown did what nature had not done, by placing the int-1 gene within a retroviral genome; the resulting virus alters the growth and morphology of cultured mammary cells. Finally, Ann Tsukamoto followed a strategy pioneered by Ralph Brinster and Richard Palmiter and by Philip Leder and introduced the int-1 gene, linked to an MMTV LTR, into the mouse germ line. All the transgenic mice, male and female, experience dramatic hyperplasia of the mammary epithelium, and most of the females develop mammary carcinoma within six months. This is about as close as we can come to fulfilling Koch's postulates for a genetic disease: by placing the virally-mutated form of the gene into the germ line—ironically, much as envisioned to occur naturally in the virogene-oncogene hypothesis—we have *recreated* the disease. (Varmus 1989a:425)[17]

Researchers used the proto-oncogene theory and molecular genetic methods to (re)create the disease, to create an explanation for disease, and to create a genealogy of tools and explanations.

6 Problems and Work Practices

Improvising on the Shop Floor

Perhaps a meander best describes the trajectory of [social] action. Yet the notion of a meander fails to characterize how purposive agents take account of multiple human and natural factors. The improvisational ways they chart their courses involve complex judgments and intricate forms of human responsiveness and cooperation.

—*Renato Rosaldo,* Culture and Truth

I turn now to researchers in their laboratories making decisions, solving daily problems, and building visions of the future. I describe various ways in which problems and laboratories were modified or transmuted in the problem-construction and -solving process. My examples present problems and solutions as the outcomes of scientific researchers maneuvering through the exigencies and vicissitudes of the daily and long-term conditions of work, the continuous reorganization of research designs, and personal and professional circumstances. This chapter provides views of the meshing and integrating of theory, problem, techniques, solutions, markets, and audiences from the shop floor. I focus on the researchers in the laboratory as they coordinated their actions with a constellation of changes in other laboratories and worlds.[1] This constellation of contingent and improvisational actions constituted, often unintentionally, the final problem and solution.[2]

In tracking problem choices, changes, and solutions, I ask, What are problems? What are solutions? How do they begin, and how do they end? These problem histories are part of the account of a cascade of changes in problem-solving activities in biological laboratories described by participants as a bandwagon. They were all in some way, large or small, connected to the proto-oncogene theory-methods package and thus provide close-up views of the changes in practices that constituted the bandwagon.

Problem-Solving as Process

For experimenters problem-solving involves asking a question about phenomena, for example, what is the process by which genes can cause normal cells to transform into cancer cells? In the cases I followed, researchers translated the question into sets of experiments with proposed outcomes. However, the technical problem structure, and sometimes the question itself, was finally reconstructed through the temporal process of "solving" the problem. This temporal process involved actions in and outside the laboratory. I refer to this temporal set of situated actions as a "problem history" or "problem path."

Experiments involve manipulation of resources to produce outcomes, which scientists use to construct fact claims.[3] Scientists translate changes in the configuration of experimental phenomena into conceptual statements about phenomena. These translations are embedded in, juxtaposed against, and adjusted to other conceptual statements or theories about the phenomena.[4] These translations are also embedded in specific work situations. For example, when we look at laboratory work we find that it includes other kinds of work beyond experiments, such as training students, publishing papers, building and maintaining careers, and securing grants, patent applications, and commercial development.[5] Researchers juggle or trade off these different tasks and commitments often on a daily basis.

Thus, while scientists often begin the research process with a plan of action, in scientific work more often than not actions are linked to developments inside and outside the laboratory. Scientists solve innovative problems in open systems.[6] That is, they solve problems within open and changing experiments, work organizations, institutions, fields of research, disciplines, national commitments, and so on. Similarly, the very process of problem-solving constitutes a changing set of conditions. For example, a solution to one part of a problem can be a resource for, or constraint on, the next part of the project depending on how researchers construct it. Thus, one can never know a priori whether an action will meet with success or not. While local and institutional conventions of work provide guidelines, what actually happens is a *juggling act:* actors improvise as they integrate changes in their routines.[7] In this setting, then, new information arising at any point in the research process will similarly have to be integrated to be useful. It will not automatically "fit" into existing problem frames and laboratory work organizations.

Contingencies also can arise at any point in the research process and can take many forms. A laboratory staff member may quit, a machine may break down, sponsors may cut the laboratory's funds, cells may get contaminated, experimental results may not fulfill expectations, peers may reject a submitted article. Changes in funding policies may open new or close extant areas of research, researchers may invent new techniques for handling difficult materials, or new findings may provide new ideas for future research, and so on.

Contingencies can be viewed as occasions for negotiation.[8] They can be perceived as either roadblocks to ongoing action or new opportunities to the work of the laboratory.[9] A contingency can also be regarded as a new opportunity by a member of the laboratory, but as a major roadblock to the laboratory's collective work. For instance, while a laboratory's collective work on a problem might have to be stopped because of an apparently insurmountable difficulty, a student, postdoctoral fellow, or technician assigned to the problem might view the difficulty as an opportunity to work on another problem in which he or she is really much more interested. In a case described in Chapter 4, Luis Parada, a first-year graduate student in Robert Weinberg's laboratory, was delighted to have found a homology between a human oncogene sequence worked on by the laboratory and a viral oncogene sequence, which translated into an important publication for him. For the rest of the researchers in that laboratory, however, the homology meant that their human oncogene was not the first of its kind, in a world where "first" means "priority" and "credit." " 'Bob [Weinberg] was upset. Everyone in the lab was upset—except me,' said Luis, grinning . . . broadly . . . 'I was overjoyed at having found the homology. It was my first big discovery. But I had to keep my gloating to myself' " (Angier 1988:115). In the examples used in this chapter, I treat contingencies, roadblocks, and opportunities as constructions of participants in situ.

Research processes often do not begin with clearly delineated research questions or problem structures. Even when initial goals are formulated (as they sometimes are in biotechnology company research projects), solutions often do not match the original goals. Researchers construct solutions or answers to their questions at points in their lines of action in conjunction with other constraints. The point at which scientists evaluate their progress and publish their experimental results is constituted of concerns of accountability (to gain tenure or merit increases, to retain and renew grants, to satisfy company demands), degree of competition, and claims to priority, as well as some technical sense of closure on a problem. Because solutions especially in basic research are not predefined, what counts as a solution is temporally and situationally defined.

> There are a lot of really bad papers that have come out in the oncogene field. They break it up. A paper should carry an idea all the way through

> to completion. But to get lots of publications, they'll break it up into eight papers. Or they'll publish little . . . bits. Part of it is they're worried that someone else will publish it first. The other reason is they want to have lots of papers [to keep their grants] . . . In the old days, people used to spend more time developing an idea fully. [Today] to keep your grants, you need to have this impressive list of publications. In the older days when you didn't have to do that, you would publish one or two papers a year. Now, "Yuzen" probably does twenty and "Chaucer's" lab probably does fifty papers a year. (Jones interview)

Solutions, as well as problems, are unpredictable, negotiated outcomes. There is no "natural" endpoint to a problem history.[10] "Nature" does not determine solutions. Instead, as the examples will demonstrate, solutions are constructed through situated collective action; scientists construct solutions and situations in interaction with each other and with other parties, within the context of simultaneous and sometimes conflicting tasks and demands. Their construction of solutions is linked to demands for accountability from their sponsors and clients and their desire for priority, as well as to their attempts to adhere to rules of "good science" as established by themselves and their peers.

Accountability to sponsors and clients requires that scientists continually evaluate their work. They evaluate the pace and success of their competitors' and colleagues' work from journal articles, press items, conference presentations, and the "grapevines" of their particular scientific subworlds. They balance those judgments against their judgments or bets on their own work, against their standards for completeness for their work, and against sponsor and client demands. However, standards of completeness are relatively subjective and variable, as the following interview demonstrates.

> *Interviewer:* How do you know when enough work has been done to publish your findings?
>
> *Researcher:* You do all the experiments you can think of to figure out what's going on. When you can't think of more experiments to do, then you publish.
>
> *Interviewer:* Do people publish unfinished work?
>
> *Researcher:* Yes, all the time.
>
> *Interviewer:* What happens?

Researcher: [Other] people get mad. Especially when you want to do a follow-up experiment, and you build on an earlier one. If the earlier one doesn't work, you have to redo it. (Field notes, Xavier lab)

In researchers' usage, the term "solution" often receives attention only for the short term, in contrast to the kind of immutability that is invested in "facts." Scientists do not wait for peers to review, reproduce, and confirm a solution before they incorporate it in the construction of their next problem. A minimum of four months (and usually much more) passes before they can write up their findings and have them reviewed, accepted, and published by their peers. By that time the researchers may have completed another set of experiments and produced new results based on their previous solution. In addition, by that time, the scientists may have decided that their previous solution was wrong or incomplete and revised or retracted it. In the meantime, however, they have created a new problem.

Problem-Solving Histories

I present here examples of several different problem-solving histories or trajectories constructed from detailed ethnographies of problem-solving efforts. I focus on the conditions under which researchers acted, their actions, and the consequences for their subsequent problem structures. I examine how beginnings and endings are decided. Some problem histories have more interruptions than others. Some end before they go very far; others continue to a form of completion. Completion takes the form of the creation of a product such as a diagnostic kit, a partial solution written up in a journal article, or a finding that provokes new questions.

The proto-oncogene theory-methods package appears in several examples as an opportunity adopted by researchers to "jump on the oncogene bandwagon." I describe the transitions in the process of moving from previous research to proto-oncogene research in several laboratories.

Some of the actions taken by the researchers in the following examples were responses to contingencies or unexpected events in the research process. Contingencies took various forms: mistakes in the performance of technical experimental procedures, recalcitrance of laboratory materials, anomalies or unexpected experimental results, inconsistent data, inadequate or unavailable instruments, questionable payoff even if the experiments do work out, loss of staff morale, loss of staff members, loss of funding, and loss of priority (that is, other researchers solve the problem first). These all sound like negative contingencies, but they could also be viewed as opportunities for making changes in research agendas and organizations. Conversely, even apparent good news can be viewed negatively.

Problem histories present rich examples of collective work and the translation of interests as researchers construct problems and solutions and respond to contingencies.

Setting the Stage

A series of local, national, and international events (discussed in Chapters 3 and 4) set the scene for changes in the problems pursued in several laboratories.[11] First, from 1973 until 1978, intense debate primarily among scientists left the application of recombinant DNA techniques in limbo. A federal moratorium on their use in several kinds of experiments, the involvement of the National Institutes of Health and Congress in the regulation of the techniques, and NIH guidelines from 1976 to 1978 restricted scientists' utilization of these powerful techniques. In 1978, however, NIH issued new guidelines that "relaxed" the regulations and that provided new opportunities for research involving the quick, exact, and large-scale manipulation of DNA.[12] This meant that slower, more time-consuming, less precise, and smaller-scale virological techniques used in the examination of genes could be replaced, and it meant that researchers could study cellular genes and compare them with viral genes.

Second, researchers began to regard the research in the field of retroviruses and viral oncogenes as essentially "mined." By 1982, there existed much data on how the retrovirus expressed itself, and sixteen viral oncogenes had been identified. However, no viral oncogenes had been implicated in the cancers of concern to everyone: human cancers. Moreover, researchers were losing interest in searching for more retroviral oncogenes; new results were becoming harder to construct. "Once the virus is found, or the tumor is found, then the techniques are clear. Then everybody jumps at it and works with it now. But to isolate a new virus in a new tumor is a very chancy thing; nobody can reproduce that. It's a totally unpredictable story . . . So there could be more [retroviral oncogenes] . . . But we have enough now to say, okay, it would be better now to understand one in depth, how it works" (Yuzen interview).

Third, beginning approximately in 1976 there was a shift in the problems being pursued in viral oncogene research laboratories. As discussed in Chapters 2 and 4, in the late 1970s and early 1980s, discoveries of "cellular oncogenes" were reported first in the laboratory of J. Michael Bishop and Harold Varmus, and later in the laboratories of Robert Weinberg, Michael Wigler, and Mariano Barbacid. These researchers proposed that there are normal cellular genes in all species that can be triggered to turn into cancer genes and to cause cells to transform and replicate. Much of the research thereafter was oriented toward locating activated proto-oncogenes in tumor tissue, pinpointing the exact mechanisms involved in the triggering of the normal genes,[13] and understanding the normal gene's structure and functions.

Fourth, as a final addition to the scenario, the connections made between proto-oncogenes and human cancer linked viral oncogene researchers to problems of human cancer and cancer funding agencies. Researchers reported finding proto-oncogenes in the normal cells of many species, including humans. A few virologists and molecular biologists then built a hypothetical bridge out of these proto-oncogenes connecting their work on the clearly cancer-causing viral oncogenes (in laboratory animals) with human cancer.

The stage is now set for telling the stories of problem construction and change in several laboratories.

Endings and Beginnings

The first example, the Xavier laboratory, was located in a private biotechnology laboratory in California. Xavier was a protein biochemist who, after completing his training in 1982 at the University of California, Berkeley, had opted to work in this company rather than move away from the San Francisco Bay Area. The biotechnology company organized its work and workers into separate departments and divisions. The divisions included basic research (Xavier's department), process and product development, diagnostics development, and preclinical and clinical development.

Xavier's earlier work during his first year at the company was on interferon, a human protein once hypothesized as a possible cancer cure. After Lab Director Xavier "solved" his basic research problem on interferon's activities, he passed his solution on to the process and product development division. "Those guys . . . take whatever ideas are developed to a certain point by research, and, if the company makes a decision [to produce it] . . . they develop a full-scale commercial process to make that product."[14] The sociotechnical solution or endpoint to Xavier's problem was shaped by the particular division of labor in his company, by his skills as a basic researcher in protein biochemistry as opposed to bioengineering technology, by his negotiations with the head of research and development, and by opportunities presented by the new and exciting field of proto-oncogene research.

Xavier then decided that he wanted to work on proto-oncogenes. In 1983 interferon was a much studied protein, while proto-oncogenes were new, "more interesting," but more risky as an investment. At that time, academic researchers had begun to study proto-oncogenes in growing numbers, and biotechnology companies were beginning to "get in on the ground floor." The oncogene bandwagon was rolling. However, the research was still not a sure bet in terms of financially profitable therapeutic products. Given the company's concerns, Xavier focused on proteins produced by several genes related to viral oncogenes that had been found in primary human tumor tissue as a more viable research strategy. He proposed to study a particular proto-oncogene, *c-ras,* and its related protein, p21, that had been found in human tumors and presented company management with a plan of action that was greeted with both caution and great interest. They negotiated and together developed a final plan that satisfied their diverse interests.

The plan represented a major commitment for company management as well as for Xavier. On the one hand, he had done no previous work on this gene or its protein product. His plan provided no immediate monetary profit for the company. In fact, the marketing division specifically evaluated the project as perhaps having no chance of producing a product at all. On the other hand, if oncogenes did finally pay off, the company would not be left out in the cold in this line of research. We see here that Xavier did not act at his own discretion in shifting from one problem to another. Nor did he have the sole right or responsibility to determine the particular shape of his new problem. It was an outcome of careful planning and negotiation among heterogeneous actors. Xavier included the production of an intermediate product, diagnostic reagents, to serve as a sign of progress to the company president and shareholders. The ultimate promise was the possibility of producing a human cancer therapeutic agent.

Xavier used his protein biochemical skills to study the new protein, p21, and also began to investigate the *ras* oncogene itself. However, since the experiments on the oncogene were not expected to yield results for a long period of time, the p21 research bought his laboratory the time needed to gain the expertise required for the molecular genetic experiments. Xavier also recruited a research technician who had been trained in the new recombinant DNA techniques at Cold Spring Harbor Laboratory. The technician, Beth, had been working with another laboratory whose director was on leave for a year. She continued her experiments in a room next to Xavier's laboratory but eventually signed on to work in Xavier's lab, in part, because of her interest in the oncogene effort and, in part, because she was without a working group.[15] Thus, Xavier's protein p21 experiments and his laboratory organization grew out of the confluence of his skills, the skills of the people available to work on his project, the company's goals, waning interest in interferon in the technical field, and growing interest in oncogenes and their proteins. Xavier lab had crafted a new problem on proto-oncogenes.

Interruptions and Negotiations

In 1984 Xavier lab researchers wanted to investigate how and where the *ras* p21 protein bound to guanine diphosphate (GDP) and guanine triphosphate (GTP). Many researchers in the field of oncogene research thought these differential binding spots and activities played important roles in controlling the oncogenic properties of proteins produced by the *ras* oncogene. Researchers in Xavier lab had produced some interesting findings on the biochemical properties of a loop of the p21 protein. They wanted to publish a paper proposing a theory of this loop's structure on the basis of the biochemical properties. However, they preferred to publish the paper jointly with an X-ray crystallography laboratory, "Lab Allan," which had produced a proposed structure for the larger stable part of their p21 protein.[16]

Researchers in Lab Allan, an academic laboratory in Scandinavia, had already written a paper about their theoretical structure for the entire protein p21. They were quite convinced that most of their proposed structure was correct. However, Lab Allan researchers, and especially "Scott," a protein chemist in that lab, had doubts about several loops that, because they were mobile, could not be clearly identified with crystallographic methods. Xavier lab researchers were convinced that their biochemical studies demonstrated that Lab Allan's guess about the structure of one of the loops was wrong. They proposed to Scott that the two laboratories publish a theoretical structure based on Lab Allan's work on most of the protein's structure and Xavier's work on the loop's structure. A joint paper proposing a complete model of the protein would make a bigger splash than a paper on the loop's structure.

The stakes were high for the Xavier lab. The biotechnology company housing it had a chance to accrue significant commercial gains from priority in publishing the structure of the entire protein. (When a highly profitable product is imminent, a company may protect itself against competition by keeping trade secrets. Patents cannot protect as well as secrecy can in the legal realm of private enterprise.)

> If their protein is the same as ours, then we can [use their model] to do computer modeling simulations of our p21 and compare our site with homologous sites of other transforming proteins. Then we could do a lot of work without ever having to actually make the crystals . . . It would take us too long. We would have to pull everyone off what they're doing now.

> Besides [Merck, Smith-Kline, and Genentech—several large pharmaceutical and biotechnological companies with more resources] are far ahead of us. Maybe a year ahead of us. This way, we could scoop them by publishing a theoretical paper showing that it's possible to design a crystal that would inhibit GDP binding and an antibody that would block that crystal's activity. And maybe get the patent on it. (Field notes, Xavier lab)

Lab Director Xavier attempted to convince members of Lab Allan that their loop and Lab Allan's structure were of one and the same protein. All but one of Lab Allan members were not willing to decide immediately on which was the "correct" loop structure and wanted to wait for more "concrete evidence." But Xavier could not afford a long delay and further extensive, time-consuming, and possibly inconclusive experiments, partly because his lab was competing in a race against another laboratory proposing a competing model of the critical binding sites on the p21 loop. Indeed, as the following statement shows, many other laboratories were working on p21 proteins and could possibly work out the structure before Xavier could provide Lab Allan with satisfactory evidence. Publishing the joint paper was a way to beat the crowd and gain priority.

> There's also a lot of politics involved. There's another group with which [Lab Allan] collaborates, who are pushing for [Lab Allan's] present model. And why should they go with our model over theirs? If we can't provide evidence that the molecule is structured this way and attaches to the GDP molecule in this way, then we can't win them over to our model. And the [competing] group is equal to us. So [Lab Allan] is just going to wait until they can get some real evidence. And they can afford to wait. They don't have a lot of competitors in their field [whereas we do] . . . Everyone is doing p21 work. (Field notes, Xavier lab)

Even if the Xavier lab published a separate paper on p21's structure, it still depended on Lab Allan's publication of its work on the structure of the larger part of the protein. However, this work was yet unpublished, as far as they knew. "[Lab Allan researchers] are inferring the structure of their molecule [in the loop]. We think that their structure is wrong and ours is right. But, in any case, we want them to publish something." Researcher Scott, the member of the Allan Lab who thought their protein was identical to Xavier's, "pulled their paper from press [because he thinks we are right and the lab at Allan is wrong]. So we're waiting on them" (field notes, Xavier lab 14:7).

Collective work also involves trust, or the lack thereof.[17] In a discussion with the company's head of research, Xavier mentioned that he was afraid that Scott would give away the lab's model to others and they would "get scooped." Scott was the only member of the Allan Lab who agreed with Xavier's model of the loop's structure and was the person with whom Xavier was negotiating for the joint work. The Xavier lab and Scott had made no formal agreements to keep the work confidential. In Xavier's words, "Scott could easily give his stuff to someone else. He owes us nothing, [there's] no commitment" (field notes). To prevent the loss of priority, the head of research advised Xavier to go ahead and write the paper before meeting with Scott and other Lab Allan members. He argued that a written paper would put Xavier "ahead of the game" and would be an incentive for Scott to continue working with the Xavier lab.

The company's legal department was also involved in the construction and elaboration of the problem. In addition to competing in the world of protein p21 researchers for scientific priority, Xavier lab scientists were also competing with other commercial biotechnology companies for patent rights and production of the protein. Xavier needed the consent of the legal counsel before he could publish his paper proposing a theoretical model of the protein. Legal counsel worried about securing the company's patent rights of ownership of the protein, should it prove to be a useful therapeutic product, before Xavier presented his proposed model to the larger scientific community. Finally, the director of research came to an agreement with the legal counsel, and he advised Xavier to go ahead with the paper and not to worry about patent rights.

To summarize this leg of the journey, the Xavier lab was competing with a large and fast-moving pool of protein biochemists as well as larger and richer laboratories working on the same protein. Xavier and his colleagues felt they could not afford to wait for further experimentation without risking falling behind. Thus, they decided to publish a theoretical structure of the protein in order to claim priority and to move the laboratory to the forefront of research on the protein's biochemical activities. The head of research, in the meantime, dealt with issues around patent rights and legal concerns.

Before publishing the theoretical paper, Xavier planned to visit Lab Allan in Scandinavia to convince researchers there of the soundness of Xavier's structural model. Lab Allan researchers were competing with a smaller group of X-ray crystallographers working at a slower pace and could afford to wait for confirming evidence. Xavier planned to make a personal "coherent presentation of all the data." In order to do so, he also had to convince management to finance an overseas trip during a spending freeze in the company.

Before his trip Xavier collected several new pieces of information that provided him with more support for his model and a possible alternate strategy if all else failed. He found an article overlooked by Lab Allan that provided new evidence supporting his proposed model. He also found an article published four years earlier by researchers in Lab Allan, in which they proposed their protein's structure, except for the mobile loops. If Xavier failed to convince Lab Allan of his structural model, his alternate and last-resort strategy was to independently publish his model and refer to Lab Allan's earlier paper as well as the paper that had been pulled from press to support their model of the protein's structure. The early paper "mentions the uncertainty about the loops even then. So they haven't moved very far since that time . . . That's four years ago! That's exactly what I'm afraid will happen this time. They'll just [fool around] and not make any decision . . . So now we can refer to the 1981 article" (field notes, Xavier lab). The head of the research division also told them to refer to the pulled paper as "in press."

In the end, these alternative plans were not used. Xavier managed to convince Lab Allan members of his model, and they published a joint paper proclaiming their model of p21's structure. In this example, we see the collective work involved in framing and constructing a scientific problem, including publishing a theoretical model to beat the laboratories working on producing the actual protein crystal, negotiating to convince another lab to work and publish jointly and thus support each other's models, writing a paper before any formal commitments were made to get ahead of the game and make oneself look like an attractive prospective collaborator, and more.

This example is an interesting case of how practical, commercial pressures were linked to questions and actions about the "content" of research.

Continuing and Beginning

The tumor virologist Yuzen had worked on viral oncogenes since his early research on the Rous sarcoma virus in the late 1960s. During the course of that work, he and collaborators had discovered and characterized the first viral oncogene, *src,* as the transforming factor. Yuzen had been elected to the prestigious National Academy of Sciences for this early work. Since that time, his laboratory at one of the most prestigious universities on the West Coast continued to pursue the question of how a particular class of viruses, retroviruses, were able to transform cells in culture and to cause tumors in laboratory animals. Yuzen also directed much of his lab's research toward the search for more viral oncogenes in other viruses and tumors including *ras.* The lab's work before 1980 was considered primarily virological research by various laboratory members. By 1982, however, Lab Director Yuzen redirected his laboratory's central research problem to the comparative study of viral oncogenes and proto-oncogenes.

The local, national, and international events discussed above set the scene for the change in the Yuzen lab's central problem frame. At the end of the 1970s and beginning of the 1980s, several of Yuzen's graduate and postdoctoral students were convinced that proto-oncogenes were "hot" and attempted to persuade Yuzen to allow them to pursue proto-oncogene experiments. Those experiments involved research into the structure and activities of the normal proto-oncogene. They were willing to bet their careers on testing the prevailing theory that normal proto-oncogenes were involved in causing human cancer. This bet was in part based on their perceptions that virology was on the wane and molecular biology was on the rise. A graduate student in Yuzen's laboratory argued, "more and more people are using the molecular biological approach, but in virology, a lot of the interesting questions have been answered. So there'll be fewer results with more effort" (field notes, Yuzen lab).

Despite his postdoctoral students' enthusiasm, Yuzen was not convinced. In the early 1980s he had argued vehemently, in both academic articles and the popular media, with proponents of the proto-oncogene theory. Yuzen claimed that there was no evidence proving that any proto-oncogene transforms normal cells in culture (in vitro), causes tumor formation in animals (in vivo), or was found in spontaneously occurring tumors. He also argued that no normal cell assay for measuring the effect of genes on normal cell activity existed at the time.

Yuzen's postdoctoral and graduate students in the end managed to persuade him to allow them to investigate what was then called a "cellular oncogene." By the end of 1984, five of the nine experiments in the laboratory were efforts to understand proto-oncogenes. The laboratory's central scientific problem compared the structural and functional aspects of viral oncogenes with their hypothesized prototypes, proto-oncogenes.

Yuzen decomposed his central problem—the comparison of the structure and functions of viral oncogenes and cellular proto-oncogenes—into various subproblems to be studied by himself and his students. Each member of the laboratory studied the structure or function of a part of a particular viral or cellular proto-oncogene. ("Cellular proto-oncogene" was the name used to refer to proto-oncogene at that time.)

Graduate student "Rob," for instance, was conducting two sets of experiments. His primary project was the cloning and characterizing a portion of cellular proto-*fes*. Rob said that this portion had never been characterized before and could prove to be important in carcinogenicity. His secondary project was to recombine a chick virus with a cat virus in order to compare the structures of two viral oncogenes, *v-fps* and *v-fes*. This problem followed on previous work Rob had done on cat and chicken viral oncogenes.

Postdoctoral researcher "Thelma" was researching two-step carcinogenesis. Some oncogene researchers had argued that altered cellular proto-*Harvey ras* and cellular proto-*myc* oncogenes were both necessary to transform normal cells in culture. Thelma's specific experiment was to prove that either the whole unmutated cellular proto-*Ha ras* or the unmutated cellular proto-*myc* gene could each alone transform normal cells if put into a virus.

Graduate student "Pei" worked on two problems. One explored the *myc* oncogenes of the MH2 retrovirus. Pei removed one oncogene at a time from the virus to determine whether the virus could continue to transform normal cells. He discovered that either one of the viral oncogenes could transform cells without the other. Pei's second problem was to study the messenger RNAs of cellular proto-*fps* in order to construct part of the structure of the proto-oncogene.

"Quint," another graduate student, also added one "brick" to the "wall of information" about the structure of cellular proto-*fps*. In a second project. Quint was studying the biology of the OK10 retrovirus and *v-myc,* an oncogene related to a particular kind of human tumor. This retrovirus was interesting because it expresses *myc* in two ways: it either overexpresses or translocates *v-myc* to a position on the chromosome next to an expressed gene like the immunoglobulin gene. Quint wanted to understand what mode of expression was being used. He took the viral genes apart, then put them together in different ways to see if position affected the expression (that is, the differences in the messenger RNAs or protein products of the genes) of one and not the other. He had spent six months taking the genes apart and was then putting them together again with one nucleotide change.

"Wanda," another graduate student, was studying the *v-myc* oncogene of the MC29 retrovirus. Using recombinant DNA techniques, she "cut out" various parts of another gene to see what effect it had on the transforming ability of *v-myc*. The theory was that the other gene might block the activity of the *v-myc* gene. Wanda said that she had chosen to work on this experiment in consultation with Yuzen. "We discussed it. You're led to think that you're planning it together, but in the end, Yuzen has the final say, because it's his lab." But that did not necessarily mean that one could not fight for a project. "[Postdoc "Wong"] who was in this lab until recently, tried to prove that Yuzen was wrong. He did a lot of experiments with that goal in mind."

Later Jürgen, a new postdoctoral student, entered the lab. He decided to study the activity of proto-*ras* gene, but with a focus different from Thelma's. On the advice of Yuzen, he decided to examine another part of the cellular gene's structure suspected to impact on its ability to transform normal cells in culture. The project produced quick results, which Yuzen, Jürgen, and journal referees deemed significant enough to publish after only seven months in the laboratory.

A fascinating aspect of this lab's problem change was its initial invisibility. From Yuzen's rhetoric, I had assumed that his laboratory's research focused primarily on retroviruses and viral oncogenes. However, despite the difference in theoretical positions, my study of the laboratory's research problems and activities through time demonstrated that their actual work was not very different from that of many proponents of the proto-oncogene theory of cancer causation, who also compared the structures and functions of viral oncogenes and their related proto-oncogenes. On the one hand, Yuzen argued with other researchers at the level of claims for the theoretical significance of their experimental results. On the other hand, in constructing his laboratory's research questions to disprove their claims, he asked questions and conducted experiments along the same lines as his opponents. For example, Thelma's research used the same methods and materials as opposing labs used (although the experiment was different) in order to prove that *either* of two genes being widely researched was adequate to transform cells in culture.

Yuzen himself acknowledged that his research choices were directly affected by the actions of his colleagues: "Particularly in cancer research, a lot of science has been done in the last couple of years with the assumption that these cellular genes are cancer genes and that everything has been checked. Papers have been written as if that were absolutely clear. Important observations have been made, changes have been found. But proof that these genes have caused the cancers (in which they were shown to be different from normal cells) is not available yet." When asked if he felt pressure to do his work in a certain way because of that, Yuzen admitted,

> It does affect me. In fact, I often do work to disprove some of these things . . . It does affect me directly and indirectly. You are forced to run with the pack to some degree to compete for grants. To publish at the same rate and same speed . . . [It also affects] the direction of your work . . . If everyone says that gene is the cancer gene, then you are more likely to look at it again and say, "is it indeed the cancer gene and how does it differ from this and how is it related to that?" (Yuzen interview)

Research comparing viral oncogenes and proto-oncogenes required Yuzen to reorganize his laboratory's work, especially the incorporation of recombinant DNA cloning and sequencing techniques. However, because the laboratory had most of the necessary equipment and materials (bacteriophage viruses, centrifuges, radioactive isotopes, liquid scintillation counters), Yuzen's incorporation effort was relatively smooth. He and his students "picked up" the techniques by working with a postdoctoral student in another laboratory and by using excellent technical manuals.

Yuzen was one of the original viral oncogene researchers. New investigators from all over the world applied to study with him and to learn his methods as graduate or postdoctoral students, even when they disagreed with his theoretical stand. Yet, despite his theoretical disagreements with his colleagues, Yuzen maintained his long-term commitments to research on viral oncogenes and assured his continuing role in the oncogene research world by redirecting his laboratory's central problem in conjunction with the shift in the field. Since proto-oncogenes were a major area of investment of cancer research funds, Yuzen simultaneously kept his work funded and in the limelight by focusing his research on the same problems. That is, he managed to be a dissenter and still maintain a viable laboratory. Nevertheless, his reference to running with the pack shows that the distributed center of authority of the theory-methods package was able to enroll even dissenters.

Problem Structures as Contexts of Problem-Solving

The actions scientists take to work around contingencies depend on the particular form of the contingency and other aspects of the laboratory's situation. Researchers may be willing to repeat technical procedures if they have made mistakes, to purchase better equipment and materials, and to troubleshoot or experiment to figure out whether an anomaly exists and is important enough to address. Even if the technology or knowledge required is unavailable, researchers might choose to construct the technology or knowledge rather than stop a particular line of research.

Scientists read and respond to contingencies within the context of their problem and project structures. If a problem is constructed of modular subproblems and pursued through projects relatively independent of each other, then researchers have more flexibility. In contrast, if scientists construct their problem as a series of steps organized as subprojects, one following upon (and based on the solutions from) another, then they have to rely on other kinds of flexibilities to work around contingencies.

For instance, the loop issue in the Xavier lab was one part of a longer story of the construction of a problem. The problem was simultaneously "scientific" and "technical," if one can make such a distinction.[18] The researchers wanted to find the structure of the protein (p21) suspected of influencing the transformation of normal cells into cancer cells. Recall that, according to the literature, this protein was produced by one of the oncogenes named *ras*. The technical problem was to produce an artifact: a series of antisera that contained the protein and that could be sold as a kit for diagnosing specific kinds of tumor tissue. Each subproject yielded a solution that was then used to construct the next subproblem. Earlier solutions became generatively entrenched in subsequent problems, and the final product was contingent on the success of each of these interlocking subprojects.[19] Thus, when three glitches—a mistake, inconsistent data, and recalcitrant materials—arose in the process, the lab director opted to reorganize the division of labor in the laboratory to work around the glitches, rather than give up his and the company's entrenched commitments. When the technician "Beth" found the data to be inconsistent and the reagent recalcitrant, Xavier assigned two other lab members to help with the work—to review the collected data, to perform shortened tests, and to repeat tests using new materials and old materials in different combinations.

In another case, when two researchers suddenly left Yuzen's lab, their problems were left hanging. No one in the laboratory was assigned to them, even with the arrival of a new postdoctoral fellow ("Jürgen"). He chose to work on another problem that Yuzen wanted to explore. In this case, Yuzen had constructed his laboratory's central problem to be more open-ended in terms of its final solution. The product did not have to take a particular shape. He had also decomposed his central problem into relatively independent subproblems, each of which was being pursued by a graduate or postdoctoral student. Yuzen then used the results of each subproblem to support his larger problem theme, thus constructing the relations among solutions. However, each subproblem and its solution(s) could stand alone, for example, as the subject of a thesis or journal article. The important point here is that the solution to the laboratory's central problem did not depend on the solution of every subproblem pursued in the laboratory. Thus, when the two researchers left, Yuzen was not compelled to replace them with other or new lab members to continue work on their problems.

In another example, graduate student Rob in Yuzen's lab dropped his first thesis problem when he discovered that this effort to clone and characterize a portion of a proto-oncogene had already been accomplished by a European laboratory. He was unaware of this work, which had been published in a European journal but was not listed in the MedLine computer bank. The requirement for original research in graduate theses made it impossible for him to continue to work on his first problem despite his heavy investment of time and interest. (Even if he had not been a graduate student, it does not pay to be second. Priority in research is assigned to the first researcher(s) to solve a problem.) Neither could Rob begin an entirely new project, since the "normal time" allotted to graduate study in his university was five years. The disciplines of molecular biology and virology also had their own version of "normal time" so that students who remained in graduate school for longer periods lost top rating. Rob had only one year left of his normal time in which to complete his thesis. He dropped his first project and put all of his efforts into what had previously been a secondary project. He was able to survive this difficult situation without having to start all over, because Yuzen regularly asked his students to work simultaneously on two different problems. Rob benefited from this organizational strategy.

I label these examples of problem structures *modular* and *generative*, respectively. In modular problem structures, it does not matter in which order scientists solve their problem.[20] Generative problem structures are necessarily serial, because the "subsequent" problem is dependent on the "previous" problem's solution. There is no subsequent problem without a previously solved problem. Lab Director Yuzen ran several concurrent projects studying the structures and activities of several genes. Yuzen integrated these projects and their solutions,[21] along with prior results, as modular pieces into his central problematic theme: the comparison of normal and cancer-triggering genes. In contrast, Xavier ran two highly intertwined and generatively entrenched projects to produce information about the structure and biochemical activities of a protein and to produce related artifacts for marketing purposes. He had constructed one problem as a series of smaller projects. However, while problems might appear more modular or more generative at one moment, they can change over time. Problem structures are set in mud, not in stone. They are rarely if ever purely generative or purely modular.

Other aspects of organizations, especially relative discretionary power and accountability, also influence problem-construction and -solving activities. Xavier's was a commercial laboratory where the researchers had less overt discretion or control over projects than do academic researchers such as Yuzen. Sponsors including the stockholders, the board of directors, and the company president (in descending order of accountability) held Lab Director Xavier accountable for the production of outcomes useful to the company. In contrast, the Yuzen lab's research sponsors, national funding agencies, were relatively separate from its organizational location, the university. Both the sponsors and the university administrators gave Director Yuzen much more discretion in his research, as compared to Xavier.

At the same time, Xavier exercised more discretionary power or authority over the division of labor in his laboratory than did Yuzen. When a senior lab technician wanted to stop assisting the antisera production and move on to another project, Xavier quietly replied, "Why don't we just leave it this way" (field notes, Xavier lab). Although both lab directors had final say in the laboratories, their relative degrees of control or discretion differed. Yuzen's was a training laboratory as well as a research laboratory. There, graduate and postdoctoral students and Yuzen himself had commitments to developing students' careers. In this situation, Yuzen chose to shape and direct the interests of each student, rather than dictate what those interests should be. Thus, students in his lab had more control over their research problems than did technicians in Xavier's lab, a commercial research and development laboratory.[22] Xavier's researchers were more compelled by their organizational constraints of product development. As one researcher stated: "We have to give them [management] something" (field notes, Xavier lab).

One could argue that these examples suggest that organizational location and demands may have an effect on the kinds of problem structures produced. For example, one set of conditions that influenced the construction of the Xavier lab's generative problem structure was the demand from the company for patentable and marketable products. In contrast, Yuzen's academic laboratory, where results were measured in terms of published research articles and graduate students, organized its research in terms of modular problem structures. However, other examples portray academic research with generatively structured problems and commercial biotechnology research with modular problem structures. Within the same organization, we can also find both types of problem structures. This flexibility might be the consequence of the blurred lines between academic and commercial biological research in the 1980s.[23] Rather than begin with assumptions of academic freedom or corporate interests governing problem choices and structures, we need to examine the actual work and conditions of problem-solving.

Problem Structures as Responses to Situations

Decomposing a problem is another way of dealing with contingencies.[24] In the following example, researchers decomposed a complex problem into subproblems that could be solved with available (or easily obtainable) resources. Decomposition, then, is an effort to adjust a problem to existing local work organization and conditions.

Lab Director Yuzen had decomposed his central problem in response to the institutional context and work requirements. In addition to the production of research and publications, his tasks included the training of undergraduate, graduate, and postdoctoral students. While students provided extra pairs of hands to do the work, they also required training in technical and intellectual skills and, for graduate students, the completion of thesis-quality projects within the time constraints of graduate student careers (at most, five years) and postdoctoral careers (generally, one to three years). To accommodate these different tasks, he decomposed his central problem of interest into subproblems that graduate and postdoctoral students could carry out and complete in relatively short time periods. Yet, the subproblems were also complex enough to require skills that these students learned in the process of solving their problems. Decomposing his problem into subproblems allowed him to use his available resources (especially his skills and students) to accomplish both his research and teaching missions within the framework of a graduate educational institution.[25] Moreover, Yuzen organized his subproblems in modular rather than generative pieces. This "recomposition" of his decomposition allowed him greater flexibility in his work organization. The failure of any one student's project did not impinge on another student's work and had limited consequences on his overall research program.

Different Strategies for Different People

I have not discussed persons and personal decisions. I chose to focus on practices and work rather than on individuals, because this focus highlights aspects of the constitution of facts that are not available or are less visible in studies where personalities take center stage. Moreover, the work of modern science, like the work of auto manufacturing and other kinds of work, is oriented toward shaping the individual to the tasks and not vice versa.[26] The work has gained an ascendancy over the people carrying it out in the late twentieth century. Although personal proclivities never disappear, people are to some extent transformed in the training process.[27] Many of my students in the sciences tell me they cannot be "themselves" when they are in the laboratory. Becoming "a scientist" means changing one's behavior. This professionalization process transforms a layperson into a professional scientist.[28]

Nevertheless, there is a danger in detaching people from the study of science-making. When we separate individuals from practices, we often cannot see the contributions—"good," "bad," and unclassified—that they bring to their scientific practice from their many other lives. People do not entirely reshape themselves when they go to work. Persons and personalities do not disappear. They too shape the laboratory's activities. Everyday understandings and experiences, say, of race, class, gender, sexual orientation, health, illness, environmental concerns, war, violence, and more are also brought to science by scientists. The literature on gender, race, and science have discussed the embedding of science in society along these lines of understandings and experiences (see Appendix).

Although this chapter presents views from the shop floor, I have omitted the dazzling and boring personalities, the intrigues and mischief, the conflicts and resentments, the warm caring and the back-stabbing, the lemon squares, chocolate pies, and raspberry-chocolate ice cream that were also quite present in the daily life of the laboratory. I wanted to avoid the stories told to me after a big fight or a party—the "you should have been there, there was lots of sociology going on" stories. We all know that scientists are people too, that they have the same boring daily routines and the same personality conflicts that every other professional suffers. The significant question is: Do these experiences affect the stuff of science? Do they affect the knowledge produced? Personalities may not affect knowledge production in the late twentieth century in the same way that they might have affected science-making in Pasteur's day. Modern molecular genetics, for example, is thoroughly a collectively organized enterprise, despite the little fiefdoms that exist across the United States (it might be different in Japan and France). But a look at some practitioners, problems, and work organizations will show how the personal and the practical were entangled in the shape of the research problem, the laboratory work organization, and knowledge production.

Consider Lab Director "Chaucer," an enthusiastic, engaging, ambitious scientist who clearly liked interacting with people.[29] A creative and accomplished scientist, he was also an excellent administrator, fund-raiser, educator, orator, politician, and manager. When Chaucer began to study viral oncogenes and cancer in his own laboratory in 1969, his staff consisted of two members, himself and a technician. In 1984, his staff consisted of himself, nineteen postdoctoral students, six graduate students, and several technical staff members. Chaucer also expanded his number of research problems from analysis of one viral oncogene to studies of several viral oncogenes, their related proto-oncogenes, and questions regarding the normal functions of the proto-oncogenes that had no clear relationship to cancer. Chaucer sought and gained new resources, including new skills and information gleaned from sharing students with laboratories in other fields of research such as developmental biology, *Drosophila* genetics, and cell biology. His story is one of expansion, translating interests, and enrolling. He and his colleagues added new problems to their existing research programs, enlarged their laboratory, and widened the network of oncogene research to include laboratories that previously had nothing to do with cancer research.

In contrast, Lab Director Yuzen concentrated his efforts in one line of research (the role of oncogenes in cancer causation) in part because of his intentionally small staff. At any given time during most of his lab's existence (about twenty years), Yuzen maintained a staff of between five and seven graduate students and postdoctoral fellows. Including support staff (technicians, dishwashers, secretary, and an undergraduate student worker), graduate and postdoctoral students, and himself, the staff totaled about ten to twelve. Expanding his laboratory staff would have required that he spend more time writing grant proposals and overseeing the work of the lab staff and less time in the laboratory doing hands-on work. Yuzen preferred the hands-on experimental work to administration. To him this choice meant that he could not expand the repertoire of problems investigated in the lab, although he could have expanded his problem frame without enlarging his laboratory staff.

Lab Director "Grant," a protein biochemist in an academic laboratory, added a set of problems that could be addressed with new recombinant DNA techniques without enlarging his staff.[30] Grant created a problem that studied the same phenomenon, protein phosphorylation, in the same protein but at a different level of analysis and using different tools, thereby introducing innovation while maintaining continuity with his lab's previous work. He wanted to understand the functional significance of protein phosphorylation for cell transformation. By deleting or inactivating the gene sequence assumed to control protein production or phosphorylation, he could assess the consequences of those alterations for the transformation process. To pursue this problem, Grant needed to acquire new skills for handling DNA.

In Grant's case, the availability and accessibility of new recombinant DNA techniques certainly influenced his decision to pursue molecular genetic analysis as the next logical step in his laboratory. In his words, "the kind of problems we are trying to understand requires some sort of a genetic analysis. That's what we're trying to develop: a molecular genetic analysis." When asked how he decided that their problems required such analysis, Grant answered that a biochemical analysis was not enough and that a functional analysis became necessary. In his definition, a functional analysis required some kind of intervention and manipulation of the genes thought to control the biochemical activities and properties.

> Well, it gradually becomes apparent, after a while, that the biochemical description of what's going on is not sufficient. The problem we're interested in is how transforming proteins transform cells and what are the cellular components which are altered in transformed cells by the viral

> transforming proteins. In some cases we know that those viral transforming proteins are enzymes which carry out a specific modification on the cellular proteins. Specifically, they add a phosphate which phosphorylate the cellular proteins. So what we've been doing, what other people have been doing, is identifying proteins which become phosphorylated in virus-transformed cells and can, in fact, identify such proteins. Now the question is: what is the functional significance of those modifications? And that's a very hard question to answer without genetics, because what [molecular] genetics allows one to do is delete a specific protein or to delete a specific phosphorylation site, and then ask what are the consequences of that alteration for the transformation process. So that, in principle, you could ask questions about, "is such-and-such a protein and is such-and-such a phosphorylation necessary for transformation, or is it not necessary?" Simply by observing what's going on with a descriptive biochemical lab work, it's very hard to determine what changes in function are significant, and just describe them. From that point onwards, it's difficult—it's not impossible, but it's difficult—to say what their functional significance is. The only way of doing it is to show that there's some functional change in the protein and to relate that functional change to the transformation process, which is a sort of long, tedious way of going about things. So, in principle, by some sort of genetic approach, one may be able to analyze the process of transformation. (Grant interview)[31]

Grant, however, was not skilled in the techniques at the molecular scale of analysis. His strategy was to import the techniques into his laboratory by apprenticing his graduate students in other laboratories on campus.

Grant chose his particular strategy because of his particular set of constraints. He had spent his sabbatical helping to care for his new baby rather than learning the new techniques. He could not take additional leave because he needed to work on ongoing research and publications in order to satisfy institutional and career requirements. Sending his students out to learn new techniques still taxed the resources that would have been put into his ongoing research, but not to the extent of redirecting all of the research in the laboratory.

The case of the cell biologist "Norma Oakdale" provides a different perspective since her work intersected with proto-oncogene research though she did not make proto-oncogene research her problem of choice. Originally trained in genetics, Oakdale was conducting basic research on cancer in university when she found a lump in her breast, which was later diagnosed as benign. "I went through the process of waiting for my results, and [I] began asking myself, 'What was I doing which had relevance to human cancer?' It turned my interest around and I began working with breast cancer."[32]

Oakdale decided to study the complexity of normal epithelial cells as they become (or do not become) cancerous. One of her primary goals was to improve early detection and successful treatment of human breast cancer. This work was based on the assumption that breast cancers differ in each person. By growing cells of each patient's tumor in culture, she could then test different treatments (various chemotherapies, prepared antibodies, hormones, and so on) in vitro and then determine the best treatment for a particular patient and tumor before the administering it in vivo.

To shift to this new research, Oakdale and her colleagues chose to build an institute. There were several reasons for this organizational and institutional relocation. First, Oakdale formulated a problem that required specimens of both normal human epithelial cells and epithelial cells in the first stages of transformation. There were two main sources of "fresh" normal human epithelial cells: cosmetic breast surgery or reduction mammoplasties and breast milk of lactating women or nipple aspirations of breast fluid from normal, nonlactating women. These procedures required working relationships with potential women donors and with those working in community clinical settings.

> I wanted to study a specific problem. I wanted to study breast cancer. And I organized this institute in order to do that. At the time I was a university employee . . . I was in the School of Public Health for a while, and then [Government Research] Labs. And I brought a couple of grants and a contract with me to fund research on breast cancer, and I wanted to be in a medical setting, . . . a community clinical setting, not a medical school. Because it's the community physician that makes the first diagnosis and does the first therapy for breast cancer. In the medical school you see the tertiary patient, the patient with advanced disease. A woman goes to her gynecologist, usually, or her internist, with a breast problem. And *that's* the area of the disease I want to study, so I have to be where it is. And so, right off the bat, I'm an oddball, because there are no research institutes in community hospital settings. They're all in medical schools. (Oakdale interview)

Oakdale had other conflicts with the university's administration.

> The university gave me a bad time, because they couldn't see focusing on one area. They didn't want to put aside a lot of space. They wanted to put

> in my basement a project that had to do with bacteria, which would destroy my tissue cultures . . . There were lots of conflicts like that . . . Then also I had a philosophical problem with the university, because I wanted to accept [private] industrial money. That's a big philosophical controversy going on throughout all of academia . . . I had a company that wanted to give me $300,000 a year for three years, with an option to continue at the end of three years . . . The university wouldn't let me take it. So I said good-bye and separated from the university, and organized a nonprofit corporation for the sole purpose of doing research in human cancer. (Oakdale interview)

Oakdale and her colleagues built their institute next to a community hospital where surgery on both normal and cancerous breast tissue was performed and where it would be easy for the institute to obtain breast fluid secretions from women on a routine basis.

In part to provide a resource for her research and in part to support the institute, Oakdale grew normal and cancerous epithelial cells in culture. Growing human breast epithelial cells in vitro was a complex job. First, fresh supplies of human epithelial cells had to be gathered constantly, since cells eventually die even when culturing is successful. Second, human epithelial cells were very difficult to grow in culture. "I grow [normal breast cells] in tissue culture in the lab. I get a very small sample from you, for example, and I put it in the laboratory, and I amplify it to a very large number. And I can then do biochemistry on them. Now that took me thirty years. I said it in one sentence, but it took me thirty years to learn how to grow those cells in culture." The few researchers who were successful at doing so all knew of one another. They constituted a very small club.

Oakdale's institute also exported its epithelial cells in vitro to experimenters in other labs, including those involved in much more highly visible research such as proto-oncogene research. For this reason, the National Institutes of Health supported their cell culture work.

> I've been funded to do something very fundamental and basic and perhaps not very exciting, because Bishop and Varmus like to call me up on the phone and say, "I need cells of such-and-such. Where can I get them?" Or, "Do you have them?" Or, "Will you make them for me?" They like to have a resource like that. So my kind of work is in their interest to see that it's done. So they'll support it . . . At some time, they need to take DNA from human cells. We've already done it on a small scale with them. (Oakdale interview)

One could classify Oakdale's work as not "hot" and even as "support" work. She and her colleagues at the institute were developing cells they supplied to other researchers who needed DNA from these rarely "domesticated" (that is, grown in vitro in large enough quantities for experimental work) epithelial cells. However, these cells were more important to them as useful tools to study breast cancer. Their vision went beyond finding the DNA involved.

> If you're really going to understand what cancer is, you have to start at the cellular level. You also have to know Bishop's area—the components of the DNA, and how they function, how they work. Then when they function and they work in a certain way, they produce a reaction in the cell, and that reaction may result in the cell becoming a tumor. Then you have to know how that tumor reacts to the whole individual. [These latter problems] are very difficult, and they don't yield neat, simple answers that you can publish and say, "I asked this question, and I did this experiment, and this is the answer." And with Bishop and Varmus's, you can. It's chemistry in the test tube. You put something in the test tube and a reaction takes place, and you analyze it, and that is it. (Oakdale interview)

DeOme, our inbred mouse cancer specialist from Chapter 2, discussed a similar kind of problem in his research that was not easily translated into experiments.

> We have strains of [inbred mice] that have quite a lot of precancerous lesions in them, even though they don't carry the virus and haven't been treated with carcinogen . . . The strange thing about one of the strains is that it has lots of these lesions, but almost no tumors . . . There has got to be some control business going on here somewhere that keeps it from doing what happens in almost every other circumstance. There are lots of things like this that intrigue me to no end . . . Not many people are interested in this [problem], because it's a real stinker to work with . . . Because it's difficult, terribly difficult. There are a lot more profitable things to do . . . I might spend ten years on this and find that you can't get the answer. Because what I want to know is what it is, not just that it happens. But proving that it happens is where you have to start . . . You have to be a brave soul to do it, or somebody my age [professor emeritus] so it doesn't matter. I don't have to prove anything anymore. I don't have to be "good enough" to be at the University of California, Berkeley . . . I can fiddle around with these things as I want to . . . This would be a high-risk thing. As you can see in my talking to you about it, what I'm saying is fuzzy and

> it's not nice and clean-cut, full of delightful experiments you can do that can give you a sure answer. It would be much more difficult. (DeOme interview)

Similarly, Oakdale could choose not to "run with the pack" of oncogene researchers because she was a senior investigator with private funding arrangements.[33] In her estimation, bandwagons developed because of the time and task constraints on new investigators.

> How do waves get started and why do they occur? I think the reason for that has to do with the way science is funded and the way young scientists are rewarded . . . [A] youngster graduates and . . . gets a job in academia, for instance. As an assistant professor, he's given all the scut work to do in the department . . . In addition, he must apply for a grant. To apply for a grant and get a grant . . . nowadays, you have to first show that you are competent to do the [research], in other words, some preliminary data. So he has to start his project on a shoestring. It has to be something he can do quickly, get data fast, and be able to use that data to support a grant application . . . that will yield results, so he can publish a paper, so that he can be advanced and maintain his job. He has to publish, he has to get grants.
>
> Therefore, he doesn't go to the fundamental problems that are very difficult, that take a long time. He's not going to study how to grow epithelial cells for thirty years. He'd be fired long before that. So he has to pick a sure-fire project that's going to yield results quickly. So he goes to the bandwagon, and takes one little piece of that and adds to that well-plowed field. That means that his science is more superficial than it should be. And that's bad for the field of science. (Oakdale interview)

These alternatives to bandwagon research were few and far between. The Oakdale example also demonstrates th
research had to submit in part to the distri
oncogene theory-methods package. Neverthe
terruptions to my "clean engine" story of on

Problems and Solutions as Sociotechn

The examples in this chapter show that problems are sociotechnical achievements. Researchers collectively constructed and solved problems within the open and changing contexts of technical experimental procedures, laboratory materials and instruments, anomalies or unexpected experimental results, work organizations, professions, institutions, fields of research, disciplines, and national commitments. Yet, these achievements were not smoothly orchestrated affairs. Instead, problems were processes that took on lives of their own. Indeed, the very process of problem-solving constituted another changing set of conditions. In some cases, a solution to one part of a problem became a resource for, or constraint on, the next part of the project, depending on how researchers constructed it. Contingent events, whether viewed as positive and negative, had to be dealt with for research to go on. The manner in which contingencies were interpreted and handled is the stuff of ethnographies and histories of problem-solving.

The problem examples presented in this chapter arose from particular confluences of events, conditions, and actions in each situation, both within and far from the laboratory. While lab researchers could and did attempt to change conditions outside the laboratory, the researchers were only one set of actors among many involved in the construction of the problem in a laboratory. Changing the conditions in other situations required tremendous efforts and resources and did not always prove successful. Indeed, these changes sometimes created another set of processes that followed their own course beyond the researchers' control.

These examples suggest that problem structures, work organization, and institutional constraints together can create or limit options for future choices in problem construction. Together, these findings challenge those analysts of science who view scientific problems as cognitive structures constituted of logical next steps, separate from work organization. Instead, a complex set of local and global conditions enter into the determination or construction of the next step. Scientific research is a linear progressive trajectory from problem design to final output only in post hoc renderings.

Problem structures are constructed within the problem frames of lines of research and disciplines and within the frames of daily and long-term vicissitudes of laboratory life in organizations and institutions. In one sense, the intersection of these various elements in a laboratory are represented and embedded in problem structures and the representations of nature they produce. We can gain some sense of how early commitments are (or are not) embedded in problem structures through observing their construction.

The world of problem-solving is constituted of seamless movements. Seams result when two parts are sewn together. Parts are created when we attempt to analyze science-crafting using our sociological eyes. Categories such as experiment, laboratory, and social worlds; inside and outside; and content and context exist because we humans create them. We constitute heterogeneous objects and actors. Social analysts like myself then take these categories and show that they sometimes obstruct our vision. With our directed vision, we may fail to see the numerous interactions that testify to the porousness of these categories. Ethnographies of practice can help us (as analysts) "see" differently sometimes. For example, the researchers I observed at work in the laboratory were dealing with stockholders, the National Cancer Institute, their professional and disciplinary colleagues, their different versions of "the economy," "the job market," reagents, protein loops, and more, as part of their daily and long-term problem-solving efforts. The neat divisions between global and national economies, experimental practices, and the development of lines of research are not as clearly circumscribed as we envision them to be, even in research that seems to be far from ideological or political battles. The view from the shop floor of the laboratory is sweeping indeed.

7 The Articulation of Doable Problems in Cancer Research

All social action is constituted by a multiplicity of discourses. Cataloging and explicating those discourses will not, in itself, enable us to understand how those discourses are articulated in people's everyday social actions. Such an understanding requires knowledge of the ways in which people connect these discourses and negtiate these connections.

—*Sylvia Yanagisako, "Sex and Gender and Other Intersections"*

Like the preceding chapter, this one uses ethnographies of practice in the laboratory but with different aims. Here I present a framework for examining the organization and practice of scientific work in the construction of problems and solutions. To examine why a strategy succeeds in contemporary science, I focus on the organization of work, including all the different kinds of work and workers involved in creating scientific facts and artifacts.

As I followed cancer research scientists at work in their laboratories over time, I came to see problem choice, problem construction, and problem-solving as intertwined processes. The key word I use to refer to this interactive process is taken from the scientists I studied who said that they chose to pursue "doable" problems. In their common refrain, "The reason that oncogene research is surging is that the work is doable; there is a productive methodology developed." A molecular biologist explained that a productive methodology "depends on the particular questions [asked] and the particular technologies [used] . . . So with recombinant DNA technology, you can ask the question, 'Are there changes in cellular proto-oncogenes in tumor cells?' and you can answer that question."

For these scientists, technological resources and capabilities generated the flurry of research activity beginning in 1978 on proto-oncogenes specifically, and on the molecular genetics of cancer generally. While scientists credit dazzling new techniques, such as those used in recombining DNA, with making problems doable, they ignore other kinds of work such as the daily, taken-for-granted tasks of washing pipettes, signing up to use the ultracentrifuge, and talking with colleagues.

Training in basic research laboratories emphasizes experimental techniques and not articulation work. Students are explicitly taught how to clone genes and grow cells in culture. They are also taught which materials are easier to manipulate for which purposes and which machines can take care of which tasks. Students may even be taught theoretical problem design, although this is usually learned by example. There are fewer explicit lessons on articulation work: how to build and run laboratories, how to cultivate sponsors, how to manage and work with students and technicians, and how to negotiate with administrators. It is this *invisible* work that I highlight in this chapter.

Work Organization

Scientific activities are carried out by different organizations or by different parts of a work organization, including the experiment station, laboratory, institution, field of research, discipline, and industry, as well as public interest activist groups and national and world organizations. Scientific activities also include the activities of "natural entities" such as restriction enzymes, scallops, and microbes.[1] I examine these activities because they help us make connections between scientific work and scientific knowledge. University administration, disciplinary entrepreneurship, and technical work at the lab bench (to name just a few) are different kinds of work that are all part of scientific practice. Convincing one's colleagues at a scientific meeting that one's findings are valid or persuading a National Science Foundation section to fund one's research are different kinds of activities than the activities involved in running experiments. All of these different kinds of work have to be done and coordinated to successfully carry out a research project. Ethnographic studies of laboratory science of the late 1970s[2] and many histories of science have celebrated the heterogeneity of elements involved in scientific work and have described science as heterogeneously engineered.[3]

I focus on the organization of work to highlight the kinds of tasks that scientists ordinarily ignore in tales of their research processes. A scientific problem is the outcome of a moving constellation of different events and actors, and their actions and interactions as they continuously articulate relationships between different kinds of work and work organizations. These include experimental work, laboratory management, and the management of social worlds critical to the laboratory work.[4] Experiments on oncogenes, for example, are situated in the scientific (basic and applied) worlds of molecular biology and cancer research. Other examples of "interested" social worlds include patients, the pharmaceutical industry, public health organizations, physicians, insurance companies, hospitals, medical ethics lobbies, medical care reformers and activists, medical insurance lobbies, national governments, and the World Health Organization.

Articulation is the more amorphous and ambiguous work of planning, organizing, monitoring, evaluating, adjusting, coordinating, and integrating activities usually considered to be "administrative" rather than "scientific."[5] This work is usually invisible and taken for granted until it is not done or not done well.

For example, "running a gel"—or gel density gradient electrophoresis—in molecular biology or protein chemistry laboratories is usually considered by scientists to be productive work. It involves a standardized set of procedures and materials for separating pieces of DNA, RNA, or proteins of different lengths and molecular weights. However, activities such as purchasing an ultracentrifuge, coordinating the experiments in the laboratory, assigning people to different projects, organizing an annual conference, and writing grant proposals are usually considered work that supports the productive work.

In my approach, articulation work is critical to the success of any project. Articulation work is not management in the sense discussed by Merton (1973) or writers in business administration. Everyone—lab director, technician, postdoc, student, bottle washer—does articulation work.[6]

Articulation work includes the sometimes mundane and often taken-for-granted activities of purchasing chemicals for reagents, negotiating with sponsors, and collecting information about the latest results of competing laboratories. As in my earlier example, to run a gel density gradient electrophoresis procedure, the laboratory technician requires various materials, skills, knowledge, pieces of equipment, and electricity. He makes sure that the gel box (or electrophoresis cell) and electrodes will be free when he wants them, that there is enough buffer prepared, that he knows how to do it or can get help from someone on the spot, and so on. A gel is run at the end of another series of tasks to prepare the DNA or protein. This means that the gel box had better be available when the DNA or protein is prepared. From the technician's standpoint, "running his gel" requires planning and coordinating with others. One alternative to coordinating gel boxes with schedules is to buy a lot of gel boxes. This converts "get a gel box" from a complex coordinating effort into a trip to the cupboard.

Consider, however, a similar solution to organizing the use of ultracentrifuges. These machines in 1986 cost from $22,000 to $32,000, not including the price of rotors, which in 1986 ran from $5,000 to $8,000 each. In contrast, gel boxes in 1986 cost from $173 to $1,600, exclusive of many necessary accessories. Given unlimited space and financial resources, new centrifuges or gel boxes could be an answer to articulation problems. Space and financial resources are always limited, however. One laboratory tried to solve the scheduling problem by having sign-up sheets next to the ultracentrifuges. But I still heard complaints about people bumping others from the schedule or leaving their materials in the ultracentrifuge so the next user had to clean it out before using it. Articulation work in science sometimes literally means dealing with messes.

The process of constructing a doable problem can be conceptualized as the art of articulation.[7] Researchers usually do not know what will be necessary to carry their research to some point of completion before they begin the research. But they do know that they need, in addition to a novel problem, some basic elements including interested audiences who will publish or use their work, sponsors or clients who will provide funding, institutional infrastructure to support their work, skilled staff to assist in getting the work done, and basic research materials and equipment and their suppliers. The specific construction of the problem and its temporal solution through the articulation among this cast of characters and their different jobs is where indeterminacy becomes locally determined. But local determination does not mean that science is done just in the laboratory. There are many participants, events, and kinds of work that go into the construction of scientific knowledge.

An Example: Oncogenes and Antibodies

I present here a case study of the process of constructing a doable problem. This example is taken from the Xavier lab, discussed earlier. Recall that the lab consisted of six members working on oncogene-antibody experiments in a private biotechnology company in the San Francisco Bay Area in California. The company focused its research and development on human health care products. When I began my fieldwork in their lab, the researchers were conducting experiments primarily aimed at characterizing the structures and functions of antibodies that seemed to counteract the transforming (cancer-forming) activity of activated oncogenes.

Articulating a Research Program

The program of research began in the fall of 1983, when the company's heads of research and development identified oncogene research as an area in which to build a project. They sought out researchers with an interest in and ideas for developing such a project. The lab director, Xavier, showed the greatest interest in the project and provided a plan of action. According to the head of research and development,

> We . . . were tracking the oncogene developments, like we track lots of other things that go on in research, and we knew that . . . our competitors . . . were cloning a number of the oncogene sequences. And we decided not to just do something "me too." It wasn't obvious what having the particular oncogene sequences might do for you in terms of the real product. So we spent, without any formal planning, a fair amount of time (over a year or so) just informally from time to time discussing the issue . . . We would call in scientists from all different parts of the company who we thought might have any ideas.
>
> [Lab Director Xavier] . . . was probably putting in the most effort on it. At one point we simply said to Xavier, "How interested in this are you? Do you want to develop some kind of oncogene program?" He said, "Yes, I really do." So we said, "Create one." So over a period of two or three months, he created—essentially himself with his contacts around the world in research labs in different places—he created a scheme which would get us into working on oncogenes. And the scheme was novel; no one else was doing it. It targeted certain things that other people had tried and been unable to do, but which it was clear, if you could do them, you could learn something that might lead you someday to an exploitation of the oncogene concept to develop a therapeutic product. (Kensington interview)

Xavier, another scientist, and management coordinated efforts to create this project. Before any experiments were begun, they worked closely together—monitoring, planning, negotiating, and coordinating—to set up a project that would respond to the demands and concerns of several audiences.[8] The team decided to begin with the development of antibodies that bind specifically to the protein product of the oncogene called *ras*. Oncogene proteins are proposed to transform the cell from its normal state into a tumor cell.[9] Laboratory-produced antibodies to the *ras* oncogene protein could be marketed as diagnostic reagents to be used as research tools in experiments and could possibly lead to the development of diagnostic tools in monitoring disease and perhaps even to some cancer therapeutic agent.[10] The head of research and development in the company described the plan to make a set of antibodies that could be used in future experiments.

> We didn't start off making these antibodies to address the problem of how the *ras* oncogene works. We thought it would be very useful to have an antibody that could detect a mutant protein. Because then one might be able to use it for screening human tumors or human cells to see if there are any activated *ras* genes in there that might have some diagnostic value in monitoring or diagnosing human cancer . . . So we started off making these antibodies in order to generate diagnostic reagents. We didn't know that they could be made, because a lot of people had failed to make these particular kinds of antibodies. We knew that if we could make them, then we could do a lot of experiments. That's the way it evolved . . . This particular set of experiments was not designed originally to address any particular problem. In fact, the [procedure] experiment was a sort of let's-see-what-happens type of experiment. (Xavier interview)

In planning their research program, the scientists weighed their research problem options against sponsor demands. They then proposed a program that incorporated their own research interests within the profit-making demands of their sponsors.[11] When asked whether he could study any question of interest to him, Xavier responded:

> Well, the general thrust [of the company] is the therapeutic area and also in diagnostics. So, when one proposes a new area of research, it should obviously be in those kinds of areas. How much support one would get for a new proposal in those areas depends on how feasible, what the kind of payoff would be, and just totally subjective assessments (out of some consensus of agreement amongst the scientific group at [the company]) that those experiments are worth doing for some reason or other—whether it is directly financial or [for] publicity [purposes], or [for] developing technology . . . There are lots of different reasons for . . . going for a certain project. (Xavier interview)

However, sponsors' demands are not necessarily hard and fast, especially when they are mediated by several agents. In this case, the scientists negotiated and even bartered to get their project accepted.

> We have this thing called the [committee name] where people with new ideas stand up there and chuck them out. [The ideas] get sort of thrashed around . . . If people like it, then there's an assessment from Marketing

> people as to what the payoff will be and from Research people [whether it is] feasible. And then some sort of bartering goes on and so on. And people decide roughly what kind of support it should get in terms of people and the layout of money, based on those facts. [Can Marketing say no?] No, they would say there's no money in it. Marketing has looked at this [oncogene] project and said, "We don't see any chance of anything we can market." But if Research says, "Okay, fine, but we think it's good for our image to do this research, and to hell with Marketing," those things will be taken into account. So Marketing is . . . only one element. The visibility, getting publications out, developing technology—there [are] a lot of other justifications for doing a project. (Xavier interview)

Thus there were extramarketing considerations, such as "image," in deciding which research problems to pursue. Ultimately, however, the oncogene project won support because management saw a possible long-term payoff in a therapeutic product and a possible short-term payoff in diagnostic products, and because the researchers succeeded in convincing the president of the potential of such payoffs.

> I guess the oncogene project that I started out here . . . was seen as a long-term thing which might have big payoff. There are some short-term elements in diagnostics that might have a shorter-term reward. [But] it was felt that getting the technology set up here would become more and more useful as more and more information comes from all over the world about oncogenes and cancer. If we're working on it and we understand what's going on and we are a part of it, we will be able to pick up on what's going on and . . . the idea is that there's always some possibility of a big payoff in therapeutics when the time is right. We don't want to just hang around and twiddle our thumbs until the time is ripe and then jump in. We want to be part of the action right from the start. (Xavier interview)[12]

Company management decided to risk resources "now and get in on the ground floor." Waiting until a clear therapeutic agent was defined would mean losing time to set up a program. Such a delay, about half a year, could be potentially costly in terms of losing patents on products in the competitive market.

Thus, to satisfy the demands of their sponsors—management, the board of directors, and private investors—the researchers constructed a problem whose short-term results could provide a quick payoff, namely, a set of antibodies specific to a set of proteins. These antibodies would be marketed as part of a kit that other scientists could use in their research and would also serve as a sign of progress toward the larger promise to produce therapeutic reagents.

The resources needed for the project were now in place. Management was willing to pay for the research and found a scientist (the lab director) who was willing and able to organize and carry out the work within existing constraints.[13] The scientist was given full support to organize his laboratory to accommodate the project. He got additional staff with the necessary skills, expanded his laboratory space, and had access to materials and supplies via organizational stock. He also had access to additional materials, skills, and knowledge through his previously established academic contacts and through the company's scientists and subcontractors.

In addition, the researchers and management constructed the project to fit into current problems in two lines of research: one, molecular biologists, protein biochemists, and X-ray crystallographers conducting basic research on oncogenes and their protein products; the other, laboratories conducting applied research in attempts to produce cancer diagnostic and therapeutic agents. Their experiments were aimed at answering questions already being pursued by these other scientists. Thus, if their answers were accepted as valid, they could be comprehended and used by these audiences.[14]

The particular oncogene the Xavier lab began to study was *ras*. Previous work showed that the oncogene *c-ras* produced a mutated form of the protein p21. The mutated p21 protein had been found in human tumor cells, whereas another "normal" or nonmutated form of the protein was expressed in normal or noncancerous cells. This observation pointed to the protein's possible importance for applied efforts to develop therapeutic agents. Both the oncogene and its protein product had been intensively studied by basic researchers. As a result, there were interested audiences and available resources (in the form of information and materials). There was also intense competition. According to Xavier,

> that was a big surprise to a lot of people, that you can actually detect an activated [p21 protein of *ras* gene] in human tumors. Because [scientists] have been working on the same [*ras*] gene in viruses and other systems for several years. So suddenly everybody realized [that] you can actually get a real handle on what's going on in human cancer—any cancer actually—at the molecular level . . . Here is a gene we're all familiar with which makes a protein we're all familiar with. And now [the protein] has been found to be involved, probably causally, in human cancer . . .[15]
>
> So [here is] a situation in which a single amino acid change in a well-known protein was enough to make it involved in human cancer. That's pretty incredible. It doesn't take much intellect to realize that that would

> be a really fantastic model for trying to test the difference between normal and cancer cells. And the gene is available. It's been cloned and sequenced. And the normal gene is [also] available . . . Now you have a situation in which you have a comparison between a normal protein that's expressed normally in cells and doesn't have any transforming effect . . . And a mutant [protein] variety of [the protein] which does transform. And it's only a single base difference, a simple amino acid difference. You couldn't ask for a nicer, cleaner system than that to try [to] understand how the protein works. (Xavier interview)

There existed, in addition, literature on some of the biochemical activities of the transforming protein p21.

After the planning, antibody production, and early testing were completed, the lab director had a good idea for a new experiment that his lab was not equipped to handle. The experiment involved blocking the biochemical activities by which the protein causes the cell to transform from its normal phenotype to its cancerous phenotype. The laboratory's in vitro tests yielded positive results.[16] "The next obvious experiment was to see what happens when the antibodies [are] introduced into living cells, [whether they are] able to turn off the [*ras* oncogene's] proteins specifically—I mean the *oncogenes* specifically." But the experiment required specialized skills and equipment to conduct in vivo tests on living cells. No one in the laboratory and organization had the necessary resources, and the lab director did not want to spend a month setting up for this one-time experiment. In this case, their strategy was to seek out and enroll the efforts of a scientist outside the organization with the necessary skills and laboratory set-up. In addition, the collaborator provided new ideas for improving the experiment.

The collaboration required articulation with worlds outside the laboratory. Who had the necessary skills? Would the expert work with Xavier's lab? Could he convince the "chiefs" and legal counsel to arrange an official collaboration? In the end, phone calls were made to this expert, and he agreed to work on the project. Collaboration contracts were drawn up by the legal counsel. Antibodies were then mailed off, and the experiment was carried out. After phone conversations, much staring at X-ray crystallographic photographs, and many decisions, the experiment was declared a success. That is, the antibody "reversed" the transformed cells in culture to normal cells. The results were presented at an important conference. They were also made public by the company president through a press release on the day of the company's annual stockholders meeting.[17] Finally, a paper reporting the research was written and accepted by a reputable journal.

The researchers were able to complete one set of experiments framed in terms of a larger umbrella problem—how do genes cause normal cells to transform into cancer cells and how can one reverse that process?—within a year and to fashion a finished product (a published paper). This product could easily be connected to other research on the transforming activity of oncogenes and work on cancer therapeutics. Aligning heterogeneous activities from contracting services at another laboratory to bottle washing had resulted in a new "fact."

This new "fact" also satisfied company management and stockholders. The company treated the experimental results as an intermediate "product." Company lawyers applied for patents on the *concept* developed from the experimental results. In fact, however, the experiments had yielded no definite or complete information about the biochemical structures and steps involved in the transformation process. Nor could the researchers use their antibody to treat whole tumors. Nevertheless, the company applied for a patent by claiming an eventual product. According to the head of research and development,

> we do know that if we inhibit [the enzymatic activity of this one protein], then the end result [tumor cell indicators] doesn't exist anymore . . . So we immediately filed patents on that. So within a year of starting the research effort, we had patents applied for on the whole concept. And what we'll do is, in each step that we work on, everything that we identify along that way, we'll file a patent application on the new concept of a potential therapeutic product. Because even if the antibody isn't a product, it's an example of a product. It's not a commercially effective product, but it's an example of a kind of product. And therefore we can claim, using that one example . . . any product that interferes with the activity of that protein. And we did. So, theoretically, if our patent was written properly, we have . . . the first claimed invention of a protein . . . a product that will inhibit the activity of the oncogenic protein. So . . . that's the way it works. (Kensington interview)

I point out at this juncture that resource commitments are continuously weighed and balanced against constraints throughout the research process. Management evaluated the laboratory's research, both formally and informally, against the state of the art in oncogene research and against profit goals. The company's head of research and development discussed his evaluation process.

> We had no idea how long [the project] was going to take [to provide some form of payoff]. My commitment to the group . . . was indefinite. I would have looked at it on a yearly basis . . . looked at the results they were getting and evaluate the progress. Is it making any sense? Is it productive? Are they doing anything? Simultaneously evaluate what is the state of the art in the rest of the world. What's going on in the other laboratories? Is this research relevant or irrelevant? Is it likely to be totally surpassed—even though relevant—. . . by much larger efforts elsewhere? You can make a huge list of factors that you have to consider to make a decision whether or not [the company]—as a single isolated little company that doesn't make any profit but with a very large research establishment—should be doing that . . . And clearly, we would not have gone five years doing research on oncogenes with no significant and interesting results. (Kensington interview)

My discussion of the negotiation among scientific and business manager partners in constructing the problem provides a different way of talking about the business aspects of research. The commercial aspects of scientific research are usually treated as something that prevents scientists from being as scientific as they might want to be. In this case, however, efforts were made to satisfy both scientific and commercial interests. Aligning their experimental work with the concerns of many audiences (in terms of proposed goals and delivered solutions) allowed the researchers to draw resources from both academic and commercial worlds. Tapping into those resources reduced development costs and provided the researchers with a ready-made set of clients for their product. Furthermore, this situation should be viewed in terms of the state of the biotechnology industry in the mid-1980s, when much of the research conducted in small research and development biotechnology companies was still very basic.

Improvisations

After the success of one of their initial experiments, laboratory members added a new set of experiments to their workload. Both new and ongoing work included experiments aimed at figuring out exactly what structures and mechanisms were involved in the antibody's and the transforming protein's effects. How did the antibodies reverse the transformation process? Conversely, how did the proteins produced by the oncogenes biochemically transform the normal cells?[18]

In addition, lab members made a major effort to use their previous success to develop a short-term product, a diagnostic kit, to satisfy management.

The diagnostic kit was based on an antibody that binds specifically to tumor cells supposedly transformed by oncogene *ras* and would be sold primarily to other scientists for further research. The scientists had conceived of the diagnostic kit product in November 1984 and had promised a July 1985 delivery date. In March, they stepped up the pace of the specificity experiments needed for diagnostic development. As one of the researchers reported, "last November [we] decided that it would be good to produce something to show that we're doing something. [It will be] good for other researchers but not really for more than that. But now it's March, and the product is due in July" (field notes, Xavier lab).

The "specificity" experimental tests entailed checking for the specific in vitro reactions between synthesized antigens related to proteins "theoretically" produced by several *ras* oncogenes and the antibodies raised against them in living animals. They were also important for linking the lab's experimental work with the oncogenic protein work of other protein biochemists. The specificity tests served as checks on their previous results and as preparatory tests for their ongoing basic research on the biochemical properties of the transforming protein itself. In other words, these tests were necessary to fulfill the technical standards of basic biochemical research—standardization, replication, quantifiability, and comparison controls.[19] In response to these standards and time constraints, Xavier asked two laboratory technicians to rearrange their workload in order to contribute some of their time to assist the technician working on the in vitro tests. Of four technicians in the laboratory, three were working on the specificity tests.

While conducting the tests, however, the researchers ran into glitches. "Nature" is recalcitrant; it does not always do what it is expected to do.[20] Experiments often fail, and company presidents get impatient when a product is not ready by established deadlines. Colleagues may disagree with interpretations of results. In sum, a project rarely runs according to plan, if there is a plan, and is usually reconstructed at least several times during the process. Scientific work is constantly *in process.*

The glitches in this case came after the initial success. The researchers changed the conditions under which they manipulated the antibodies in the effort to develop a marketable product and a more useful tool for their own experiments. They then conducted new in vitro tests for specificity. Unfortunately, during several months of this work, one laboratory technician had generated many experimental results in a haphazard manner. Xavier finally realized that something was wrong and asked another technician to take on the job of organizing and making sense of the results. The second technician's final report was "bad news."

> [We had a] bunch of random experiments which [didn't] mean much and we [had] to redo them. (Field notes, Xavier lab)

> [The first technician] had been doing separate pieces of work every day; new [tests] every day without looking at the larger picture of the project. [The second technician is] now in charge of [the specificity tests]; she did a chart of the results from the many [tests]. She also asked [the first technician] to run some [tests] with a sample from an old batch. They got very different numbers. So the variations among numbers makes it impossible to make any conclusions. No standardization. So they have to run the whole set of [tests] again. A lot of work. But they can't do anything else. (Field notes, Xavier lab)

A second glitch occurred when new test results showed no specificity. The antibody that had originally bound specifically in vivo no longer bound specifically to the synthetic protein in these new in vitro tests. Neither did their other experimental samples bind. Researchers came up with two possible explanations: (1) The original experiment had been a fluke, and they had never had specificity; (2) The change from strong to mild detergent buffer conditions had affected binding. Lab members commiserated together about this second difficulty. "It's binding all kinds of proteins. We decreased the detergent [in the buffer], so the other proteins aren't controlled, so they can [now] bind. Think that's why no specificity. Don't know. Or there's no affinity difference between the specific and the nonspecific."[21] When the department head entered the room, Xavier reported the bad news.

> *Xavier:* The bottom line is, in the new mild detergent conditions (of the buffer), we have no specificity anymore . . . [We just get] streaks of black across the film [autoradiograph].

> *Department head:* Did we ever have specificity?

> *Xavier:* [Yes], for X, Y, and, Z. Never tested for P, Q, and R . . . There are [also] differences [between] the early and late bleeds of [animals]. The [P] pool that [laboratory assistant] is [now] working with [shows] no specificity. (Field notes, Xavier lab)

In theory, this should have been no major problem. The researchers would have redone the tests or attempted new ones. However, they were working under many constraints. By that time (about a year and a half into the project), the president was becoming impatient for more results; he already felt that he had pumped a lot of resources into this laboratory without adequate progress. The president "was giving me a lot of flak during the [meetings] for spending so much on the oncogene research," reported Xavier. "He wants it both ways. He uses [the success] to appease the stockholders. But when he sees the deficit on the balance sheet, he gets upset. It also depends on the time of year. Things are great at the beginning of the year, but as it gets closer to the stockholders meeting and the end of the year, there's more pressure" (field notes, Xavier lab). The president had also issued a company-wide freeze on spending to boost the company's profit profile before the July fiscal report. So repeating tests using expensive materials was problematic. Moreover, some researchers in other laboratories resented the resources going into this laboratory when they saw very little good coming from it. (At lunch, these researchers from another lab sarcastically asked me whether "my" lab did any work.) In addition, the three technicians were getting tired of the specificity experiments and wanted to move on to new experiments. On top of all this, the original results had been announced to colleagues and the public, and another company had already sent a letter requesting the rights to market the product. Thus, the "outside" is always "inside" the laboratory.

The laboratory members held several impromptu emergency meetings as well as a formal laboratory meeting to figure out how to deal with these contingencies. Lab members suggested reasons for the loss of specificity and strategies for technical manipulations to satisfy management.

> So we cannot sort by serum, but we might be able to get around it by using monoclonals.[22]
>
> We have to give [management] something. Can we give them the antisera (the fluid containing the antibody) as a diagnostic probe with the restriction that they have to use [a particular buffer]?
>
> You can sell it once, but you can't sell it twice.
>
> We said we'd have a tool for specifically diagnosing tumor tissue. [But we don't.] But what can we salvage? What can we give [management]?
>
> [One lab technician's] going to try two different experiments to see if she can salvage it. (Field notes, Xavier lab)

The idea of marketing the antisera with restrictions on its use was dropped, since other researchers would certainly want to modify the conditions under which they used the antisera. Thus, both high specificity and the purified antibody were necessary to develop a marketable product. Researchers finally decided to try to purify and market the whole antibody. In this way they hoped to satisfy the president in the short run and convince him to keep putting resources into the laboratory.

However, no one in the lab was highly skilled in purification or product development. Theirs was a basic research laboratory. They wanted to pass the purification work to a specialist from another department and the product development work to a third department to save the lab members time and effort. This division of labor required new negotiations between the lab director and the heads of these two other divisions in the company. Meetings soon expanded to include several members from each participating laboratory. These meetings took time and effort away from experimental work and raised tempers.

Back in the lab, the division of labor had to be coordinated with experimental work. One constraint was the work schedule of the "expert" at antibody purification. His skills were in high demand. While he could fit the Xavier lab's purification work in between two other scheduled projects, the timing had to be right. The lab had to be ready with results from its part of the work before the expert was scheduled to start on another project.

During these negotiations, Xavier researchers continued experimental work. Improvising, a technician attempted to salvage the experiments. Before testing for specificity again, she inserted an extra step; she filtered the antisera to remove "junk" antibodies (those not related to their question). She then retested for specificity and got a better result. Two other lab technicians worked on tests with other samples of antisera. They changed conditions (especially the detergent strength) again and repeated the tests on old and new samples of antisera. One of these tests yielded even more positive results.

> [The lab technician] got indicators of specificity (color changes) between the antigens and antibodies using the antisera [in an immune-precipitation test] . . . [Xavier] got really excited, because he hadn't expected it. He started to sing "Oh, We've Got Specificity," while [the technician] sang "We're in the money . . ." One of the reactions barely showed up on the autoradiograph but definitely showed up in [the immune-precipitation] test. [Xavier] said that maybe they can make a specific antisera or antibody to p21 and sell it, based on this test, at least. (Field notes, Xavier lab)

But even positive results were only bits and pieces of information that had to be coordinated with one another and with early, highly varied results. Only then could the researchers make any consistent and careful conclusions either for publication (claims made to their basic research audiences) or for product development. In addition, new tests required decisions about costly new antisera samples. Lab members deliberated about what they could "get away with"—that is, what tests were absolutely required to fulfil standards of "good science" and what tests could be safely eliminated. Which samples were worth the time and expense for additional tests and which were not?

Every day during the next three weeks and squeezed among their other activities, the researchers continued to discuss and work on the specificity tests. And they continued to devise possible strategies for dealing with the problematic results.

Thus, the laboratory members faced an array of contingencies and possibilities. As contingencies arose, they constructed, considered, and attempted strategies to adjust their experimental frame to satisfy the demands primarily from two specific audiences—their sponsors and other molecular biologists and protein biochemists working on oncogenic proteins. Adjusting the experimental frame, or rewriting the rules of the game in midcourse, in turn required rearrangements in the work organization in the laboratory. When the demands from the two social worlds conflicted—for example, when a product deadline interfered with standards of good science—lab members tried to devise strategies that met both demands "well enough." Their main strategy was to divide up the labor of product development and pass some work on to other departments. This system allowed them to continue their biochemical tests. Note the tension of juggling laboratory commitments to the conflicting demands.

Satisfying conflicting demands in this case might perhaps have been easier had mistakes not been made and had the experimental results been unproblematic. But *mistakes at work are routine,* and experimental results are problematic more often than not.[23] When these contingencies arise, researchers adjust. In this case, constructing doable problems took the form of creating strategies that allowed the scientists to juggle and balance multiple simultaneous demands in multiple aspects of their work process and organization.

Scientists make adjustments in work organization as they construct, reconstruct, and solve their problem. Just as unexpected results could lead to the development of an entirely new problem (given adequate resources), they can also lead to difficulties if demands in other parts of the work require the researchers to stick to the current problem structure. Before I leave this example, let me emphasize that work process and organization are temporal and local. What works in a laboratory at one time might not work in another laboratory or in the same lab at another time. The uncertainty in scientific work, the complex division of labor, and the multiple audiences keep researchers constantly alert to problems and to the articulation necessary to make their work "doable."

Articulation and the Standardization of Practices and Technologies

Through articulation work, researchers juggle and balance multiple, simultaneous demands in multiple aspects of the work process and organization. The example above of a short time in a laboratory's process of constructing a doable problem shows the links and translations constructed between different events, conditions, practices, demands, and interests as they emerged in the work site. The final problem structure arose out of the confluence of these elements and the scientists' handling of them.

This portrait of problem-solving is one of ad hoc tinkering. Basic science requires facing uncertainty on a daily basis. Uncertainty inhibits researchers' abilities to plan ahead, which means that much of the work is carried out on an ad hoc basis. When constructing and solving problems, however, scientists sometimes attempt to create other conditions that reduce uncertainty or alleviate its consequences.

As discussed in Chapters 2 and 3, one strategy is to standardize materials, instruments, and techniques. More generally, much of the experimental work in molecular genetic research on oncogenes has been packaged into standardized protocols and problem structures.

> The molecular biologists [looking at oncogene regulation] are literally marching along the DNA . . . They call it that, because the technique they use to find these regulatory elements is "walking down the DNA." If you know a piece of DNA is an oncogene, you can make a probe to it, and you can cut it with restriction enzymes. So you have the known piece, then you have a little tag on it that you don't know, but you have a probe for the known piece. So you take the known piece, and then you sequence the piece downstream, and then you make a probe for that. Then you take it and cut it so that it's a little bit longer. And then you "walk down." You take a known oncogene, and you walk down the DNA until you find the regulatory sequence. You walk down the other way and find the regulatory sequence. (DeVita interview)

As also discussed in Chapters 2 and 3, the materials and other tools used in the experimental work have been packaged and distributed for use in varied sites; witness cell lines, in-bred strains of mice, and air-mailed antibodies and

DNA samples. This packaging and distribution accounts for the similarities found in the hands-on oncogene experimental work carried out in laboratories in different physical and institutional locations.

For example, the development of gel density gradient electrophoresis systems in molecular biology in the 1970s made it possible for researchers in many laboratories to participate in biological problems that required skills in DNA separation and measurement. Before then, only a few researchers were skilled enough to use ultracentrifuges to separate DNA fragments. They were said to have "golden hands." The situation changed when slow and difficult DNA measurement techniques were simplified, standardized, and packaged into instruments. These standardized technologies were then easily adopted by many laboratories and became the standard procedures for separating DNA fragments.

The publication of standard cloning techniques or cookbooks noted in Chapter 3 helped eliminate a lot of direct person-to-person teaching.[24] Standardized procedures have also replaced many more expensive trial-and-error strategies. In contrast to the standardized procedures for cutting, cloning, tagging, and sequencing genes, working with proteins elicits language like "black magic," "troubleshooting," "sensitive," and "trial and error." Proteins become "actors" in the researcher's frame of reference.

> If you ask me why I do certain techniques, [use] certain solution mixtures, certain things in my buffer, I won't be able to tell you. You try some things and your proteins like it, so you keep doing it that way. I can only say that you have to keep your proteins happy. There's a lot of black magic in protein biochemistry. (Field notes, Xavier lab)[25]

> There's lots of trial and error. You have to . . . fiddle with so many parameters . . . So you have to have a broader range of knowledge. And you won't find the techniques in a textbook or a manual like that [pointing to the Cold Spring Harbor Laboratory's *Molecular Cloning: A Laboratory Manual*] . . . You [really] have to learn them from other people. You pick up things from different people. You have to know how to deal with a whole range of possibilities, what is likely to happen. (Field notes, Xavier lab)

Compared with molecular biological technologies for working with DNA, technology for handling proteins was viewed as less standardized. A telling comment by a trained protein chemist studying both oncogenes and their proteins demonstrates the differences between the two sets of work processes.

> DNA does very consistent things. Proteins are much more changeable. There are more different kinds of proteins, and they move around. They do different things at different times. They're very sensitive. So you never know what's going to happen, what you're going to find. I like to do both kinds of work.
>
> It's fun to do DNA work after you've worked on a protein for a while and are frustrated with it. You can work out the problem and experiment pretty much in your head. The experiment is pretty much confirming, and the results are usually what you expect. So you can go quicker. And it works! (Field notes, Xavier lab)

This researcher stated that he could "work out the problem and experiment pretty much in [his] head" because the techniques, knowledge, and even the experimental designs (that is, the kinds of things one can do with DNA) have been packaged into standardized protocols.

The standardized technologies and protocols described in Chapter 2 package up work that would otherwise require more organization and coordination. Machines also "package" previously complex work and reduce articulation needs. Besides centrifuges, molecular biologists employ incubators, photographic equipment for autoradiography, automatic cell counters, liquid scintillation spectrometers, computers, automatic DNA and amino acid synthesizers and sequencers, and PCR thermocyclers. Researchers often have no knowledge of the hands-on process of an experimental task before its instrumental incarnation. For example, when using an automatic X-ray film processor to produce autoradiographs of protein sequences, one research assistant said, "I feel totally tied to this machine. I never learned to process it by hand" (field notes, Xavier lab). And when researchers thought about it, they assumed that the by-hand process resulted in more human errors. That machines also produce errors was acknowledged but considered less problematic than human error. Researchers treated machines as dumb but accurate and humans as intelligent and creative but flawed.

The Daily Detail

Clearly, science is constructed of "the daily detail," the activities often taken for granted by scientists. Problem construction is embedded in these activities as well as in the particular confluences of events and conditions of work in each situation. This chapter has provided an extended example of researchers' efforts to deal with the daily detail and confluences of events as they construct doable research problems.

Articulation work involves the discretionary work necessary to plan and coordinate activities. The small team working for a biotechnology company and researching oncogenic antibodies attempted to adjust and mesh its experimental work with interests and practices of several audiences. We saw the attempts, both successful and unsuccessful, to mesh materials, experiments, laboratory staff assignments, collaborations with other laboratories, funding, and audiences for the research. From fiddling with a protein's reagent to "keep it happy" to persuading the National Cancer Institute or the company stockholders to fund their research, cancer molecular biologists align all parts of their work in order to take a project to its conclusion. In other words, in addition to handling matters of technique, they also negotiate, persuade, market, organize, and reorganize their practices.

Most internalist views of problem-solving assume that it consists of hypotheses and experiments whose form, content, and results are framed in terms of established knowledge in the field (Laudan 1990). I have added marketing, debating, publishing, politicking, administering, and addressing all their attendant strategies and considerations to experimentation in the activities of constructing and solving problems.

My approach also differs from that of Peter Medawar, the biologist and historian of biology whose discussion (1982) of the "art of the soluble" emphasizes the "art," intuition, and tacit knowledge of constructing problems that can be solved in a given situation. As I exemplify above, there is much art, intuition, and tacit knowledge in cancer molecular biology. Researchers have to "know which reagent my protein likes," as well as which proteins to study in the first place. But that unacknowledged work, articulation, also involves art, intuition, and tacit and local knowledges in the making of doable problems in contemporary science.

Medawar also does not attend to funding issues, which are critical in determining which problems become doable. This is an example of worlds other than the technical that are involved and implicated in laboratory research.

When we look at each work site, we see that actors shift their attention from national economies to nonspecificity of rabbit antibodies in a seamless fashion. To take an example from my ethnography, shifts (or rumors of shifts) in the American and world economies were part of the laboratory's daily work life. Sitting in the laboratory in a biotechnology research and development company, I listened to lab members and administrators discuss the company's stockholders report during an economic slow-down. The stockholders and the president wanted the laboratory to speed up its research schedule. In the next hour, I listened to concerns about the lack of specificity of rabbit antibodies that would delay the research schedule. In their efforts to deal with these two events and sets of demands, researchers changed their experiments and problem. There is no outside and inside in the research laboratory. Both university administrations and biotechnology stockholders were often present, often in spirit and sometimes in embodied forms. The world is in the laboratory, and the laboratory is in the world minute by minute. There is no micro-, meso-, or macro-sociology in the laboratory.

8 Conclusion

Crafting Oncogenes

> *According to [the] new kind of materialism that Marx proposes, the object is not only transformed but in some significant sense, the object is transformative activity itself and further, its materiality is established through this temporal movement from a prior to a latter state. In other words, the object materializes to the extent that it is a site of temporal transformation. The materiality of objects, then, is in no sense static, spatial, or given, but is constituted in and as transformative activity.*
>
> —*Judith Butler*

From a Small Doable Problem to a Thriving Enterprise

In the late 1970s and early 1980s in the United States, participants from many different lines of work came to agreement on how best to study cancer. A new representation of cancer, the proto-oncogene theory, initially constructed by a few molecular biologists and tumor virologists, is today accepted as a fact by other researchers, sponsors, suppliers, and diverse participants in the cancer research arena. Cancer has become defined for many people as a genetic disease to be studied in terms framed by the proto-oncogene theory, recombinant DNA, and other molecular biological theories and technologies. Proto-oncogene research became the state of the art. Today it is established cancer research.

The story of the development of proto-oncogene research is the story of how the definition of the situation constructed by molecular biologists became the accepted definition. I have tried to understand this major change in science by borrowing Everett Hughes's (1971) view of the workplace as a place "where [diverse] peoples meet." Hughes asked how members of different social worlds cooperate in the workplace while holding different "definitions of the situation."[1] He concluded that work gets done in these places only through the conflict, struggle, and negotiations over the set of conventions that should guide action and interaction at this meeting of worlds. Hughes's question was framed around the situations in factories in Chicago in the 1940s and 1950s, when new immigrant groups from diverse cultures came together and were obliged to work together. In my case, the "invisible college" of cancer research is the "extended workplace," a "factory floor" that is "located" in a complicated web of interconnected places.[2] This web of interconnections is the arena where members of different worlds meet and negotiations ensue about how one should approach solving the problem of cancer.[3] This meeting of different worlds created a new world, the hybrid world of proto-oncogene research. Moreover, the diverse worlds involved in the interactions changed in the process.

However, there is a difference between the two situations. In the case of cancer research, participants from different social worlds were not forced to work together. Until the early 1980s, when oncogene funding commitments attracted many new and established researchers to the field, scientists and medical researchers in different lines of research could have continued in their own trajectories without engaging in or using the findings of new proto-oncogene research. Recent histories and sociologies of science have also shown that agreement and consensus are not taken for granted in science; they are not based on reality's guiding hand. To understand how proto-oncogene research was so quickly accepted by scientists, I have presented a framework for understanding scientific knowledge production. I have described the processes involved in the construction of knowledge by focusing on how tools, practices, and theories are co-constructed, incorporated, and refashioned in a continuous process of negotiation.

Callon, Latour, and Law framed their actor-network theory to explain the construction and stabilization of facts and artifacts as the enrolling and disciplining of many heterogeneous actors into a network of stable linkages. The story I present here is one of co-enrollment. Bishop and Varmus were enrolled into the ongoing enterprises of scientists working in normal growth and development, just as these scientists became enrolled in Bishop and Varmus's proto-oncogene project. I focus on the linking of practice, work routines, experimental systems, and theory in diverse situations across time in order to tell a story of proto-oncogene mutual enrollment. Mine is a story of collective history and situated action.

I give an account of a package of theory and methods that became a *distributed* center of authority. Through the proto-oncogene theory and molecular genetic technologies, players in the worlds of science transformed practices in many different situations and linked the activities of research laboratories and communities, sponsors, and DNA sequences in a network of practices. Through its satisfaction of the "desires" of researchers, funding agencies, biotechnology companies, venture capitalists, and others, this package gained and then wielded an authority that went beyond any individual person or scientific institution.

Co-Constructing Practices, Problems, Laboratories, and Worlds

Researchers crafted science by shaping and adjusting materials, instruments, problems, theories and other representations, and social worlds as well as themselves and their laboratories in the process of co-construction.

In Chapter 2 I discussed how standardized experimental systems were used to construct representations that were comparable between laboratories. These systems included inbred animals, tissue culture materials and methods, and tumor viruses, which gave materiality to "genes" and "cancer." They were technical-natural objects that reinvented nature. These standardized experimental systems transformed practices in multiple sites. From this transformation of practices emerged a new technical (as versus abstract) definition of genetics and cancer. The animals and the collective work behind them defined a new technical work space for "realizing" the concept of genetics in cancer research. These animals represented and embodied the commitments of the cancer research community to a particular definition of genetics. They were cyborgs, artifacts created under artificial conditions according to a formula that combined a commitment to rational scientific principles with a particular conception of cancer as a genetically caused or transmitted disease.

Moreover, standardized experimental systems were used to reconstruct laboratory work practices and, in turn, experimentally produced representations. These practices and representations were assumed to be homogeneous and comparable across laboratories and through time. Thus, scientific technologies are highly elaborated symbolic systems. They are meaning-laden and meaning-generating systems, just as language and writing are. In the case of standardized experimental systems, they represent values and commitments to the ideals of invariability, homogeneity, standardization, and reproducibility. These values are also real laboratory (arti)facts embodied in these standardized experimental systems.

In Chapters 4 and 5, I used the concept of packages and the linking of practices and interests to show how different social worlds interacted through time and space to collectively craft science. In the construction and adoption of the standardized package of proto-oncogene theory and recombinant DNA technologies (Chapter 3), each world changed in some manner, yet each also maintained its uniqueness and integrity. The package provided both dynamic opportunities for divergent meanings and uses as well as stability. In crafting and recrafting the theory, Bishop and Varmus, colleagues, and supporters constructed multiple translations between oncogene research, on the one hand, and evolutionary biology, developmental biology, cell biology, carcinogenesis research, and more, on the other hand. As they drew on concepts and arguments from these lines of research, the theory, practices, inscriptions and materials became part of these ongoing lines of research. The package created "homologies" between laboratories as well as between representations of phenomena. As the proto-oncogene theory and the package of theory and technologies were being constructed, they were also used to reconstruct laboratory work organizations as well as experimentally produced representations.[4] This process created a productive balance between standardization and novelty.

In Chapters 6 and 7, I described problem-solving in basic science as a dynamic, shifting, uncertain process. For any project to succeed, scientists must negotiate tasks ranging from convincing funding agencies of the project's worth to making or buying necessary supplies to conducting experimental manipulations of DNA. Uncertainty and ambiguity reign at every turn in research processes as scientists adjust, shape, and align elements from recalcitrant proteins to conservative company presidents in order to construct doable problems. In Chapter 7 I focused on researchers' efforts to construct and solve problems by meshing the practices of experiment, laboratory organization, and social worlds in an interactive work process I call "articulation work." The scientists' (and my) term for successful articulation is "doable" problem construction. My definition of doable problems incorporates experimental, theoretical, social, and cultural practices. These problems include audience demands, sets of conventions for organizing work in specific disciplines and fields, and cultural and historical practices, in addition to more commonly understood technical and theoretical practices. These practices interact with each other in diverse ways in every situation. Thus, while all are significant in defining each situation, we cannot predict how practices will interact and play out in each situation. In Chapter 6, in views from the shop floor, I examined how scientists negotiated their various daily as well as long-term and often conflicting demands from many audiences in their efforts to construct and solve problems. Scientists juggle and fiddle with one of the practices in order to adjust it to changes in other practices in a continuous interactive process. Throughout this continuous process, scientists patch together some coherence among scientific tools, scientific representations of nature, and audience demands.

Coherence

These portraits of problem-solving in specific situations present another problem for this study and for the social studies of science. Granting the contingency and variability of problem-solving, and the many kinds of coherent descriptions of nature that can be constructed, how does science move from local constructions to community agreements? Ethnographic studies of science have shown us the contingencies, the mistakes, the blind alleys, the fiddling with and coddling of materials, the magic and black art and tacit knowledge, the self-styled and adjusted and shortened versions of "cookbook" techniques, most of which are deleted from or buried in the final representations. However, even granting this variability of scientific practice, there still are stabilizations and continuities across situations and through time. Scientific knowledge about cancer is constructed by participants in many different worlds, including the biotechnology industry, the National Cancer Institute, molecular biology, embryology, developmental biology, oncology, epidemiology, cell biology, genetics, and, more recently, affected communities (breast cancer activists). There is no one world that owns the problem or the solutions. The problem of cancer is distributed among the worlds, each with its own agenda, concerns, responsibilities, and ways of working.

Given this diversity of scientific practice, how did the proto-oncogene approach gain so many adherents across situations, laboratories, fields of research, granting agencies and other worlds? How did these situations come to at least some common practices, if not agreement on proto-oncogenes? I next address these questions with several accounts of coherence in scientific work across situations and through time.

Traces of Continuity in Practices and Concepts across Time

In Chapter 4, I presented an account of researchers crafting science in a historical context through an analysis of situated action; the construction of new research programs is a case of situated action, built on past practices, concepts, and representations. That is, scientific actors consider new proposals, theories, and methods in light of earlier routines and representations and in light of the conditions of the new situation. In cancer research, for example, the development of proto-oncogene research provides an example of historical continuity. For years, as part of the war on cancer, tumor virologists had studied particular sequences of RNA in retroviruses, a class of viruses whose genetic material is RNA, not DNA.[5] Their research showed that these sequences caused cells in culture to turn cancerous and sometimes caused tumors in laboratory animals. Researchers developed and stored viral strains in order to study whether viruses or at least these viral oncogenes were possibly involved in human cancers. However, large investments of effort, time, and finances provided no indications of a significant involvement in human cancers. Instead they provided grounds for criticism of this research from many quarters.

Into this scene, Bishop and Varmus introduced a new hypothesis that reversed the direction of the relationship from gene insertion to gene pilfering by the virus. They claimed that their research found DNA sequences "homologous" (similar in structure, function, and evolutionary ancestry) to the RNA tumor-causing sequences in normal cells of life forms from yeast to humans. They argued that, contrary to the existing hypothesis that viruses had transferred their cancer-causing sequences to humans, the viruses had gained their cancer-causing sequences from normal cellular DNA.[6] This reversal, they argued, meant that cancer-causing genetic elements are conserved through many species and thus must be related to some essential function in all organisms. Their new research program thus aimed at finding and studying the properties and expression of these normal DNA, "proto-oncogene," sequences in cells of different species, but especially in humans.

Bishop and Varmus had turned the earlier virus studies on their heads. The movement of genetic elements reversed directions literally and figuratively, but the same or homologous gene sequence continued through time and space. Driving this "gene conservation" were molecular probes created using molecular hybridization techniques. Critically, however, their work made use of the same (or similar) DNA sequences that previously had been studied extensively and that had gained enormous material and sentimental support from many researchers, organizations, and institutions. As discussed in Chapters 2 and 4, the construction of the proto-oncogene theory was situated within a network of historical and contemporary commitments to particular representations and representational practices in cancer research, such as inbred animals, cell lines, Temin's protovirus ideas, the provirus theory, viral cancer research and collected materials, new molecular genetic methods, and prior commitments of funding agencies.[7]

From Local Contingency to Traces of Continuity across Situations

I have discussed the co-construction of proto-oncogene theory, problems, technologies including instruments and materials, data, and practices. These elements are shaped and adjusted to mesh with one another. In each situation, a problem is constructed and solved through the interactions that articulate and weave together practices in experiments, laboratory organizations, and multiple social worlds including laboratory, field of research, discipline, profession, regulatory agency, industry, and research tool suppliers. In addition to performing experiments and negotiating with relevant audiences, scientists fiddle with one of the elements in order to adjust it to changes in other elements in a continuous process of problem-solving. Throughout this process, scientists patch together some coherence among scientific tools and resources (materials, instruments, techniques, data, skills, labor, money, time), scientific representations of nature (conceptual, theoretical, material models), and audience demands (or jobs). This local construction of coherence in science is *craftwork,* done in constant interaction with the rest of the world.[8]

This description of coherence construction ignores a problem to which I now turn: If we have coherence within local situations, and if local situations differ, how then do we account for the coherence between different situations that is the basis of a stable theory? In this case, how do we account for the apparent stability of the proto-oncogene theory across laboratories given different local contingencies? To answer this question, I must first discuss the issue of local contingencies, before analyzing two kinds of coherence that are conflated in proto-oncogene science.

Coherence constructed from the contingencies within local situations has been the preoccupation of recent social studies of science. Laboratory ethnographies have provided us with understandings of the bricolage, tinkering, discourse, tacit knowledge, and situated actions that build local understandings and agreements. These studies point to various tacit knowledges employed by scientists in specific situations and their different interpretations (Collins and Pinch 1982, Collins 1985, Pinch 1986). Studies of "experimenter's regress" problematize established ideas about "replication" and alert us to the issue of how "the same" is established on each occasion of reproducing an experiment (Collins 1985). We have observations of "the occasioned character of research" as exemplified by tinkering and patching things together within the contingencies of the local situation (Knorr-Cetina 1979, 1981). The local laboratory discourses and negotiations among people involved in constructing and evaluating representations (traces), phenomena, and instruments are elaborated in great detail (Gilbert and Mulkay 1984; Gooding, Pinch, and Schaffer 1989; Lynch 1985; Lynch and Woolgar 1990; Traweek 1988). Other studies, as mentioned above, present science-making as the accumulation and arraying of "inscriptions" to enroll "allies" (Callon, Law, and Rip 1986; Latour 1987; Law 1986) and the elaborate processes involved in constructing and using instruments in experimentation (Latour and Woolgar 1986, Traweek 1988; see also Suchman 1987).[9]

Historical cases have also provided material for the detailed examination of attempts to achieve local coherence. For instance, in his historical studies of high-energy physics, Pickering (1984, 1990, 1993) proposes that researchers like Giacomo Morpugo adjusted their material procedures and instruments and their phenomenal models to "cohere" with one another. Holmes (1985) and Schaffer (1989) also examine laboratory activities in historical cases to trace the production of local coherence.

These studies have opened the daily contingent, collective, and collaborative practices in scientific laboratories to sociological analysis. Drawing variously on the ethnomethodological tradition, discourse analysis, semiotics, phenomenology, and detailed historical analysis of documents, many of these scholars are interested in the collective performance of actions in specific situations—for instance, the coordination of local actions around a technical object—as the means to study collective production of scientific knowledge and artifacts.

However, despite the local and situated nature of scientific work, stabilities and continuities (if not agreements) across situations and through time are often reported by scientists and engineers. Their reports open up questions about definitions as well as productions of continuity, agreement, and stability.

Philosophers and sociologists of science have attempted to explain continuities and stabilities across situations in terms of replication, triangulation, and black boxes. For example, Collins (1985) has rendered replication problematic by demonstrating that conditions are usually changed in some way in efforts to replicate experiments. Using the case of gravity waves, he demonstrates that each repetition of the original experiment incorporated some change in equipment or protocol that made the experiment arguably different and would therefore allow the experimenters to claim a new finding or to debunk the original finding. For another example, Star (1989) has discussed the efforts of turn-of-the-century brain researchers and physicians to triangulate[10]—that is, to use each other's work to produce a theory of localization of behaviors and functions in brain tissue—despite differences in their local units of analysis, levels of abstraction, temporal orientations, and orientations toward anomalies. She argues that one process by which triangulation occurs is the transformation of uncertainties in local situations into abstract global certainties through the deletion of the work of adjusting the local situation to fit the global claim.

Oncogene research is not turn-of-the-century brain research. And despite experimenter's regress, there is continuity in oncogene research. How is this continuity produced in contemporary science? In my study of oncogene research, I have found the concept of standardization germane to the production of reliability, reproducibility, replicability, triangulation, and authentication. What is standardization? How is standardization achieved across situations?

Difference, Equivalence, and Standardization

Using molecular biological techniques, oncogene scientists proposed that they found similar (technically "homologous") gene sequences in cancer cells and in the cells thought to be involved in normal growth and development. Based on this and related research findings, they argued that cancer is the end result of a series of genetic changes (qualitative, quantitative, or time-frame changes) in which once-normal growth genes are transformed into tumor-producing genes. This book raises the sociological and epistemological question of how experimental outcomes are determined to be similar. We cannot assume that assertions of similarity are true in any metaphysical sense. Instead, we need to investigate how similarities are produced and understood. This book examines the conditions under which and the processes through which these particular experiments, problem structures, and outcomes came to be regarded as similar in laboratories across the country.[11] For example, how does a technique *become* reliable, to produce stable, similar representations across spatial and temporal situations?

A related set of studies focusing on continuities of representations argues that continuities across situations are shaped by the black-boxing of skills and techniques in "inscription devices" or instruments (such as the printing press, scientific instruments, and cameras) that reliably reproduce results or "tracings" (such as texts, maps, diagrams, photographs, autoradiographs, and computer readouts) in each laboratory (Callon and Latour 1981; Callon, Law, and Rip 1986; Latour and Woolgar 1986; Latour 1990a; Schaffer 1989).[12] The term "black box" refers to a tool that is no longer questioned, examined, or viewed as problematic, but is taken for granted. The more elements one can place in black boxes—modes of thoughts, habits, forces, and objects—the broader the construction one can raise. Latour calls the inscriptions produced by such black boxes "immutable mobiles" because they can travel without withering away, they do not fundamentally change along the way, and they can be sent to and interpreted by others as well as linked to a variety of other elements.

Latour and Woolgar (1986:238) attribute this production of stability to the strategy of "materialization."

> One cannot take for granted the difference between "material" equipment and "intellectual" components of laboratory activity: the same set of intellectual components can be shown to become incorporated as a piece of furniture a few years later. In the same way, the long and controversial construction of TRF [a hormone] was eventually superseded by the appearance of TRF as a noncontroversial material component in other assays. Similarly . . . investments made within the laboratory were eventually realized in clinical studies and in drug industries . . . We . . . refer to the above process as *materialisation,* or *reification.*

An interesting example of the production of stable knowledge through visual inscriptions is provided by James Griesemer's (1991) discussion of August Weismann's theory of the continuity of the germ-plasm (genes) and the discontinuity of the soma. Griesemer shows that Weismann's theory as inscribed in written text and diagrams was later distorted in a diagrammatic simplification by E. B. Wilson in his widely used biological textbook, *The Cell in Development and Inheritance,* published in 1896. Wilson's version emphasized the continuity of germ cells (an individual's complement of genes). However, Weismann specifically "expressed his theory as the continuity of the germ-*plasm* and discontinuity of the soma, not as the continuity of the germ-*cells.* Weismann took great pains to distinguish his theory from the one Wilson's diagram depicts; several of Weismann's contemporaries held the theory of the continuity of germ-cells and Weismann clearly wanted to distinguish his developmental view from it" (Griesemer 1991:4, Griesemer and Wimsatt 1989).[13]

Nevertheless, it was Wilson's (re)presentation, not Weismann's diagram, that educated several generations of biologists and has carried on into twentieth-century molecular biological views of causality and agency attributed to the molecular DNA form of genes in Crick's central dogma (that DNA makes RNA, which makes proteins). Griesemer argues that this shift from the "genetics" of Weismann to the "genetics" of "Weismannism" (the distorted version of Weismann's theory) helps to explain the change from social hereditarianism to eugenics in the turn of the century's efforts to "improve" society.[14] By historical analogy, Griesemer's concern is that the simplicity and portability of Wilson's version of Weismann's diagram has extended itself into current ideas and practices in biology and society to influence ideas of genetic causality in contemporary efforts to understand the molecular mechanisms of human genomes. For my purposes here, this case demonstrates that scientific representations have far-ranging symbolic, historical, and geographical (time and space) consequences, when packaged in durable inscriptions and black boxes.[15]

Once black boxes and their inscriptions are put out into the world, they can be destabilized, either through different uses in other situations or through changes in situations (see Bennett 1989). They are more mutable than Latour (1988) indicates. However, this book (and Griesemer's example) argues that the more portable a tool is, the more reliably it will be reproduced in other situations. Portability refers to the qualities of simplicity and ease of movement and use. For example, research protocols with clear directions and standardized and commercially produced reagents are highly portable tools. Griesemer's discussion is an interesting illustration of this point. He attributes the "invisibility" of Weismann's own diagrams and texts in part to their complexity as compared with Wilson's simple diagram of Weismannism.

Jordan and Lynch (1992) challenge the idea of portable inscriptions and black boxes. In their study of "mundane" laboratory techniques, they show that even a simple "plasmid prep" is often performed in different ways. "Plasmid prep," alias plasmid purification and isolation, has been a basic procedure in molecular biology for producing large quantities of genetic sequences of interest for further study. Jordan and Lynch raise questions about what "sameness" means in practice (compare Collins 1985). They ask what makes one performance "the same" as another, through all the changes in ingredients, protocols, personal techniques, purposes, and circumstances. In ethnomethodological fashion, they conclude that sameness is decided through the course of (re)producing laboratory practices, not in any "outside" observer's decision of sameness.

I want to appropriate Jordan and Lynch's conclusions to argue that their study confirms my point about portability and black boxes. Indeed, I have argued that the success of a technique in producing "reliable" results is not only due to its clear protocols or its ability to (sometimes) produce the "same" results. It is also constituted by researchers' assumptions, understandings, and agreements that they produce the "same" or similar results.[16] In the case of oncogene research, researchers treat molecular biological tools and protocols as standardized tools producing results reliably consistent across situations. When they read their experiments as not producing the expected results, the researchers might (or might not) repeat the experiment until they "obtain" the expected results. "Replication" and "reliability" depend on what researchers *agree* to be the same results.

Thus, standards and black boxes are the consequences of collective understandings and practices developed in the laboratory and through history, as well as the tools for reconstructing practices in laboratories (and across generations!). As they give form to each new situation, they simultaneously are shaped by the situation. In each situation, these new tools or commitments have to be integrated with ongoing commitments, conventions, and contingencies. This integration is part of the job of articulating alignment, fiddling with things to get them to "work" together to achieve situational coherence. Situational coherence is achieved in part *because* standards, black-boxed technologies, and concepts are not rigid and inflexible as they might seem.

Another example might help to clarify this point. In their study of the constitution of continuity, Keating, Cambrosio, and Mackenzie (1992) show that concepts like "avidity" and "affinity" held through a hundred years of immunological research despite changes in the techniques for measuring and defining them. They attribute this conceptual continuity, despite differences in the definitions and measurement techniques attached to the concepts, to "disciplinary stakes" or "disciplinary interventions." They argue that affinity and avidity were retained through time and across situations because they were tied to "specificity," a concept that was vital for the maintenance of the discipline of immunology as an institutional entity separate from other units. For another historical example, we need only to look at the long-term debates about the nature, structure, function, and units of hereditary "factors" in the history of genetics and population biology to remind ourselves that dissension in science can exist even as debating groups use the same concept, "gene," in their different theoretical constructions.[17]

These examples demonstrate that standardization, continuity, and stability do not necessarily mean equivalence in an ontological sense. Concepts, practices, and even material entities (such as instruments and biological organisms) can be treated as the same or continuous or stable, and yet still be divergent in some fashion. Thus, sameness is something produced through experimenters' collective efforts, understandings, expectations, and agreements.

Constructing Coherence, Circumventing Realism

In this book I adhere to the basic assumption that the similarities produced by oncogene researchers are the results of situated work and not of ontological reality. I want to address the question of realism in terms of the relationship between the confirmation of the oncogene hypothesis and the experimental findings. With respect to the confirmation of the hypothesis, I find philosopher of science Ian Hacking's concept of the self-vindication of theories in laboratory sciences useful for dealing with the question of realism.

Informed by work in the social studies of science,[18] Hacking attempts to explain continuities across situations. In his framework, the relationship between observation and theory in mature laboratory sciences is mediated by a number of elements and activities whose particular way of cohering are not preordained. Instead, theory and observation are connected to each other via processes of *meshing* these mediating elements including ideas (questions, background knowledge, systematic theory, topical hypotheses, modeling of the apparatus), things (target, source of modification, detectors, tools, data generators), and marks and the manipulation of marks (data, data assessment, data reduction, data analysis, and data interpretation). Time is a critical element in his model of laboratory science. "As a laboratory science matures, it develops a body of types of theory and types of apparatus and types of analysis that are mutually adjusted to each other . . . [These theories, apparatuses, and analyses] are self-vindicating in the sense that any test of theory is against apparatus that has evolved in conjunction with it—and in conjunction with modes of data analysis" (1992:2). These elements and their "mesh" with one another are constructed through time, through a "fitting" process. In other words, referring to Callon and Latour, Hacking argues that they are co-produced.[19]

A theory, then, is vindicated by or "true to" its own set of ideas, activities, markers, and marks. Hacking (1992:31) argues that "our preserved theories and the world fit together snugly less because we have found out how the world is than because we have tailored each to the other."

> Theories are not checked by comparison with a passive world with which we hope they correspond. We do not formulate conjectures and then just look to see if they are true. We invent devices that produce data and isolate or create phenomena, and a network of different levels of theory is true to those phenomena. Conversely we may in the end count them as phenomena only when the data can be interpreted by theory. Thus there evolves curious tailor-made fit between our ideas, our apparatus, and our observations. A coherence theory of truth? No, a coherence theory of thought, action, materials and marks. (Hacking 1992:57–58)

Hacking specifically refers to the meshing of practices, beliefs, and tools with one another through the developmental history of a mature laboratory science. He confines his discussion (except for a few qualifications) to the "internal" workings of a laboratory science. His goal is to explain the stability of theory within a particular laboratory science, and he refrains from arguing that a stable theory extends outside that science to explain the "real world" in any metaphysical sense. His is a self-consistent system within which theory is "vindicated."

The charm of the self-vindication framework is that it allows us to discuss the construction of stable theories without appealing to realism. For example, I can argue that the theory that various oncogenes cause cancer has been vindicated in the manner indicated by Hacking. Oncogene researchers argue that the hypothesis is consistent with the data generated in their laboratories: that is, oncogenes can produce cell transformation and tumor growths in laboratory experiments. At the moment, oncogene theory is "true to" the many kinds of data and data interpretation, laboratory procedures, materials, instruments, and historically preceding hypotheses and background knowledge. However, this brief statement masks an enormous amount of careful, detailed work that went into, and continues to go into, the continual adjustments made by researchers as they study cancer at the molecular level. This "coherence work" continues in response to new data produced in laboratories and to criticisms and questions about anomalies. New entities, such as anti-oncogenes or tumor suppressor genes, appear on the scene as researchers attempt to fill the gaps in their data.

However, in the study of diseases and other such practical problems addressed by laboratory sciences, it seems implausible that a single laboratory science can generate and vindicate a theory (Fujimura and Chou 1994). The case of cancer is no different. The production of knowledge about cancer spans many biomedical research and medical communities and practices as well as patients, government institutions and agencies (NCI, Congress), and private industry (biotechnology companies, biology supply firms, health insurance companies). The meshing of theory and observation becomes much more complicated to achieve when, for example, many basic and clinical research sciences as well as clinical practices participate in producing knowledge about cancer. Adjudicating between and attempting to mesh together their different kinds of data, different forms of argument, different units and levels of analysis, different temporal orientations, and different orientations toward anomalies require, among other things, time, data processing and interpreting efforts, research funds, and collective work.

In fact, oncogenes have yet to achieve solid links to the worlds of biomedical treatment.[20] They have been used in diagnostic research, but even this work is still in the experimental stage. Yet, in basic science, oncogene theory enjoys the kind of vindication discussed by Hacking through the links established by experimental work with problems of normal growth and development, cell biology, and evolutionary theory. Oncogene researchers argue that the hypothesis that oncogenes cause cancers meshes with the data generated in these other fields of biological research. More complex and messy processes of meshing occur because different worlds are attempting to mesh their different sets of practices and products. How do we explain these cross-links with other biological sciences using Hacking's scheme?

In the example of the proto-oncogene theory, my explanation for its obduracy relies on a history of oncogene science that goes beyond recent proto-oncogene research per se. I suggest that the proto-oncogene theory of the 1980s was vindicated through a long process of coherence-making activities extending back to the search for cancer viruses in the 1960s and viral oncogene research of the 1970s. Viral oncogene research, of course, built on previous efforts to make experimental work on viruses "doable" in the laboratory. For example, Renato Dulbecco created the tissue culture assays that were crucial to experimental research on animal viruses which, in turn, were used to experiment on DNA and RNA tumor viruses in the 1960s and 1970s. However, Dulbecco's work itself grew out of the questions and goals of patrons such as the National Foundation for Infantile Paralysis and James Boswell, a wealthy and shingles-stricken benefactor (Kevles 1992). Similarly, research on enzymology and bacteriology, along with DNA tumor virus research (also funded by the NCI), were critical to the construction of recombinant DNA technologies. These technologies were used by proto-oncogene researchers to demonstrate their theory.

The obduracy of oncogene laboratory science also resulted from the reconstruction of other situations in which different kinds of science had been practiced. Even when attempting to triangulate results from other laboratory sciences (such as normal growth and development, *Drosophila* genetics) to argue that those findings support their theory, proto-oncogene researchers may have produced a kind of "truth in applications." That is, although proto-oncogene theory appears to be stable across worlds of practice, this appearance is in part an artifact of the reconstruction of other laboratories in the image of oncogene science.

Along the same lines, Hacking argues that success in such situations "external" to a laboratory science is the outcome of a great deal of work done to transform the world. He borrows the example of Pasteur, as discussed by Latour (1988a), to point to the hard work involved in changing the world, in remaking it into a quasi-Pasteurian laboratory, to reproduce the same events on the farm and elsewhere that Pasteur produced in his Parisian laboratory.[21] Similarly, what appeared in the 1980s in oncogene research to be rapid interworld collective action was the consequence of the laborious work of reconstructing laboratories and problems in the image (practices and materials) of proto-oncogene research.

Thus far, the self-vindication framework provides a useful answer to the question often posed to sociologists and anthropologists of science "lumped" into the "constructivist" approach category: If science is merely constructed, why does it work? Although my approach is not strictly constructivist, I am also often asked this question. My answer is that things constructed in laboratories often *do* work. However, they usually work only under laboratory conditions. NIH 3T3 cells can reliably be transformed into cancer cells upon the introduction of certain chemicals. Laboratory-produced OncoMice reliably produced tumors within months. Indeed, a major task of science is to make things work in the laboratory, and many sciences are quite successful at this job. That oncogene science works vindicates oncogene theory. However, it does not "prove" in some metaphysical sense that oncogenes cause naturally occurring human cancers. The self-vindication framework is useful for understanding the stability and continuity of scientific theories over time without having to resort to the metaphysics of truth and reality as explanatory variables.

Hacking's framework borrows from sociologists of science such as Callon, Latour, and Law, who have attempted to deal with such heterogeneous instances of science-making by arguing for a more encompassing framework of "heterogeneous engineering." Like most post-1970s sociologists of science, they have argued that sociologists and philosophers of science must include many other categories of things and activities beyond internalist analyses. Latour's telling of the Pasteur story emphasizes Pasteur's enrollment of many different audiences to vindicate his theory of the microbe.

Given this commitment to seriously deal with heterogeneity of the world, actor-network analysis promotes a very dynamic and plastic framework for examining scientific fact production. The framework celebrates the heterogeneity and dynamics of science in a fashion unprecedented in science studies (except for Feyerabend 1987; see also Dupré 1993). In actor-network theory, with the perpetual *diversity* of interests among heterogeneous parties, the *stability* of facts is more difficult to explain than is scientific change.

I propose that stabilities and continuities in molecular biology are the extension of molecular biological laboratories out into the world. Molecular biological apparatuses, theoretical frameworks and descriptions, markers and markings, and procedures remake the world in ways that reproduce selected features of the molecular biology laboratory in other biological research laboratories, hospital laboratories, and forensic laboratories, and install the language of molecular genetics in patient charts, insurance policies, federal health funding priorities, school curricula, television and film *(Jurassic Park)*, novels, and so on. Put more dramatically, molecular biology is being used to reproduce and proliferate its tools and environments.

Oncogenes and Collective Action

Hacking's framework does not provide us with an understanding of collective action or of how standardization contributes to the creation of stability. But science is collective action, and common conventions and coordinated collective action can be produced when actors come together. Chaos and discontinuity are also possible outcomes. I want to expand Hacking's conception to examine an interworld situation where one would have expected instability yet one finds stability. I increase the number of elements and complexity of interrelations between elements discussed by Hacking. Indeed, I argue that divergences and commonalties in practices and concepts are critical to understanding the success of the proto-oncogene research and theory.

In the case studied here, the package of proto-oncogene theory and molecular genetic methods accommodated heterogeneous practices and conventions because it was both abstract and concrete. The proto-oncogene theory, a tool for handling heterogeneous situations, and molecular genetic technologies, tools for creating homogeneity, allowed for points of commonality between worlds to accommodate collective action but also yielded points of difference to allow for different interpretations.

In each situation, researchers used the assemblage of oncogene concepts (genes, cancer, cancer genes and viral genes in the proto-oncogene theory), their material representations, and recombinant DNA technologies to further restrict and define these concepts, materials, and technologies. Such co-definition and co-restriction gradually narrowed the range of practicable actions, or potential practices, but also left a certain degree of flexibility in the choice of specific materials (there are millions of genes open to study), research protocols, technologies, interpretations, and so on.

Through these different interpretations and concrete (re)presentations and through reorganization of local practice, linkages among researchers, laboratories, research institutions, and even biotechnology companies and diagnostic clinics were created. Conversely, the interactions among these different worlds created the package of theory and methods. Through this dialectical process, cancer and cancer research worlds have been reconstructed.

Thus, proto-oncogene theory and recombinant DNA technologies were part of the transformation of diverse situations and the creation of a new state of the art in cancer research. In a relatively short period of time, the combination of these oncogene concepts, their material representations, and recombinant DNA technologies "solidified" into a package of standard practices—that is, practices and understandings shared by several social worlds. Simultaneously, because the package encompassed a range of flexibility in details, researchers were able to maintain continuity with past commitments to particular methods, materials, and theories. A dynamic *interface* between multiple social worlds, the package represents the flow of resources (concepts, skills, materials, techniques, instruments) along multiple lines of work.

Criticism and Controversy

Although most scientists think that the evidence proved that Bishop and Varmus were correct in their claims about proto-oncogenes, a few dissenting voices have challenged the hegemony of the proto-oncogene theory of cancer causation. While there has been much support for the theory in the late 1980s and work continues within this frame in the 1990s, a few cancer researchers and molecular biologists have openly criticized the proto-oncogene theory.[22]

One set of criticisms launched at the proto-oncogene theory was based on the theme "if you have a hammer, every problem is a nail." According to these critics, the introduction of new technology (including knowledge) in a particular line of work results in a corresponding theory of cancer causation.

> I believe the current rush to accept cellular oncogenes as the origin of human cancer . . . is at best premature. I have discussed previously some of the problems inherent in a simple genetic interpretation of cancer . . . and others . . . have pointed out flaws in experimental design which raise serious questions of interpretation of the results that engendered the present excitement. Further detailed criticism is unlikely to have much effect. It should be pointed out, however, that explanations for the origin of cancer have been varied and plentiful in this century. A limited list would include early theories of chromosomal alterations, virus infection, high glycolytic rates, damaged grana (mitochondria), enzyme deletion, and reduced immunological surveillance. Each of these was carried to the fore by developments in a corresponding area of basic biology or biochemistry, and each time many were convinced that a final answer had been found. In retrospect, the supporting evidence always was strong, but it later turned out to be inadequate to establish causality. I believe we have confused advances in molecular biology and its attendant technology with deepened understanding of the nature of malignancy. (Rubin 1983:1170)

In a similar vein, Peter Duesberg and J. Schwartz argued that small mutations noticeable only through the development of better detection and signaling techniques are "amplified" by researchers into a theory of cancer-causing proto-oncogenes.[23]

> The scientific community has been virtually unanimous in admiring its recent triumphs in biotechnology—above all the detection and amplification of minute amounts of materials into workable and marketable products. However, in clinical diagnostic applications the new detection methods have become a mixed blessing that benefits medical scientists but not necessarily their clients . . . Cellular mutations have become detectable that do not, or just barely, affect the function and activity of genes. Yet when the affected genes are structurally related to retroviral oncogenes, they are assumed to be just as oncogenic as highly active retroviral oncogenes. However, the evidence for these hypotheses is only circumstantial—based on structural similarities to classical pathogenic viruses and viral oncogenes. (Duesberg and Schwartz 1992:136)

Another set of critics had previously argued against drawing conclusions from the use of assays such as NIH 3T3 cells, which were used to determine transformation of cells from normal to malignant. NIH 3T3 cells are called "preneoplastic" and do not function like normal cells. Neither are they examples of cells progressing from normal to malignant, according to critic G. B. Dermer.

> Progression . . . suggests that the cells have acquired features of the malignant phenotype . . . [If so,] should they not exhibit a high degree of variability, since much recent work indicates that variability or heterogeneity is a fundamental characteristic of the malignant phenotype? . . . Even human tumor cells are not immortal when taken from the patient and put in culture. Only after a period of adaptation and selection does an immortal clone sometimes arise from a primary cell culture. The relationship between these immortal tumor cell lines and the original tumor cells is in my opinion distant. To my knowledge, none of the human tumor cell lines derived from adenocarcinomas retain the glandular organization characteristic of adenocarcinomas or the normal cell types from which they arose. Furthermore, human tumors do not appear to exhibit the alteration of a cellular oncogene that has been demonstrated in long-term carcinoma cell lines. Since a basic concept in biology is that structure and function are related, we can conclude that human tumor cell lines do not function like human tumors. Perhaps cancer researchers who experiment with long-term cultures should pause for the purpose of imaginative thought so that appropriate experi-

> mental systems for studying human cancer can be developed. (Dermer 1983:318)[24]

Some of Dermer's arguments also take the form of a sociology of error. He asserts that particular aspects of the organization of scientific work produce errors in scientific knowledge. His question of the relevance of the NIH 3T3 research to human disease is reminiscent of similar criticism against the Virus Cancer Program (Chapter 4).

> I would . . . say that some cancer research may actually have set back our understanding because the data produced are not relevant to the human disease. The pressures of decreasing research funds, intense competition, and furthering careers often make publishing more important than meaningfulness of data. Thus it is not surprising that long-term "normal" and "transformed" cell lines are favored systems for study, as data can be quickly obtained. The cells grow rapidly, producing large numbers in a short period. (Dermer 1983: 318)

Why Oncogenes?

These critiques and my discussions with scientists show that initially not everyone agreed that oncogenes played a significant role in human cancer. Despite this initial disagreement, many researchers and organizations made commitments to Bishop and Varmus's new proposal to study the relations between proto-oncogenes and established ideas, routines, and practices in many different situations. As the commitments grew, support for the thesis was created through the reconstruction of laboratories, tools, and skills. I have described the interactions, the processes, the symbolic meanings, and the practices involved in the construction of the current situation.

That this particular representation became "the ruler over the realm" of cancer research is a historical event which cannot be explained in terms of simple cause-and-effect relationships and therefore could not be predicted. Some of my respondents explained the proto-oncogene bandwagon as a response to time and funding pressures. In contrast, I do not "explain" proto-oncogene research as an outcome of a unilinear cause-and-effect relationship. Instead, I seek explanations that retain the complexity of scientific work and life. The story of this bandwagon is a process of co-production. The theory-methods package made it possible for researchers to produce results and papers in short time frames and therefore, partly because of this productivity, expectations of short time frames were co-created. The package made this possible by including a definition of what problems and results were. Granting institutions could not expect results in short time frames if they had not accepted the definitions of results prescribed by the package.

The commonalities in representations—for example, from cancer as caused by viruses to cancer as caused by proto-oncogenes—do not demonstrate any ultimate progression in the history of science and scientific development. Neither do they demonstrate any necessary connections between the pasts, presents, and futures, or between contemporary situations. Instead, in each situation actors construct new hybrid representations, theories, and methods, by constructing networks of common practices. My description of historical continuity does not infer causality. It is instead an account of association between the past and the future. Histories and social studies of science have painstakingly demonstrated that inference and prediction are not the stuff of writing histories. Ethnographic studies of science have emphasized the situated, contingent, tacit, and "magical" aspects of scientific work. In this picture of science coherence is cobbled together from local contingencies and practices. In a related vein, I have examined the success of a particular representation and explanation of cancer etiology as part of a particular construction of history and local practice. The Nobel Prize Committee's decision to award its 1989 prize in physiology and medicine to Bishop and Varmus is part of this historical construction and construction of history.

A few respondents have suggested that the success of the proto-oncogene hypothesis was due to the current historical privileging of genetics in Euro-American society. I agree in part. However, I do not want to separate society's interest in genetics from that of scientists. Why does "society" find genes so fascinating? Could it be because scientists promise that genetic research will produce a cure for cancer? Or do scientists favor genetic studies because that is what "society" desires? Or are society and science indivisible?

A Molecular Biological Bandwagon in Cancer Research

My story describes a confluence of situations, events, and actions into which Bishop, Varmus, Weinberg, and researchers in several laboratories introduced a theory that, together with new molecular genetic technologies, captured the interests of several social worlds and could be reconciled with the situations and practices of those worlds. Researchers in various research settings adapted and adopted the package as a tool for getting their work done. Researchers and university and science funding administrators constructed associations with their past and existing commitments, with their futures (for example, individual and organizational career concerns), and with their audiences. The linkages thus created, in turn, shaped the commitments and work organizations of researchers, laboratories, research institutions, biotechnology companies, and more recently clinical diagnosis in hospital laboratories. This set of multiple cascading commitments to molecular genetic cancer research constituted a bandwagon. As defined earlier, a scientific bandwagon exists when large numbers of people, laboratories, and organizations commit their resources to one approach to a problem because *others* are doing so.[25]

The proto-oncogene bandwagon represents a particular configuration of events, actions, and situations through which cancer and cancer research worlds have been reconstructed. The multitude of commitments to oncogene research does not establish the fact that proto-oncogenes play significant roles in causing human cancers. Excitement and enthusiasm over a particular research program do not necessarily mean that a theory is accepted, as the case of polywater and cold fusion clearly exemplify (Franks 1981, Gieryn and Figert 1990, respectively). However, the sustained commitments and the continued momentum of oncogene research long past its initial emergence is one sign of the stabilization of the oncogene theory. Proto-oncogenes are now facts in undergraduate biology textbooks and the building blocks of new research programs, theories, and diagnostics. Whether they will continue as textbook facts is also not assured. (As writers of biological textbooks and historians of science understand, many one-time facts have been deleted from the history of biology as well as from contemporary biology.) But their use in diagnostics has already begun. Consider N-*myc,* an oncogene used as a tool in prognosis to supplement the other signals of the virulence of cancer cases. Indeed, according to Bishop (1989:14), "The *New England Journal of Medicine* has provided its imprimatur by arguing that in neuroblastoma, amplification of the N-*myc* gene is of greater prognostic value than the clinical stage of the disease. Thirty years after deserting the bedside, I have found clinical relevance in my research." In addition, cellular oncogenes moved into cardiovascular specialties and are being studied for their possible action on cardiac muscle cells (Mulvagh, Roberts, and Schneider 1988, Mulvagh et al. 1987a,b).

My focus has been on the practices and activities, both routine and extraordinary, through which such coalescences of commitments occur. Much of scientific work is simultaneously routine and risky. The work is full of detailed minutiae and repetition that might not pay off in great intellectual or technological contributions, that might lead nowhere or to minuscule contributions of scientific facts and technologies. Within these daily and long-term exigencies of work, scientists attempt to manage as best they can to get their work done. While intellectual activities—for example, creating theories and technologies—play important roles in the story, their significance is manifested through their roles as managers, organizers, and translators of the interests and practices of participants in scientific work.

I have presented a history of the development of the molecular biological turn in cancer research framed in terms of the standardization of research tools and practices and the flexibility of an abstract concept. The molecular biological turn in cancer research emerged from the coalescences of two major developments in biology during the late 1970s and early 1980s. First was the development of materials, instruments, and techniques that standardized and streamlined recombinant DNA and other molecular genetic tools for studying cancer in the eukaryotic cell. Second was the construction of a theoretical framework that was shaped by and adapted to the interests of several social worlds involved in cancer research and molecular biology.

After a few years, the proto-oncogene theory and molecular genetic protocols themselves had been reconstructed by proto-oncogene researchers. The proto-oncogene theory has been adjusted many times over to fit new experimental findings. Single point mutations are no longer considered sufficient to cause a cell to become cancerous. "Catalogues of genetic damage within individual tumors are taking shape, showing us how the malfunction of several different genes might combine to produce the malignant phenotype: for example, carcinomas of the colon contain no less than five different yet prevalent lesions—some genetically dominant, others recessive; carcinoma of the breast, at least five lesions; carcinoma of the lung, at least four; and neuroblastoma, at least three" (Bishop 1989:15). Another modification of the original proto-oncogene theory is the addition of "anti-oncogenes" or tumor suppressor genes—for example, the retinoblastoma and p53 anti-oncogenes that are thought to cause retinoblastomas and colon cancers if they are inactivated by point mutations, truncations, or deletions (Benz and Liu 1993, Cooper 1995, Stanbridge 1990, Stanley 1995, Vogelstein and Kinsler 1993, Weinberg 1991).[26]

Collective Work or Fashion?

My use of the term "scientific bandwagon" is not equivalent to Crane's (1969) term "scientific fashion," which she defines as the situation where there are no theoretical reasons for larger numbers of scientists to simultaneously choose to work on a particular problem. When asked, oncogene scientists and sponsors present good theoretical and experimental reasons for choosing to engage in oncogene research. They argue that the package of proto-oncogene theory and recombinant methods provides them with opportunities to create novel findings and to construct doable problems.

Sperber (1990) demonstrates that particular scientific paradigms, approaches, theories, or facts are the outcomes of the "fashion process." Widely accepted scientific explanations are accepted because they are promoted by established opinion leaders. These leaders in turn "crystallize and reinforce the vaguely expressed collective tastes of the public" (1990:13). However, whereas the persuasiveness and the political savvy of particular oncogene researchers certainly contributed to the growth of oncogene research, political persuasiveness and tapping into the sentiments of the time cannot explain the development of a bandwagon. Indeed, some of the original oncogene researchers did not want a bandwagon of efforts around their work. Using his later work on the retinoblastoma gene as his example, Weinberg vividly described the *disadvantages* of having too many people working in the same field.

> My lab discovered a gene called the retinoblastoma gene in 1986. Within months other people had jumped on it and had begun doing interesting experiments and using reagents we did not have available at the time, which made them able to rush ahead of us, taking out of our mouths, as it were, our bread, in the sense that they could do things with our discovery that we couldn't. That's not my own peculiar history, that's the history of everybody who has made interesting discoveries. And now the field rapidly, within a matter of months, became very competitive, indeed so much that we were often frequently at a great competitive disadvantage. And I was not at all pleased, nor were the people in my lab, that it had become a bandwagon. It's not my own selfish or paranoid reaction to such events. It makes perfect common or simple sense that the involvement of many, many other groups rapidly working on the same problem we were working on made life much harder for us.[27]

I suggested to Weinberg that he might want many other laboratories to participate in this research in order to corroborate his facts and to enhance his laboratory's work. His response was negative; indeed he saw it as a threat to his laboratory's work.

> The fact is, if you have one or two other groups working [on the same problem], that's just perfect, because that's enough to already independently reproduce your work . . . That single other independent reproduction is all you want. You don't want 20 or 30 or 40 groups working on it . . . Because now all of a sudden many of the things you had wanted to do yourself,

> much of the trajectory of future experimental plans that you had laid out for yourself, becomes impossible because they are constantly leap-frogging ahead of you doing things that you yourself had wanted to do. (Weinberg interview)

In 1994 Terrence Rabbitts of the United Kingdom Medical Research Council echoed Weinberg's sentiment in an interview with the science writer Jean Marx (1994:1944). "The size of the field also means it's easier to be scooped. When he started, 'you knew who your rivals were and the kinds of things they were doing,' and so could avoid duplicating their efforts. 'Nowadays, everything you are doing replicates what someone else is doing.' "

Harold Varmus has also expressed concern about the effects of the oncogene bandwagon. As quoted by Jean Marx (1994:1944), he "worries that the sheer size of the field now could inhibit researchers, especially young ones, from taking a chance on exploring uncharted waters. In his talk this year at the Tenth Annual Oncogene Meeting in Frederick, Maryland, he said: 'In the areas of light, there is so much to do, and with so much light, it's difficult to be attracted to areas of darkness—areas where the most remarkable discoveries are waiting to be made.' "

According to Weinberg, bandwagons are not the result of active encouragement or enrollment of the initiating scientists.

> There is a bandwagon effect in the sense that if there is a rapidly expanding field of science which seems to have a large number of successes, then lots of people may eventually jump on it, because they see that other people are having success with it. But that is not due to the active encouragement of those who are already in the field necessarily; rather, the reason they are doing it is that they see right away that lots of interesting questions can be attacked. (Weinberg interview)

Even if Weinberg had encouraged others to study oncogenes in the same manner in which Bishop and Varmus promoted their proto-oncogene theory, the articulation required between parts of complex work organization in modern science cannot be accomplished purely through the persuasion of charismatic or politically astute opinion leaders or of highly skilled experimenters. This particular scientific bandwagon involved many practices and activities in many locales, between locales and across time, between people and objects in a continuous process of co-construction. Through this process, the package of the proto-oncogene theory and recombinant DNA technologies emerged to play a significant role. It facilitated and oiled the gears of the organization of collective scientific work, while changing that organization; in short, it facilitated the construction of doable problems.

Bandwagons, Research Programs, and Paradigms

My accounts of representations and the proto-oncogene bandwagon raise the issue of similarities and differences among scientific bandwagons, "normal science," and "research programs." Is a bandwagon a revolutionary vehicle or event or is it consistent with normal science? Do the package of proto-oncogene theory and molecular genetic methods, and the problems constructed using it, exemplify paradigmatic normal science (Kuhn 1970) or research programs (Lakatos 1978, Lakatos and Musgrave 1970)?

Lakatos attempted to salvage Karl Popper's rational reconstruction of scientific growth in his model of research programs. Briefly, Lakatos argued that a research program is a cohesive series of theories with certain methodological rules that determine which paths of research to reject (negative heuristic) and which to follow (positive heuristic).

The negative heuristic consists of the "hard core" of central principles that cannot be questioned and is protected by a belt of auxiliary hypotheses that can be continually adjusted or replaced to defend the core.[28] A research program is successful if these defensive adjustments bring about progress, that is, novel facts and "progressive problemshifts." Otherwise these adjustments become degenerating. A successful research program can be considered continuously "content-increasing" with periodic reviews to corroborate the newly discovered content.

The positive heuristic allows scientists to avoid a continuous struggle to explain anomalies. According to Lakatos, the positive heuristic creates a program that seeks to build ever more complex models to simulate reality. Under this framework, scientists expect their model to encounter anomalies that it cannot explain. This serves only to force scientists to improve their model. The aspects that the model verifies become the "contact points with reality." The verifications keep the research program going. Thus, scientists recognize anomalies but often do not deal with them in hopes that future models will explain them. When scientists start to address specific anomalies in a trial-and-error fashion, it becomes readily apparent that the research program is degenerating (not predicting novel facts or not corroborating predicted facts).

A successful research program, then, requires continuous growth: it continuously predicts novel facts that are periodically corroborated. A research program may also have rival programs, even if it is not degenerating. Although competing programs can exist, one program is generally more accepted than others. The more generally accepted program will be replaced only if there is a competing theory that explains those facts that the old research program explained and predicts novel facts that are later corroborated. For Lakatos, "mature science consists of research programs in which not only novel facts but, in an important sense, also novel auxiliary theories, are anticipated; mature science—unlike pedestrian trial-and-error—has heuristic power." Thus, Lakatos's research program has a strongly rational character (see also Laudan 1990).

Lakatos argues that "continuity in science, the tenacity of some theories, and the rationality of a certain amount of dogmatism" is explained by his model of science as a "battleground of research programmes," where the research program with the most "heuristic power" (that is, yielding the most corroborated factual novelty) wins. For Lakatos, the history of science has been or should be a history of competing research programs in which heuristic power rules.

The proto-oncogene bandwagon in some ways resembles Lakatos's research program.[29] Oncogene research as thus far manifested exhibits all the properties prescribed by Lakatos for a thriving research program. However, I do not subscribe to Lakatos's agenda to rationally reconstruct the growth of science. More important, I disagree with Lakatos's singular definition of rationality. He limits his discussion of research programs to an analysis of aspects of the particular features of the program that make the decisions of adherents appear rational. He sees the "good reasons" for joining the research program in the very properties of the research. But there may exist many kinds of rationales, many kinds of "reasonable" reasons why participants choose to join in oncogene research. Although Lakatosian "good" reasons for studying oncogenes might exist, these are not necessarily the reasons why researchers, students, and funding agencies joined in proto-oncogene research. They might be good, even excellent, post hoc justifications for participants' actions, but they do not "explain" participants' actions at the moments of commitment. Indeed, there may be no decision to join at all; some participants might fall or back into oncogene research. Events can occur or emerge without individuals planning for them.

A difference between my bandwagon and Lakatos's research program, then, is that the proto-oncogene bandwagon does not singularly depend on theory as its driving force the way Lakatos's model of a research program does. In my discussion of the oncogene bandwagon, I focus on theories, methods, practices, and work.

In contrast to Lakatos, Kuhn (1970) views the history of science as a succession of periods of normal science interrupted by revolutions that, according to some readings, are ruled by "irrational" social psychological factors. There can be no rational reconstruction of the growth of science for Kuhn, at least within the framework of his *Structure of Scientific Revolutions*. Instead, for Kuhn each new paradigm brings an incommensurably new rationality.

Do the package of proto-oncogene theory and molecular genetic methods, and the problems constructed using it, exemplify paradigmatic normal science? Proto-oncogene research certainly has some of the properties Kuhn ascribes to a paradigm. The proto-oncogene theory appears to be a significant framework governing a way of doing biological research since 1980, and probably for the next few years. However, the theory is not revolutionary in the Kuhnian sense. If we read Kuhn to say that revolutionary science fundamentally and incommensurably shifts the frame of scientific understanding, then proto-oncogene research is definitely not revolutionary.[30] Indeed, as seen in Chapter 4, scientists claim that the proto-oncogene theory provides a unified way of looking at a disparate array of facts in different fields, while simultaneously not conflicting with any of this existing science.

Furthermore, if we limit our frame of discussion to molecular biology, the proto-oncogene theory currently appears to provide a coherent theoretical framework for organizing the many detailed mechanisms constructed in molecular biology. The philosopher of science Richard Burian (1991:2) has argued that "molecular biology is not unified by a central theory, but instead by an approach to explaining and altering organismic function by reference to, and use of, an omnium gatherum of detailed molecular mechanisms . . . Indeed, the subject matter of molecular biology is detailed mechanism—and what it studies is mechanisms all the way down (but also . . . all the way up)." I argue that the proto-oncogene theory provides a model of the kinds of theories that could possibly link these mechanisms in ways previously unavailable to molecular cell biology as well as to other biological disciplines. Oncogene theory and oncogene protocols are the constructions that theoretically and methodologically currently link many of these molecular mechanisms studied by different biological subdisciplines to produce a working program or paradigm (at least for a while). As such, the proto-oncogene theory is very much a part of "the molecular biological paradigm."

The package of the theory and molecular genetic methods also is one of the links being constructed by current institutional rearrangements of biological subdisciplines. For example, the University of California, Berkeley, has undergone an extensive reorganization of its biological departments into two new divisions, Molecular and Cellular Biology, on the one hand, and Integrated Biology, on the other. These institutional rearrangements represent more than mere institutional changes. They also represent changes in the disciplinary work done within the new institutional boundaries. Walter Gilbert and James Watson provide another instance of the use of proto-oncogene theory for building institutions, programs, and technologies. Both Watson and Gilbert have noted the theoretical promise represented by oncogene research in their use of proto-oncogene theory as an example of the kind of possible payoff the human genome project could yield. Like oncogene research, they argue, the project to map and sequence the human and several model animal genomes could yield many more homologies between gene sequences that would provide clues for solving important problems.[31] Thus, if proto-oncogene theory is to be revolutionary, it will be as one part of a set of theories about many detailed molecular mechanisms, several technologies, and many institutional linkages now under construction. Simultaneously, it will remain normal science, a part of the general molecular biological paradigm.

Are representations "merely" the products of reigning thought styles, paradigms, or research programs (Fleck 1979, Kuhn 1970, Lakatos and Musgrave 1970, respectively), or are they "true"?[32] I discussed my position on the realist-constructivist debate in Chapter 1. To briefly recapitulate, I assume that truth claims are real(ized) in their consequences, to paraphrase Thomas and Thomas (1928). Therefore, we need to look at the changes in practices and organization of work that are the consequences of new representations like the proto-oncogene theory. Scientists might well be creating representations that "work." However, studies of representational practices tell us that they too are the products of fiddling and tinkering to make things fit together in ways that "work" under particular conditions. In that sense, they are constructed.

My focus has been on the interplay of the establishment of a theory as fact and the processes of articulating, tinkering, gaining allies, aligning practices, and making things work.

The tinkering and adjusting in local situations using standardized molecular genetic technologies have transformed biology and much of cancer research. Standardization is generally viewed as antithetical to basic science. "State-of-the art" science is popularly viewed as ambiguous, novel, creative, imaginative, and certainly not standardized. But most scientific practice resembles Kuhn's notion of normal science. Normal science is the "place" where theories and models become robust. One definition of robustness is Richard Levins's statement that "our truth is the intersection of independent lies." I have focused on the work that goes into creating robustness, and I argue that truth is the intersection of *dependent* assertions. In this case, the communication back and forth and the transfer of theories, visions, and standardized technologies across laboratories transformed work practices and representations. In an incremental fashion, representations interactively constructed in multiple laboratories have been used to create a new truth based on dependent assertions. In this way, standards are used to construct the world. But standards are not neutral. They are the outcomes of particular desires, positioned rationalities, and historical circumstances.

Appendix
Notes
References
Index

Appendix
Social and Cultural Studies of Science

Social and cultural studies of science have radically changed in the last twenty years. I here present a brief idiosyncratic discussion of the ideas and writings in this multidisciplinary field that have been useful to my research, some as oppressive shadows, others as inspiring and suggestive lights.[1] Before the 1970s, sociology of science in the United States was dominated by the Mertonian school, named after the sociologist Robert K. Merton.[2] In contrast to the Mertonian functionalist approach, studies since the 1970s have been heavily influenced by Thomas Kuhn's book *The Structure of Scientific Revolutions* (1970 [1962]). A physicist and historian, Kuhn proposed that scientific knowledge changes through "paradigmatic shifts" and scientific revolutions, and not by the gradual, incremental growth of knowledge. His ideas were a major catalyst for much of the subsequent work in the social studies of science, which since the 1970s has emphasized the extent to which knowledge produced by scientists is subject to social and cultural influences. Some writers have also raised concerns about how one should write about or "tell" science. Their studies have provided us with an extensive body of historical and contemporary examples of how science is locally practiced and produces knowledge. Often called "constructivists," they represent in fact a diverse group of perspectives that share some commonalties and many differences. These differences have been the source of creative debates in the field.

Inspired by the works of Feyerabend (1975), Fleck (1979), and Kuhn (1970), some sociologists, historians, and philosophers of science explored the processes through which scientists make judgments about what is a better or worse representation of nature.[3] These studies questioned the assumption that judgments are grounded in some absolute truth or reality and compared scientific representations against one another by examining the methods and negotiations among heterogeneous elements used in the production of representations. They argued that science analysts should be agnostic and symmetric in their treatment of the cognitive elements of scientific representations.

One group of British sociologists, historians, and philosophers who began writing in the early 1970s called their work a "sociology of scientific knowledge" or "SSK." Among these writers were those who identified themselves as the Strong Program tradition of sociology of science and the Relativist Program.[4] In their studies of the social construction of scientific knowledge, these authors studied the social and cultural contingencies under which knowledge claims and artifacts were produced. They argued that both authorized and unauthorized science should be treated symmetrically in sociological analyses (Bloor 1976). They used sociological analyses to explain scientific facts and artifacts and argued that social interests could be used to explain the construction of particular theories about nature (Barnes 1977).

These British sociologists of science presented linguistic analyses, discourse analyses, ethnographies, and histories of science to examine the social construction of science. Many of these studies paid attention to controversies and debates in science (Collins and Pinch 1982). Others examined how anomalies are managed in science (Woolgar 1976, Mulkay and Gilbert 1982); the development of specialties and disciplines (Collins 1975, 1981; Law 1973); the expression of scientific knowledge (Barnes and Law 1976); the interaction of contexts and ideas (Barnes and Law 1976); the construction of "deviant" versus "accepted" science (Collins and Pincer 1982, Wallis 1979).

Ethnographies of Practice

The sociology of scientific knowledge generated strong interest and reactions from sociologists and anthropologists conducting contemporary and subsequent ethnographic studies of laboratories. A primary criticism was that, in trying to demonstrate the embeddedness of science in society, some of the SSK studies constituted a separation between scientific cognition and the work arrangements and processes through which the ideas were produced. To relate scientific ideas and technologies "back to" their "social context," these studies constructed visions of the "social" using traditional sociological theories of society.[5]

In contrast, ethnographers of science coming from other sociological and anthropological traditions chose to study scientific work without first instituting this division between the content of science and the material and social practices and contexts of scientific work. Three overlapping collections of studies emerged. The first drew on ethnomethodological criticisms of the constitution of the social as an explanation of the content of science by sociologists. Work incorporating lessons from ethnomethodological studies of society examined the practices of scientists (Lynch 1985; Garfinkel 1982; Garfinkel, Lynch, and Livingston 1981; Suchman 1987, Woolgar 1981),[6] the production and construction of scientific objects in the laboratory (Latour and Woolgar 1986; Callon, Law, and Rip, 1986), and the production of facts (Latour and Woolgar 1986, Knorr-Cetina 1981).

A second collection of studies drew on symbolic interactionist studies in sociology. These studies explored through ethnographies and histories of the laboratory how ideas became separated from the contexts and processes of scientific work via practices of reification, abstraction, and representation (Star 1983, Fujimura 1987, Clarke 1987).

A third group of scientific ethnographies came from anthropology. Sharon Traweek (1988) studied physics communities and their co-production of scientists and science within these communities. Emily Martin (1987) compared medical and scientific textual representations of women's bodies with interviews of women's experiences of their bodies.

Actor-Networks

My mode of inquiring about science has some commonalities with what has been called the "actor-network" approach associated with researchers such as Callon (1986a,b), Cambrosio and Keating (1992), Latour (1983, 1987, 1988a, 1993), Law (1986, 1994), Callon, Law, and Rip (1986). Callon and his colleagues framed an actor-network theory that focused on strategies for building networks to make findings into facts and small actions into Tolstoyan feats (Callon and Latour 1981). Their perspective partly derives from an ethnographic approach to the study of science. Their methodological imperative is to follow scientists as they go about doing their work. To understand how scientists manage to produce knowledge *in* the laboratory, one must follow the scientists as they move *out* of the laboratory.

The actor-network theory perhaps too strongly portrays the scientific playing field as a field of battle where the "fact" that proves more "factual" depends on which actors succeed in enrolling allies, much as leaders enroll armies and armories and politicians enroll sponsors.[7] To accomplish this, scientists employ strategies such as translating others' interests into their own interests and making their laboratories "obligatory passage points" or "centers of calculation." The enrolled and disciplined heterogeneous actors have then to be maintained as a network of stable linkages by constant and careful work.[8] In more general terms, translation is the way in which certain actors gain control over the organization of society and nature, by which "a few obtain the right to express and to represent the many silent actors of the social and natural worlds they have mobilized" (Callon 1986a:224). However, as Latour (1987) pointed out, others are simultaneously trying to enroll allies to support their fact. In this view, science is a battle over which fact—that is, which network—will become and remain stabilized.

Callon and Latour have also accredited actor status to nonhumans (see also Haraway 1985). Microbes, scallops, batteries, doors, acid rain, and other nonhumans are part of the (re)construction of nature, society, and science. For example, Latour described how microbes, once introduced into the world, reconstructed social and cultural practices and our "natural" landscapes. Microbes have become part of our everyday maps of the world as we go about our lives. Latour insisted that "society is not only made up just of [people], for everywhere microbes intervene and act . . . We cannot reduce the action of the microbe to a sociological explanation, since the action of the microbe redefined not only society but also nature and the whole caboodle" (Latour 1988a:35–38).[9] Latour (1990c:156) wants to redistribute "politics . . ., will, liability, respect, humanity, soul as well" to nonhumans. According to actor-network theory, we should return to a whole new (or ancient) ontology where things and people are both actors in the (re)construction of the world.

Although this is a radical concept in one sense, it does not differ from most sociologies and histories of science as much as one might think. Many sociologists and historians of science have treated "scientific objects" as representations rendered through human perspectives and actions. They do not ascribe to nonhumans the kind of agency they ascribe to humans. Callon and Latour similarly do not ascribe conscious agency to nonhumans. In their scheme, nonhumans are elevated to a status equal with human actors only in the sense of "actants."[10] That is, both humans and nonhumans can be elements in a network, and both humans and nonhumans can be "macro-actors" when supported by strong networks; they become actors that command immense resources, "sovereigns" in their domains.

A good example of the actor-network approach is Latour's version of the success of Pasteur's germ theory in eighteenth-century France. Latour tells the story of Pasteur simultaneously building his science, his microbe, and French society by constructing strong associations between what Callon, Latour, and Law call "heterogeneous elements." He argues that Pasteur through Herculean efforts linked together microbes, the hygienist movement, army doctors, farmers, surgeons, brewers—to define and redefine microbes, diseases, fermentation processes, the practices of army doctors, hygienic measures used throughout the country, theology, and French society. Post-Pasteurian France was entirely different from pre-Pasteurian France. According to Latour, Pasteur's genius was located not so much in his microbiology as in his strategy of "placing his weak forces in all the places where immense social movements showed passionate interest in a problem" (Latour 1988b:70–71). For example, the passion and the power of the nineteenth-century French hygiene movement was used to promote Pasteur's still little known microbe. Pasteur later discounted the contribution of these powerful allies to his effort and instead claimed that the success of his microbe came from fundamental research in his laboratory, from "Science." Thus, he recruited allies in these social movements and then negated the role of this network in his scientific success.

Latour's version of Pasteur's success in the eighteenth century demonstrates the political skills involved in producing "scientific" facts. Latour and Callon view "science" and "society" as artifices, as historically contingent creations of the seventeenth century. Latour (1993) agrees with Shapin and Schaffer (1985) that the authority to rule over "science," to decide matters of nature, was captured by "scientists" of the Royal Society while the political authority to rule over people was captured by sovereign leaders.[11] For Callon and Latour, through the seventeenth-century work of Boyle and Hobbes, science and technology became autonomous realms of production. Only after Boyle and Hobbes was it possible to separate technoscientific projects from their inventors. However, in Latour's view, this division is a myth. Latour (1993) argues that the line between inventor and product, between subject and object, does not exist, and in fact never existed. For him, the world has never been modern. He explicitly asks us to abandon the myth of the modern world and to realize that we are still living in an amodern or premodern world.

Callon and Latour thus propose a new ontology (Callon and Latour 1981; Callon 1986a; Latour 1990c, 1993). Using actor-networks, they wish to explain not only science but also the world. They do not want to stand in society and explain science as if it were in a separate category. Callon and Latour want to explain the construction and reconstruction of science, nature, technology, society, polity, and history. They argue that science and society are the things to be explained, not the explanations. Indeed, humans, scientific ideas, technological objects, and other nonhuman entities are not easily divisible categories (see Latour's *Aramis*). The conclusion they draw from their view of seventeenth-century history is that technosciences construct humans and nonhumans alike, and that humans and nonhumans construct technoscience. Society and humanity are objects as well as subjects in the science game.

As part of this reordering of the world, then, Callon and Latour agree with other ethnographers of science that the social context cannot be used to explain scientific knowledge. Since neither "science" nor "society" existed before Boyle and Hobbes and their fellows created them as separate entities, we cannot use one to explain the other. Instead, we should question equally some social scientists' categories and modes of explanation (for instance, single flows of sociological cause to scientific effect) as well as some natural scientists' categories and explanations (single flows from nature to scientific representations).[12]

Using an actor-network framing of the study of science, one does not demarcate which set of actors has which set of interests. Rather than try to fix group membership and group interests, the actor-network approach asks us to follow the actors and describe their actions in terms of network-building. This approach also avoids classifying actors into categories and fixing future interests.

Comparative Cultural Studies of Science

The actor-network approach has congruences (and differences) with ideas raised in debates raging in anthropology and feminist studies, where this is a time of simultaneously liberating and disconcerting mutability and flexibility. In these fields of inquiry, the familiar sources of collective and personal identity—cultures, customs, selves, bodies, communities, and places or spaces—have become radically contingent categories. Scholars working on problems that have become known as interdisciplinary "comparative cultural studies" especially have had to confront the rejection of universalizing schemes for narrating the histories and cultures of represented "Others."[13] Many anthropologists have written about the perverse and continual slippage in categories of identity that have traditionally designated the fixed "object" of ethnographic inquiry.[14] These include those who find themselves both subjects and objects of study and those who wish to avoid further imposition of their versions of reality onto Others.

Represented Others have begun to speak against their representations and to write their own versions of themselves. Within the confusing mutability of the once familiar landmarks, there is an opposed tendency to map these disappearing markers so we can still find our way. Some authors attempt to fix identities in political economic terms of material relations of domination and subordination, of master and slave. Other authors preserve the identity of communities by mourning their disappearance. And yet another set of authors attempts to authenticate essences (sexuality, race, and gender) through the technologies of science. National and international projects to map and sequence "our" human genome are examples of such efforts to fix the reality of persons and identities. It is these concerns that are at the heart of a newly developing comparative cultural studies of science.

Positionings

Although it is useful to recognize the historically contingent construction of the science-society dichotomy, the consequences of that construction are real and have, at times, been devastating for some groups of people and nonhumans. The feminist and cultural historian of science Donna Haraway best expresses the imperative of our positions as students of science when she argues that we should study the "collective subject position" of actors despite their constructed and contingent nature. She impresses on us the understanding that collective subject positions have rationalities or ways of knowing, which she calls "situated knowledges" and "partial perspectives" and which together can produce a kind of communal "objectivity" without transcendence.

> We seek [knowledges] ruled by partial sight and limited voice—not partiality for its own sake but, rather, for the sake of connections and unexpected openings situated knowledges make possible. Situated knowledges are about communities, not about isolated individuals. The only way to find a larger vision is to be somewhere in particular. The science question in feminism is about objectivity as positioned rationality. Its images are not the products of escape and transcendence of limits (the view from above) but about the joining of partial views and halting voices into a collective subject position that promises a vision of the means of ongoing finite embodiment, of living within limits and contradictions—of views from somewhere. (Haraway 1988:196)

For Haraway, "our problem [as science analysts] is how to have simultaneously an account of radical historical contingency for all knowledge claims and knowing subjects, a critical practice for recognizing our own 'semiotic technologies' for making meanings, and a no-nonsense commitment to faithful accounts of a 'real' world" (Haraway 1988:187). The semiotic technologies and positioned rationalities used in this study have been constructed in dialogue with, among others, the authors and literatures discussed in this appendix.

Notes

1. Introduction

1. Bishop and Varmus are professors at the University of California, San Francisco. In 1993 Varmus was appointed director of the National Institutes of Health.
2. See, for example, the histories and reviews of oncogene research noted in Chapter 4.
3. However, conventional forms and their organized institutional frameworks are in turn the outcomes of conflicts as well as coordinated efforts among actors, and therefore are open to change. See the section "Science as Collective Work and Practice" later in this chapter.
4. MedLine is a computerized data base from which scientists can quickly retrieve up-to-date information on current biomedical research. It is supported and maintained by the National Library of Medicine, an organization under the umbrella of the National Institutes of Health.
5. In the United States alone, in addition to biomedical researchers, other participants range from the National Cancer Institute to Congress, the American Cancer Society, patients, health activists, hospices, hospitals, alternative cancer treatment research, pharmaceutical companies, and health insurance companies.
6. See Shimkin 1977.
7. The appearance of the universality of scientific entities and objects is achieved through expensive and labor-intensive efforts to standardize and regulate measurement practices, technologies, and laboratories. The quotation marks around the term "technical" indicate such "social" practices of standardization.
8. See Chapter 7 for a discussion of "language," semiotics, and "forms of life."
9. Pragmatist philosophers include Arthur Bentley (1926, 1954), John Dewey (1920, 1925, 1929, 1938), William James (1907, 1914), and George Herbert Mead (1932, 1934, 1938). See also Mills 1966 on sociology and pragmatism, Shalin 1986 on pragmatism and symbolic interactionism, Kuklick 1977 on the history of pragmatism, and West 1989 on the genealogy of pragmatism.
10. I especially rely on the work of Howard S. Becker (1960, 1967, 1982, 1986), Howard S. Becker and Michael McCall (1990), Herbert Blumer (1969), Rue Bucher (1962, 1988), Everett C. Hughes (1971), Tamotsu Shibutani (1955, 1962, 1986), and Anselm Strauss (1978a,b, 1985; Strauss et al. 1985).

11. For reviews of this work, see Fujimura 1991a, and Fujimura, Star, and Gerson 1987. For work using this approach in the study of science, see Clarke 1987, 1990a,b, 1991; Clarke and Fujimura 1992; Fujimura 1987, 1988, 1992; Fujimura and Chou 1994; Gerson 1983; Gregory 1984; Star 1983, 1989, 1992; Star and Griesemer 1989; and Volberg 1983.
12. See Clarke 1990a for a review and discussion of the social worlds literature.
13. Perspectives on science similar to the symbolic interactionist tradition and to recent work in the social and cultural studies of science include Bourdieu 1990 and Foucault 1975, 1970, 1978, 1980a,b.
14. Note the resurgence of these ideas in contemporary postmodernist writings. I discuss issues in postmodernist literature such as "truth" and "representation" elsewhere (Fujimura 1991c). Postmodern pragmatist writings include Rorty 1982 and 1989.
15. Viral oncogenes are RNA sequences (packaged in retroviruses) that cause "normal" cells in a culture to be transformed to tumor cells and sometimes cause tumorigenesis in animals.
16. Their actions and interactions in the construction of science have been studied in great detail by historians and sociologists of science.
17. Kranakis 1989 presents an amazing example of entrenched commitments to theoretical approaches to bridge engineering in nineteenth-century France (rather than to hands-on empirical research such as model building and testing), which resulted in the collapse of an entire structure.
18. See, e.g., Latour and Woolgar 1986, Restivo 1983, and Star 1983.
19. See, e.g., Amann and Knorr-Cetina 1990, Latour 1990a, Lynch 1991b, Taylor 1992, and Griesemer and Wimsatt 1989.
20. For example, Duster argues that there is no such thing as universality in the application of regulations on genetic screening procedures when local conditions differ. "While federal and state laws are clear and unequivocal on [the matter of voluntarism and informed consent in genetic screening programs], the implementation of the ideal is fraught with almost insurmountable problems for the administering agency, the clinic, and the physician. Regulations call for standardization, yet the populations being screened are infinitely variable . . . Individuals from certain groups come to the bureaucratic setting with widely varying expectations of what will ensue once one rejects or refuses what are described as routine procedures" (Duster 1981:131).
21. For studies of the labor of technicians, see Haraway 1989, Shapin 1989, and Star 1992.

2. Tools of the Trade

1. However, this statement is made in almost every presentation on clinical cancer research.
2. A neoplasm or tumor was defined by the pathologist Rupert A. Willis (1960) as an abnormal mass of tissue the growth of which exceeds and is uncoordinated

with that of the normal tissue and persists in the same excessive manner after cessation of the stimuli that evoked the change. Tumors have been described as irreversible, progressive, uncoordinated, and autonomous. But many tumors beyond neuroblastomas do not exhibit all of these properties. Some tumor growths are not autonomous and are controlled by the endocrine system. Tumors are difficult to define because they come in many varieties, because tumor cells change, and because hosts, which influence tumor cell growth, differ in many ways.

3. For a history of cancer research before this period, see Rather 1978. My history is based on primary and secondary sources, on interviews with scientists, and on documentary sources. Cancer research became an experimental science at the turn of the century, along with most of the biological sciences. For some histories of this experimental cancer research, see Braun 1969, Bud 1978, Cairns 1978, Goodfield 1975, Patterson 1987, Rettig 1977, Russell 1981, Shimkin 1977, and Strickland 1972. These studies focus on different issues and problems in cancer research histories. For a political history of cancer theories, see Proctor 1995. There is no "comprehensive" history of cancer research in the traditional style of history of science.
4. Indeed, disciplines and disciplinary boundaries are political entities used and maintained for specific purposes independent of scientific and technical work (Cambrosio and Keating 1983, Gieryn 1983).
5. Although I do not discuss the history of eugenics in this book, it must be noted that ideas about genetics in the 1920s and 1930s (and some argue even today) also incorporated ideas about eugenics, that is, the application of knowledge about heredity to the engineering and control of selective reproduction of human communities. For literature on the eugenics of the late nineteenth and early twentieth centuries in both the United States and Europe, see Adams 1990; Allen 1975; Duster 1990; Haller 1963; Kevles 1985; Ludmerer 1972; Proctor 1988; Rolls-Hansen 1980, 1988; Schneider 1990; Searle 1976; Weindling 1990; and Weiss 1987.
6. Attributions of credit for particular theories and "discoveries" are marked by journal or book citations in histories of science. I adopt this textual device but do not intend these references to refer to unproblematic attributions of credit. In my text, these markers identify the authors who are credited—by themselves, by other scientists, or by historians of science—with the introduction of these ideas. I am using these attributions as the basis for writing my story. Whether these authors were the first to propose these theories, whether their original statements really said what they now say they said, and many other issues are left to be explored by historians of science. For example, Löwy (1990) has compared retrospective discovery accounts with original statements in three cases to show that memories often distort the meanings of key terms in the original statements. My purpose here is not to present a "correct" history of science, if that is even possible, but instead to write the story in the way that scientists in my oncogene case study understood the history of earlier work in cancer research. These are the tales, the myths, that fuel their own work. These references

have meaning for them, and that is the significance I give to these references in my present "history." (See the last note to this chapter for literature on the issue of writing "true" histories.)

7. Transplantation biology also helped to create current knowledge of tumor immunology, especially understandings of tumor specific antigens. For a history of that line of work, see Löwy 1993. I am here more interested in the technologies produced from this early research that have influenced all of biology and specifically genetic research on cancer.
8. A more detailed study of the roles of inbred mouse colonies and established cell lines in cancer research would add to our knowledge of how institutionalized commitments to particular research materials affect the kinds of knowledge produced.
9. Why genetics was *the* mode of explanation studied in the 1920s and 1930s is a subject well examined by historians of science. See, for example, the histories of eugenics (Allen 1975, Kevles 1985, Kimmelman 1987, Hubbard 1990).
10. Interview with Kenneth DeOme, former director of the Cancer Research Laboratory and emeritus professor, University of California, Berkeley, February 1, 1984.
11. These results were later reinterpreted in terms of oncogenes. See Prehn and Main 1975.
12. See also Clarke 1987 on the work involved in developing new research materials in reproductive biology in the first half of the twentieth century.
13. See Chapter 1 of this book and the introduction to *The Right Tools for the Job* (Clarke and Fujimura 1992) for further discussion of the concept "situation."
14. See, for example, Rainger, Benson, and Maienschein 1988.
15. See, for example, Wolgom 1913.
16. See also Fujimura 1991b on the universalization of representations via the standardization of practices in contemporary molecular bioinformatics. See Star 1989 for another universalizing modality, where abstraction was the means for deleting local variance in nineteenth-century neuroscience.
17. See Chapters 1 and 7 (and Fujimura 1991c) for discussions of the relationship between language and thought or action. For recent work on the relationship between language and molecular genetics, see Haraway 1981–82; Hayles 1984, 1990; Keller 1992b, 1994; and Nelkin and Lindee 1995.
18. Some scientists and historians of science have represented the products of recent laboratory sciences primarily in terms of narrowly defined and localized experiments (cf. Gooding, Pinch, and Schaffer 1989). For example, Rheinberger (1992), following Ludwik Fleck (1979), also uses the term "experimental system" but defines it as "this smallest unit of scientific research." He argues that if one wants to know how scientific research actually works, "one must begin with the characterization of an experimental system, its structure and its dynamics" (p. 3). Using Latour's definition of experiment as an event, he asserts that the history of science should be constructed as a history of "experimental events." An event creates a situation that did not exist before the event. For example, he shows us the different technical conditions (in this case, different

centrifugal forces) used to produce different scientific objects (in this case, cellular components), which are created by and defined in terms of these different centrifugal forces. However, in all this detail Rheinberger's close analysis sacrifices the collective processes and actors involved in constructing any scientific object. As just one example, Burian points out that "mRNA was found by a kind of triangulation involving the work of many groups working with different tools and in different traditions" (Burian 1991:11). We do not see these different groups in Rheinberger's story, except perhaps in the footnotes. In Rheinberger's system, the collective processes and actors involved in constructing these specific experimental technologies and protocols that produced the "scientific object" are lost or at least not clearly identified. While his framework has advantages, it also makes faceless many of the other elements in science-making that are of interest to me. In contrast, Griesemer's (1992) "analytic system" is more collective and historically processual in conception and covers a broader set of actors, time, and spaces (cf. Star and Griesemer 1989). Compare these two conceptions of experimental systems with Haraway's even more comprehensive list of constituents and participants.

19. Other early-twentieth-century pioneers involved in creating these new techniques included, in the United States, Alexis Carrel and Montrose Burrows (both at the Rockefeller Institute for Medical Research in New York), W. H. Lewis and M. R. Lewis in Baltimore, and "Strangeways and his co-workers in England, Fischer in Denmark, von Mollendorff in Germany, Champy and Ephrussi in France, Chlopin in Russia and [Giuseppi] Levi in Italy" (Willmer 1965:7). Wilton Earle at the NCI and George O. Gey in Baltimore participated in later contributions (respectively, cell culture media and the roller tube). Kevles (1992) adds John Enders to this cast of characters. See Willmer 1965 and Witkowski 1979 for literature on tissue culture methods. See below for Peyton Rous's and Renato Dulbecco's contributions to this research.
20. Willmer (p. 1) used the term "Tissue Culture" to refer to three kinds of cultures: ". . . 'tissue culture' proper, in which small fragments of tissue are explanted into a suitable medium and encouraged to grow in isolation, to form colonies, and perhaps to continue some of their normal functions. In such cultures, the original organization of the tissues may be lost, but the constituent cells emerge into the zone of outgrowth where their activities may be directly observed. Secondly there is 'cell culture,' in which the cells of a tissue or even individual cells, are made to grow in much the same way as bacteria are grown: all the organization of the original tissues in such cultures is discarded as irrelevant and cell multiplication and growth in uniform populations are the dominant interests. The cells of different tissues tend to become 'dedifferentiated' and relatively alike in appearance. Thirdly, in 'organ culture,' growth is only of minor interest, but embryological development and the maintenance of normal physiological functions are the chief aim and object. The outward growth and migration of 'dedifferentiated' cells is positively suppressed and the maintenance of the normal organization of the tissue is of first importance."
21. This table covers historical landmarks as understood by the authors in the de-

velopment of tissue culture techniques critical for the cross-laboratory comparisons of cell research, including oncogene research. It ends with the mid-1970s, when oncogene research as we know it today began.

22. Similar tales are told in contemporary science about the transformation from "art" to "science." Before the development of standardized electrophoresis procedures and equipment, one molecular biologist was known to have "golden hands" for separating DNA lengths of different molecular weights using the centrifuge. Other researchers found the centrifuge too clumsy a tool for such delicate work. Thus, few researchers performed experiments requiring separation techniques until the invention and marketing of the electrophoresis equipment. With the development of standardization, experiments requiring the technique flourished. More recently, similar statements have been made about the transformation of immunology art to science via the new monoclonal antibody technologies (Cambrosio and Keating 1988). However, in neither case can one say that standardization eliminates local knowledge and skill.
23. Hematopoietic cells are blood cells that differ depending on their location in different organs; e.g., the spleen produces lymphocytes, while bone marrow produces myeloid and erythroid cells. Fibroblast and other mesenchymal cells form the connective tissues in animals. Epithelial cells form the skin of bodies and the linings of many important organs, including the colon, uterus, lungs, and mammary ducts. A discussion of cell lines follows.
24. We will see in Chapter 4 that Robert Weinberg's laboratory used NIH 3T3 cells in their transfection experiments, to experimentally transfer what they proposed was a transformed gene from one cell to another. See Pollack 1981 for further discussion of NIH 3T3 cells. See Chapter 7 for my summary of critiques of NIH 3T3 cells as a research tool.
25. An example of an established tumor cell line, as opposed to an established "normal" cell line, that has been through many passages is the notorious HeLa cells. HeLa cells were grown in artificial laboratory conditions for experimental research from the "immature glandular" tumor cells of an African-American woman named Henrietta Lacks who died of cancer in 1951. Her cells have lived on to gain notoriety as especially adaptable and virulent cells for experimental research. They are adaptable in that they manage to thrive in laboratories around the world, despite differing conditions. They are virulent in that they manage to thrive often to the detriment of other kinds of tumor cells under investigation. Researchers have had to adhere to strict sterilization methods in order to avoid contamination or edging out of the tumor cells they were studying by HeLa cells. Some researchers tell tales of airborne HeLa cells transported from one petri dish into their culture dishes of other kinds of tumor cells. Covering one's culture dishes became crucial to prevent such airborne "invasion." This virulence became the subject of investigations first by a Seattle geneticist named Stan Gartler in 1966 and in the 1970s by Walter Nelson-Rees, who maintained tumor cell cultures from laboratories around the world in his University of California, Berkeley, laboratory. Both researchers conducted investigations to confirm the pedigrees of purported cultured tumor cells. Both found

that many samples had been contaminated by HeLa cells. Both researchers suffered castigation by cell culturists and researchers for their public announcements of contaminations. Nelson-Rees managed to publish articles reporting the contaminated samples, along with the names of the researchers and laboratories from which they had been sent, in the journal *Science* (see Nelson-Rees, Flandermeyer, and Hawthorne 1974; Cullinton 1974). The articles, both in the publication process and in their reception, created an uproar of angry responses from the named scientists, who accused Nelson-Rees of publicity-seeking and defamation. Science, we see again, is a human enterprise. See Gold 1986 for an entertaining if disturbing tale of Gartler's and Nelson-Rees's efforts, tenacity, and torments.

26. For more discussion of markers of differences between normal and transformed cells, see Pollack 1981.
27. This deletion of the context occurs despite the frequent qualification by practicing scientists today of their results with concerns about whether one can extrapolate from their data to theories about nature. Similarly, many of the articles on tissue culture methods by scientists practicing in the 1960s in Willmer 1965 provide testimonies to this concern during that period of major developments in tissue culture methods. This process of generalization and universalization is not avoided even in the situations where scientists hold the best of intentions, in part because they cannot control the downstream interpretations and applications of their research results. See, for example, Hornstein and Star 1990.
28. See Kevles 1992 for a history of Renato Dulbecco's tissue culture studies and research. Kevles provides detailed descriptions of various tissue culture techniques that Dulbecco used to develop his own method of quantitative assay of viruses in cell culture. Kevles argues that Dulbecco developed his techniques of cell culture and methods of assay in the context of his quantitative (and not reductionist) approach to viral behavior. Dulbecco's commitment was to "quantitative positivism," or the need to have precise mathematical and physical measurements for valid experimental research on viruses. Therefore, argues Kevles, Dulbecco's contribution to the development of molecular biology via animal virology was guided by his dedication to a particular approach to laboratory practices and not to the philosophical stance of reducing biology to physics and chemistry. In response, one could argue that a commitment to mathematical quantitation is a commitment to reductionism to mathematics and statistics, which sat even higher on the totem pole of scientific philosophies "above and beyond" physics and chemistry.
29. This statement was made in a course on tumor biology taught by Martin in 1984.
30. Cairns was then director of the Cold Spring Harbor Laboratory. Cairns was also a major figure in the development of bacterial molecular genetics.
31. "Expression" refers to an actual phenotypic display of the function or feature "coded for" by a segment of DNA. In this case, the existence of the DNA in the genotype was not enough.
32. A virus is constituted primarily of its genetic material and its protein casing.

Human Immunodeficiency Virus (HIV), the virus thought to cause Acquired Immune Deficiency Syndrome (AIDS) in humans, is also a retrovirus. HIV is not currently understood to cause cancer.

33. Later MacPherson and Montagnier (1964), Dulbecco and Vogt (1954), and Dulbecco and Stoker (1970) made similar contributions to the work on DNA tumor viruses.
34. The heroic stories of the discovery of the structure of DNA by James Watson and Francis Crick, as well as the human stories of the contests for credit and races for priority in their competitions and collaborations with Linus Pauling, Rosalind Franklin and Maurice Wilkins, Max Perutz, Matthew Meselson, François Jacob, Salvador Luria, Max Delbrück, and Jacques Monod have been told by, among others, Watson (1968), Judson (1979), and Ann Sayre (1975).
35. In Foucault's (1970, 1978) terms, this is a *normalization* of biological activities and their attendant entities. See discussion in the conclusion to this chapter.
36. Howard Temin died in February 1994 in Madison, Wisconsin, where he had taught for thirty-four years.
37. See Baltimore 1970; Baltimore, Huang, and Stampfer 1970; Mizutani, Boettiger, and Temin 1970; Temin and Mizutani 1970; and Temin and Baltimore 1972. Temin and Mizutani used low concentrations of detergent to "open up" the virus particle and showed that the RNA produced cDNA (complementary DNA).
38. Again, it is within the framing of "normal action" that their finding was so unusual. It was an unexpected and therefore amazing discovery. This framing, according to those involved, is what caused the discovery to be delayed.
39. Many of these researchers call themselves virologists; some also call themselves molecular biologists or protein chemists. Harry Rubin, an iconoclast, only partly in jest called himself a veterinarian because his Ph.D. was from the school of veterinarian science at Cornell University. Rubin is widely considered the father of modern retrovirus research, for which he has received the Lasker Prize and membership in the National Academy of Sciences. Duesberg's early work was also instrumental in the delineation of reverse transcriptase. Duesberg has gone on to criticize the proto-oncogene thesis (see Chapter 7) and the current view that HIV causes AIDS (see Fujimura and Chou 1994). When G. Steven Martin entered Rubin's laboratory, Rubin was "working at that time on regulation of cell growth and virus transforming cells, just as a way of getting into work with animal cells. I decided at that time to work on trying to isolate viral mutants, which had defects in transformation, and came up with some interesting mutants, and so I just stuck with it. So I guess at the end of my graduate student career at that time, I didn't think that I was necessarily going to be working on cancer and tumor viruses, and it basically just took off during my postdoc, and I've stuck with it ever since. So it was a historical accident and luck. It wasn't a rational and considered choice." Interview on September 28, 1984, with G. Steven Martin, professor of biochemistry and molecular biology at the University of California, Berkeley, and a noted researcher on viral oncogenes and protein biochemistry. I also interviewed Professor Martin on January 27, 1984, and July 20, 1990, and was a student in two of his courses on tumor biology.

40. This is a simplified summation of much work to demonstrate the transforming properties of the Rous sarcoma virus. For more details of the experiments, see Duesberg 1968; Duesberg, Martin, and Vogt 1970; Duesberg and Vogt 1970; Hanafusa and Hanafusa 1971; Lai and Duesberg 1972a,b; Lai et al. 1973; Martin 1970; Martin and Duesberg 1972.
41. One reason why Rous sarcoma virus was the first and for a long time the most well studied RNA tumor virus was that it usually transformed all primary chick embryo cells in culture. In contrast to other, later defined oncogenic viruses, the *src* tumor virus can transform cells in culture without the assistance of a helper virus. Most other oncogenic viruses require the assistance of a helper virus in cell transformation.
42. See Fujimura and Chou 1994 for a reading of the debate on HIV as the cause of AIDS.
43. This case is very much like Thomas Hunt Morgan's affinity for *Drosophila* and Charles Manning Child's choice of *Planaria* for their respective research tools as presented by Mittman and Fausto-Sterling 1992.
44. Chemotherapy research prior to World War II was disreputable owing to accusations of quackery and fraudulent claims. This situation changed after the war with the introduction of war-related chemicals (Bud 1978).
45. The reader should note that I am not especially concerned about the lack of coherence in cancer research. What interests me is that researchers such as Cairns, who *are* involved in the work, are very concerned about coherence. Indeed, it is the construction and definition of coherence by these researchers that fascinates me.
46. For example, some small tumors had normal diploid karyotypes. The frequency of transformation was much higher than the frequency of mutation.
47. Friedewald and Rous may have first framed a two-stage model of an initiation-promotion process of carcinogenesis, but pathologists had long considered tumorigenesis to be a multistep process as represented in their analyses and classifications of tumors and tumor cells into benign versus malignant and into different stages of malignancy. Cytological factors such as loss of differentiation and tumorilogical factors such as invasiveness were their properties of classification.
48. However, even promotion can be a multistep process.
49. There has been some debate about this. Ames has since argued that many naturally occurring food products ingested by humans are also mutagens, and that chemical compounds are therefore not the only carcinogens.
50. On the basis of experimental animal studies, hormones were also considered to be promoters, if not initiators, of carcinogenesis. In the 1980s epidemiological studies also provided data linking artificially introduced hormones with tumorigenesis. I have chosen to focus on the chemical carcinogenesis research because of the extensive history of that research and because of its links to proto-oncogene research. Yet these studies of other environmental carcinogens before the molecular genetic period should not be ignored. For literature on this topic, see Farmer and Walker 1985. On radiation studies, see especially Bloom's (1972)

study of carcinogenesis in atomic bomb survivors. I have included very little about hormonal studies in this chapter because of time and space limitations. Hormonal cancer studies deserve a chapter, if not a book, of their own. See Oudshoorn 1990 for a history of hormonal research in reproductive biology.

51. Transfection is the uptake and expression of DNA into mammalian cells. See Chapter 5 for more on these developments and for a detailed delineation of the differences between the viral and molecular biological roots of oncogene research.
52. These experimental systems also incorporated sociocultural understandings as they reinvented nature. See the concluding chapter for further discussion.
53. My argument here has some similarities with Foucault's *normalization* of social forms of organization as "natural." In some ways, Foucault's normalization parallels the concept of standardization I discuss here. See n. 35 above.
54. For literature on the construction of history, see e.g. Borofsky 1987, Burke 1991, and Dening 1988.

3. Molecular Genetic Technologies

1. See also Rubenstein et al. 1977, Stein and Stein 1984, and Nagley et al. 1983 for more discussion of the use of recombinant DNA technologies in exploring biological problems on eukaryotic organisms during the early 1980s.
2. I do not reify disciplines as stable entities in society. Disciplinary boundaries are also constructed and therefore can be destabilized. Indeed, much of the politics within universities centers on turf battles over the maintenance or destabilization of departmental and disciplinary boundaries. What is molecular biology, for instance, has changed ever since its "birth" through its movement into other realms of biological research and biological institutions. A similar warning goes for "cancer research," which is an umbrella term that applies to an even larger and more heterogeneous group of research and researchers.
3. Financial profits, in this case, accrue mostly to commercial enterprises. An important exception here are the Cohen-Boyer patents granted in 1980 and 1984, which by 1989 were estimated to be worth $1 billion. Apart from product profits from royalties of between 0.1 and 0.5 percent of net annual sales, commercial firms pay annual license fees of $10,000 to Stanford University and the University of California for use of the recombinant techniques. Cetus, the second largest biotechnology company in the United States in the mid-1980s, voiced objections to this license fee. (Cetus has since been acquired by Roche, a multinational corporation.) Academic researchers do not have to pay to use the techniques. For further reading on the recombinant DNA patents, see "Cetus in Revolt over Gene-Splicing Royalties" (Budiansky 1985), Cherfas 1982, Juma 1989, Lappe 1984, Office of Technology Assessment 1984, and Yoxen 1983.

4. However, what constitutes a species and the boundaries of a species is still under debate by philosophers of science and by scientists.
5. For histories of the development of these tools other than those discussed in this section, see Research and Education Association 1982 and Watson and Tooze 1981. For concerned or critical accounts focused on the controversies about physical, social, and ethical hazards of recombinant DNA research, see Duster 1990, Jackson and Stich 1979, Judson 1992, Juma 1989, Kenney 1986, Kevles 1985, Krimsky 1982, Lappe 1984, Yoxen 1983, and Zimmerman 1984.
6. Eukaryotic DNA also appears to be very different in both structure and function from prokaryotic DNA. For example, eukaryotes have "intervening sequences" or "introns," which are "silent" regions of DNA, whose functions are not understood by molecular biologists, while other regions called "exons" do code for amino acids, the constituents of proteins. Prokaryotes have no introns in their DNA. Molecular biologists report that many assumptions based on the simpler systems do not appear to hold true for eukaryotic systems. In the late 1970s, these findings conflicted with earlier understandings of DNA structure and function, and thus posed new problems for molecular biologists. The purpose(s), or lack of purpose, of introns is still being debated. More recently, several researchers have argued that introns, or so-called junk DNA, play critical roles in the organism's activities (see Britten 1994, Koop and Hood 1994). See Fujimura 1995a for further discussion.
7. Judson (1992) divides the development of recombinant DNA technologies into three time periods. Using Judson's benchmarks, the technologies that played significant roles in the development and growth of oncogene research would be included in Judson's first two stages. His last "wave" of gene technology development occurred between the mid-1980s and 1991. The later tools were valuable for the massive, large-scale mapping and sequencing of the human and model animal genomes presently under way in the United States, Japan, and Europe.
8. I confine my discussion in this chapter to DNA because DNA manipulation has played a key role in recent cancer research on cellular oncogenes (see Chapters 4 and 5).
9. This particular piece of DNA was chemically synthesized for the experiment and not isolated from human cells.
10. Molecular biologists have also developed gene transfer techniques for introducing foreign DNA into germ cells of mammalian organisms. For example, Palmiter, Chen, and Brinster (1982) transferred the growth hormone gene from a rat species into fertilized mouse eggs. Most of the mouse progeny that incorporated the gene were much larger (in some cases, up to 50 percent larger) in size than was normal for that species. Palmiter, and colleagues (1983) later microinjected a human growth hormone gene into fertilized eggs with similar results. In 1983 the journal *Science* displayed this "transgenic mouse" as its "cover organism" (see the November 18 issue). Gene transfer experiments today aim at future human genetic therapies even to germ line cells. To say that this is

currently a highly controversial issue would be an understatement. Today, gene transfer is being experimentally used in the new "pharming" industry, where human genes coding for "useful" proteins (hemoglobin; alpha-1-antitrypsin, a hormone that helps to fight emphysema; and lactoferrin, which is found in breast milk and provides iron and immunities to babies) are transferred into domesticated farm animals, which are then bred to produce these proteins in large quantities.

11. Many scientists and members of the lay public were concerned about the issues of biohazard, evolutionary interference, and possible military appropriation of these technologies (Juma 1989, Krimsky 1982, Lappe 1984, Piller and Yamamoto 1988, Wright 1986, Yoxen 1983; cf. Watson and Tooze 1981). From 1973 until 1978, intense debate primarily among scientists left the application of recombinant DNA techniques in limbo. A federal moratorium on their use in several kinds of experiments, the involvement of the National Institutes of Health (NIH) and Congress in the regulation of the techniques, and NIH guidelines from 1976 to 1978 restricted scientists' utilization of these powerful techniques. In response to public protest, NIH established recommended containment standards, running from P1 to P4 in increasing order of security, for new experimental recombinant DNA research. For example, a P4 laboratory was even more carefully secured. In the United States only one P4 facility existed; it was located at the Frederick Cancer Research Facility, whose facilities were inherited from the former Biological Defense Research Laboratories at Fort Detrick, Maryland.

    Many biotechnological research and development companies constructed P3 facilities to contain biological materials used in the original recombinant DNA experiments in the mid-1970s when fear of biohazards was high and NIH guidelines for recombinant DNA research were very strict. A P3 laboratory, for example, looked like a miniature submarine. It was sealed from the outer environment. Air was pumped in through special ducts and then pumped out through special filtering systems that kept harmful materials from entering the environment. Workers sterilized all waste before removing it from the facility. In the late 1970s, a P3 lab cost $150,000 or more.

    Private industrial compliance with NIH biohazard guidelines was purely voluntary. NIH primarily funds researchers in academia and nonprofit research institutes. These researchers were required to comply with the guidelines to continue receiving funds. Although commercial biotechnology companies were officially exempt from NIH guidelines, most did comply. Since then, however, NIH has relaxed its restrictions for various reasons. One reason given for the relaxed guidelines was the development and use of "weakened" organisms like the *Escherichia coli* K12 bacteria, supposedly less able to survive outside carefully constructed and maintained laboratory conditions. Researchers still conducted some research, especially those employing human viruses, in P3 facilities. However, since 1978 most recombinant DNA research has not required such containment. (These facilities are now being used to study feared highly infectious viruses.) See Research and Education Association 1982 for more information on NIH guidelines for recombinant DNA research during the early 1980s. See Berg et al. 1974 and Krimsky 1982 for the specific kinds of recombinant DNA research that were put on hold during the moratorium. These NIH guidelines were considered to be extremely strict by the participating scientists and very loose by critics (e.g., the Council for Responsible Genetics) of recombinant DNA research.

Duster (1990) and Kevles (1985) have discussed the possible impacts of these technologies specifically in terms of a new eugenics. More recently, similar concerns have focused on the Human Genome Initiative. These concerns range from the privacy of information and the efficacy of genetic screening and its social impact, to the "geneticization" of social and behavioral disorders and the social construction of definitions of disease and health. See Holtzman 1989, Hubbard 1990, Hubbard and Wald 1993, Keller 1992a, Lewontin 1992, Nelkin and Tancredi 1989, Suzuki and Knudtson 1989. See also Evelyn Fox Keller (Casalino 1991) and Donna Haraway (Darnovsky 1991, Penley and Ross 1990) on the human genome research.

12. When they did decide to invest later, they made substantial commitments. See Table 3.2 for a list of the investments in biotechnology made by multinational corporations in the early 1980s.
13. This is not a new situation. Histories of science show us that academic science and even the often quoted exception of biology have been influenced by industrial concerns throughout the twentieth century. See Hughes 1986; Kimmelman 1983, 1987; Kranakis 1992; Latour 1987; Lenoir and Lécuyer 1995; Paul and Kimmelman 1988; and Wise 1985 for historical examples of and debates on the public-private and science-technology divides.
14. The divisions between government, university, and business appear simultaneously porous and definite. For example, the commercial enterprise IntelliGenetics, Inc., was created to run GenBank (a computational resource containing DNA, RNA, and some protein sequence information) in September 1987. The T-10 group, which had built the data base under Walter Goad's direction at the federally run Los Alamos National Laboratory, was subcontracted to maintain the data base. The decision to contract out the management of the data base was based on the perception of some of the decision makers that private industry could run such an enormous effort better than academics could. Whether right or wrong, the perceptions of different styles of work organization in these different arenas help to maintain the divides.
15. The boundaries between university and industry are even fuzzier in the 1990s. The University of California recently established a biological research institute that is simultaneously quasi-public and quasi-private. Its aim was to retain the future profits of inventions devised within university laboratories for the financially strapped institution. (Researchers had previously created their potentially profitable ideas and inventions using university facilities and public funds and later, after waiting the required eighteen months, moved to private industrial laboratories to develop products for market.)
16. Hine 1994 notes an interesting divide between "human" and "honorary human"

model organisms in human genome research. In order for nonhuman organisms such as *Drosophila,* mouse, *E. coli,* and *C. elegans* to warrant significant roles in genomic research, they must take on the role of "tool" or "model" rather than "organisms" in their own right.

17. Perhaps it is in this messiness that alternatives and possibilities of change can arise.
18. According to Wright (1986:320), improvements in three areas were required to make the procedures more easily reproducible in other laboratories: (1) "the means to define and control precisely the piece of DNA to be inserted into a living cell and the manner of its insertion"; (2) "the means to increase the efficiency of replication of a piece of foreign DNA"; and (3) "the provision of a wide variety of sources of pure DNA so that gene-splicing techniques might be applied to any gene from any source." (Wright gives a summary of technical developments that solved these difficulties between 1974 and 1976. See especially pp. 320–324.) Moreover, although recombinant DNA methodologists had resolved these technical difficulties by 1976, it was not until the fall of 1977 that they managed to refine a technique for expressing the genes of "higher" organisms by bacteria (Itakura et al. 1977). As finalized in their techniques, expression took the form of bacteria first incorporating a gene from another organism (such as humans) and then producing the proteins of that human gene. According to molecular biologists, this step was necessary before they could apply the set of techniques to problems involving the genes of "higher" organisms. But the techniques were still not refined and standardized enough for non–recombinant DNA methodologists to efficiently (in terms of time, skills, materials, and money) use them in substantive experiments.
19. The discussion of incorporating nonhuman actors into analyses of the craft of science remains a heated debate. For example, Collins and Yearley (1992) argue that this move takes the word of the technoscientists about the activities of nonhuman actors and therefore conveys double agency to scientists by allowing them to speak as scientists and as nonhumans. For Latour and Callon, the term "actor" is similar to the semiotic term "actant" and emphasizes the changes in the world produced by the association of entities. Their "actor" status does not refer to the kind of agency attributed only to sentient beings by traditional philosophy. Lynch (1993) and Lenoir and Lécuyer (1995) argue with Callon and Latour's equation of actor and actant.
20. This language—of "invasion by foreign" elements, "genetic purity," and "degrading foreign but not self" DNA—is fascinating from a cultural anthropological perspective. Concepts of invasion and contamination, of genetic purity, and of "self" as separate from "other" are themes present in Euro-American (and other) cultures (Kondo 1990; Martin 1994; Traweek 1989, 1995). Anthropologists remind us that these concepts are not universal, that they are culturally specific, that they are not limited to science. Indeed, these concepts are often read from one domain of cultural life onto others (Yanagisako and Delaney 1995).
21. I summarize and simplify the current molecular theory of heredity here for the

reader's convenience. Sequence information for genes is in terms of deoxyribonucleic acids (DNA) and ribonucleic acids (RNA) and, for proteins, in terms of amino acids. DNA is currently (and has been for about forty-four years) biologists' candidate for the hereditary material. DNA is envisioned as a long, double-helixed molecule composed of chains of smaller molecules of nucleic acids. According to current molecular biology, a gene sequence is the reading out of the nucleic acid bases that constitute the genes on an organism's genome. The human genome, for example, consists of genes located on twenty-three paired chromosomes. It contains 3 billion nucleotide base pairs (constituting perhaps 50,000–100,000 genes). A chromosome consists of a single, long DNA molecule plus many proteins. A gene is an ordered sequence of nucleic acids, or pairs of nucleotides. The genetically significant part of each nucleotide is another smaller molecule. In DNA, this smaller molecule is one of four types—adenine (A), guanine (G), thymine (T), and cytosine (C). Nucleotides containing A pair specifically (by way of hydrogen bonds) with those containing T, while nucleotides containing C pair with those containing G. These four base molecules are organized in different permutations to form the DNA. The particular sequence of these base pairs is currently considered to be the organism's genetic code, that is, the particular sequence determines which proteins are manufactured in the cell. More specifically, a set of three consecutive base pairs of DNA, via messenger RNA, code for one of the 20 amino acids (plus a terminator) that constitute a protein. A protein is a chain of linked amino acids. A protein's specific "nature," or activity, is determined by the specific type and sequence of amino acids in the chain. An average-sized protein consists of 400 amino acids, which means that the gene coding for the protein is at least three times 400, or 1,200, base pairs long, plus a few repetitive and (at this time thought to be) noncoding sequences. Finally, a cell's properties are determined by the activities of various proteins, each protein exhibiting one or several particular function(s). This is a very simplistic description of the current dominant view in current molecular biology.

For more extensive discussions of DNA and molecular genetic theory for lay readers, see any basic molecular biological textbook, e.g., Suzuki et al. 1989 and Watson et al. 1987. For more critical renderings and alternatives to this unidirectional, "blueprint" view of DNA, see Fausto-Sterling 1985, Hubbard 1990, Keller 1991, Oyama 1985, and Yoxen 1983. They argue that words such as "determine" and "code for" obscure the many other elements within the cell, the organism, and the larger environment that play interactive roles with the genome in "determining" events in the organism's development and activities.

22. See Cherfas 1982 and Watson, Tooze, and Kurtz 1983 for literature on early research on restriction and early efforts to find enzymes that cut at specific points on DNA.
23. See Dixon 1970 and Allen 1975 for an introduction to the history of enzymology.
24. For different perspectives and histories of molecular biology, see, for example, Abir-Am 1982; Bartels 1981; Cairns, Stent, and Watson 1966; Fleming 1962;

Judson 1979; Luria 1984; McCarty 1985; Muller 1937; Olby 1974, 1984; Portugal and Cohen 1977; Stent 1968; Stokes 1982; Waddington 1969; and Watson 1968. Many of these texts discuss the history of the battle over the nature of the fundamental hereditary material between molecular biologists who promoted DNA and biochemists who argued for nucleoproteins. Jan Sapp presents yet another claim to the fundamental unit of inheritance in his history of research on cytoplasmic inheritance (1987). After the "birth" of molecular biology, many biochemists began to study DNA structures and mechanisms as well as protein biochemistry. See also Abir-Am 1991; Burian 1991; S. Gilbert 1982, 1991; Olby 1991; and Zallen 1991. For more critical work, see Haraway 1989; Hubbard 1990; Hubbard and Wald 1993; Kay 1993; Keller 1985, 1992b; and Lewontin 1992.

25. Note the similarity to computers as black boxes. Note also that just as computers are black boxes for their users but not for their builders and hackers, so too are various standard technologies in recombinant DNA work black boxes for their users but not for their builders and tinkerers.
26. Cherfas quotes from Singer's "Introduction and Historical Background," in J. K. Setlow and A. Hollaender, eds., *Genetic Engineering,* vol. 1, *Principles and Methods,* pp. 2–3, New York: Plenum, 1979. Cherfas also includes a readable (by laypersons) description of many of the enzymes, other than restriction enzymes, used in recombinant DNA technologies.
27. Other plasmids included DNA from viruses and cosmids (artificially constructed hybrids of plasmids and viruses).
28. In the late 1980s, a new technique for increasing the speed and quantities of amplification of recombinants was invented by Kary Mullis at what was then Cetus Corporation in Emeryville, California. This method, called "polymerase chain reaction," or PCR, used the associated coming apart and annealing properties of the hydrogen bonds of DNA molecules during heating and cooling, respectively, to create a situation where multiple copies of a strand of DNA could be made without recourse to a cell's "machinery." PCR quickly spread throughout the molecular biology community and today is an indispensable part of molecular biology laboratories. See Jordan and Lynch 1993 and Rabinow 1996 for social and historical analyses of this method.
29. For more information on cloning techniques, see Darnell, Lodish, and Baltimore 1990:214–223 and Suzuki et al. 1989:405–409. Identification and isolation of the gene of interest did not mean that the exact sequence of the DNA in the gene of interest was known.
30. Many new techniques for mapping the genome have been developed recently under the auspices of the Human Genome Initiative, formally begun in 1990. The allocation of $3 billion to this project to map and sequence the human genome (as well as the genomes of other model organisms) has engendered much heated debate (Boffey 1986, Dulbecco 1986, Lewin 1986c). See also Fortun 1993 for a history of the American human genome project.
31. I am indebted to Mike Fortun and Tom Maniatis for allowing me to use some

of Maniatis's comments from Fortun's interview with him in Cambridge, Mass., February 1992.

32. For the purposes of comparison, in another laboratory where protein biochemistry was the primary disciplinary background, one copy of the same manual sat on a bookshelf in the scientists' office (not on the bench) and was referred to infrequently primarily by two of the six lab members.
33. See Keller 1995, Shapin 1989, and Star 1992 on invisible female and technical labor.
34. In 1993 Richard Roberts of the Cold Spring Harbor Laboratory and New England Biolabs won the Nobel Prize in medicine together with Philip Sharp of the Whitehead Institute at the Massachusetts Institute of Technology. Roberts won the prize for his laboratory's work on splicing messenger RNAs, which showed the baroque architectures of RNAs in eukaryotes.
35. For other sources on university-industry relationships during this period, see Krimsky 1982, Office of Technology Assessment 1984, and Yoxen 1983.
36. I thank Scott Gilbert for this story.
37. Chromosomes X and Y (considered to be the genes that produce female and male, respectively, humans) are counted as two different types of chromosomes, bringing the number of different types to twenty-four.
38. Hood's laboratory also produced an automatic protein synthesizer and an automatic protein sequenator. In the fall of 1992 Hood's laboratory was designated and funded as a technology development center for the human genome project by the National Science Foundation. Hood has since moved his laboratory and technology center to the University of Washington in Seattle, where William Gates, the head of Microsoft, has heavily funded Hood's new department of molecular biotechnology.
39. There are many other data bases currently available. See Fujimura 1996 on molecular biology data bases and computational biology.
40. GenBank was officially under the auspices of the Department of Energy until 1992. Until 1987, GenBank was organized and administered by the Theoretical Biology (T-10) group at Los Alamos National Laboratory. From September 1987 (10 million base pairs) to September 1992 (100 million base pairs), it was also administered by IntelliGenetics, Inc., a commercial venture in Palo Alto, Calif., which won an operational contract from the Energy Department.
41. Control of and access to GenBank has now moved to the National Center for Biological Information in Washington, D.C. There currently are many debates about how the GenBank information should be organized and accessed (Fujimura 1994).
42. Today even more search and analysis procedures have been computerized. Data bases and their attendant software are an important subject for social study, because they are the basic informational resources *in and with which* biological and biomedical knowledge is currently being collectively constructed. See Fujimura 1996 and Fujimura and Fortun 1996.
43. For molecular biologists, "homology" means functional identity or at least simi-

larity. They do not distinguish between homology and analogy, whereas scientists working in evolutionary biology do distinguish between the two. For evolutionary biologists, "homology" means both similarity and common ancestry. See Fujimura 1996 for discussion of the use of homologies in constructing meaning of sequence information in data bases and in the human genome projects.

44. Interview on November 2, 1984, with a professor of biology at a major Californian University. Hereafter I use the pseudonym "Sander" because this oncogene researcher requested that I not use his name in this text.
45. See Darnell, Lodish, and Baltimore 1990:212–214 and Suzuki et al. 1989:410–414 for descriptions of both methods. Judson 1992:65–66 presents a less technical description of the Sanger-Coulson method.
46. This quote is from Cherfas 1982:124, which quotes Gilbert's lecture, "DNA Sequencing and Gene Structure," published in *Science* 214 (1981): 1305–1312.
47. A debate on whether and where technology replaces expertise was held at the conference "Rediscovering Skills in Science, Medicine, and Technology," organized by David Good and Harry Collins, Bath, United Kingdom, September 1990.
48. Yet, even this advanced design instrument was inexpensive when compared with a protein synthesizer, which cost $250,000.
49. See Becker 1982 for a similar exposition on the role of conventions in the "logic" in music and art worlds. Molecular biologists have constructed the set of recombinant DNA technological packages that cancer researchers are using to order their work.
50. I thank John Law for the term "geography of forces."
51. See Chapter 8 for a discussion of Harry Rubin's critique of oncogene research. Rubin has pointed to the dominance of proto-oncogene research in explaining the loss of other paths to be followed in pursuing answers to cancer.
52. Although standards can be resisted or transformed (Gilroy 1991, 1993; Pratt 1992), resistance and transformation in scientific laboratories require much more effort, time, and money than most researchers have at their disposal.
53. Scott Gilbert runs his embryology laboratory at Swarthmore College. He was also chair of the Biology Department and has published both a well-circulated developmental textbook and many articles in the history of biology.

4. Crafting Theory

1. Oncogenes were first called "cellular oncogenes" by Bishop and Varmus (1982) to distinguish them from viral oncogenes. Bishop (1989:9) later explained this nomenclature: "As transduction by retroviruses came into common discourse, the need arose for a generic term to describe the cellular progenitor of *src* and other retroviral oncogenes. The first to find general usage was 'cellular oncogene.' Although I was a nominal member of the responsible committee on nomenclature, I was uncomfortable with this term because of its unwarranted

implication that the native cellular genes carried intrinsic tumorigenic potential, that they tended not to be changed to cause trouble. So in playful homage to Howard Temin, I began to use the term 'proto-sarc.' The generic 'proto-oncogene' followed in short order." For the early literature on proto-oncogenes, see Bishop 1982, 1983; Cooper 1982; Duesberg 1983, 1985; Hunter 1984; Land, Parada, and Weinberg 1984; Robertson 1983; and Weinberg 1983. For other reviews, see Varmus 1989b, Weinberg 1989, and the numerous review articles on proto-oncogenes and oncogenes that have been published in journals such as *Advances in Cancer Research, American Journal of Reproductive Immunology, Annual Review of Genetics, Biochemistry Journal, Biochimica et Biophysica Acta, Cancer Genetics and Cytogenetics, Cancer Investigation, Carcinogenesis, Cell, Cell Tissue Kinetics, Cytometry, FASEB Journal, Human Genetics, Journal of the American Academy of Dermatology, Journal of Cancer Research, Journal of Cell Biology, Journal of Molecular and Cell Cardiology, Nature, Journal of Pathology, Pharmaceutical Therapeutics, Reviews in Oncology,* and *Science.* One can also find review articles in books, and there are new journals (e.g., *Oncogenes, Oncogenes and Growth Factors Abstracts,* etc.) and many books devoted to oncogenes. Recent internalist accounts of oncogene research include Morange 1993 and Vecchio 1993.

Proto-oncogenes differ from cancer genes designated as having a familial link, such as the breast cancer gene (BRCA 1) located by Mark Skolnick's laboratories at the University of Utah and Myriad Genetics, Inc., of Salt Lake City in September 1994. In contrast to proto-oncogenes, which are somatic cell genes, the latter genes are considered to be heritable through the germ line. BRCA 1 is thought to be a tumor suppressor gene. See Skolnick et al. 1994, King et al. 1990, and Chapter 8.

2. Compare the proto-oncogene theory with Duesberg and Schwartz's (1992) "rare recombinants" theory of cancer causation.
3. RNA tumor viruses are retroviruses that have genes constituted of RNA sequences rather than DNA. They replicate by producing a strand of DNA sequences through the activities of an enzyme called "reverse transcriptase." See Studer and Chubin 1980 and Watson et al. 1987.
4. Researchers did report suspected links between some human cancers and retroviruses. See especially Gallo 1986.
5. The work that led to this announcement began much earlier. According to Bishop, "the oncogene we studied was not discovered until 1970. By 1972 I had several people in the lab thinking about the oncogene itself and what it does. Our first and still most important experiments were done within that year. By 1974 we had found that that gene was a normal cellular gene as well. And that is the work for which we are still best known. That work was conceived of and done within that first year of beginning to think about oncogenes." Interview on February 11, 1984, with J. Michael Bishop, professor of microbiology and immunology, biochemistry, and biophysics, and director of the George Williams Hooper Foundation at the University of California, San Francisco, Medical Center. I also interviewed Professor Bishop on July 20, 1990.

6. At the time recombinant DNA techniques were not available owing to the moratorium on their use imposed by the NIH at the urging of several key molecular biologists developing and using the techniques. Bishop and Varmus constructed molecular probes using molecular hybridization techniques.
7. Molecular biologists claim that since a gene is constructed of a specific sequence of nucleotide bases along a continuous strand of DNA, simply locating a particular gene of interest is akin to searching for the proverbial needle in a haystack. Molecular biologists argue that constructing DNA probes is one way of locating homologous DNA sequences. In 1978 probes were still relatively difficult to construct. In 1990 most probes were constructed by automated DNA synthesizers. Today the procedure is routine.
8. By 1995 over fifty possible proto-oncogenes had been reported in the literature.
9. See below and Studer and Chubin 1980 for more details on the effort to link viruses to human cancer by the National Cancer Institute's Virus Cancer Program.
10. See Bishop 1982 and 1989 and Varmus 1989a,b for accounts of the genesis of their laboratory's research. Temin (1971) had earlier proposed a related theory. His protovirus hypothesis argued that cells contained endogenous viruses (RNA to DNA transcription) whose task it was to move around the cell genome and generate genetic variations through incorrect transcription or by cobbling together new genes using parts of cellular genes. According to Bishop (1989), he and his colleagues had been unaware of Temin's proposal at the time of their initial research. By the time they formulated their own theory, however, they certainly acknowledged Temin's work. "The genesis of retroviral oncogenes by recombination with cellular genes had been postulated by several observers. Although it may sound self-serving, I confess my ignor[ance] of those speculations when we began our work on cellular *src*. I was motivated by a desire to test the Virogene-Oncogene Hypothesis of Huebner and Todaro, not by an interest in the origins of oncogenes. But in due course, Howard Temin provided a useful inspiration with his suggestion that all retroviruses arose by the cobbling together of disparate genetic elements in the cell, with intermediates that he called proto-viruses. [However,] the inspiration was for taxonomy, not experiment" (Bishop 1989:8).
11. There was dispute at the time over the Nobel committee's exclusion of Dominique Stehelin, the postdoctoral student who had done most of the hands-on research on the project. In another vein, Peter Duesberg wrote a letter to the journal *Science* arguing that the prize should not have gone to Bishop and Varmus, since in his perspective proto-oncogenes did not cause cancer in the manner suggested by the prize recipients.
12. Evolutionary biology, and especially evolutionary genetics, is so embroiled in debates that oncogene researchers may succeed in this effort to propose a role for oncogenes in evolutionary biology. The units-of-selection debates so closely studied by philosophers of science are just one indication of the lack of consensus about the unit, levels, and processes by which selection and evolution occur. See,

for example, Lloyd 1988 and Brandon 1990 for overviews and analyses of the units-of-selection debates.

13. Other suggestions of oncogenes as a source of genetic variation and as an indication of the course of evolution were made by Temin (1971, 1980) and by Walter Gilbert's research group (Schwartz, Tizard, and Gilbert 1983), respectively.
14. It is often difficult to distinguish between molecular biologists and tumor virologists because, by now, their technologies are very similar and overlapping, if not identical. These researchers often call themselves by several field or disciplinary titles. I make the distinction here based on the techniques used and problems framed by the researchers at the beginning of their research programs. I base my distinction on the early training of the researchers studied.
15. See also Chapter 2 and Robertson 1983 for more general discussions of gene transfection experiments.
16. See Der, Krontiris, and Cooper 1982; Goldfarb et al. 1982; Land, Parada, and Weinberg 1983; Parada et al. 1982; Santos et al. 1982; Shih et al. 1979; Shimizu et al. 1983; Tabin et al. 1982; and Weinberg 1983 for literature on gene transfection oncogene experiments in the laboratories of Barbacid, Cooper, Weinberg, and Wigler.
17. These "normal" cells, called NIH 3T3 cells, are somewhat ambiguous cells. They are not entirely normal, since they have been "passaged" so many times in the laboratory. That is, the original cells taken from normal mouse tissue in the early 1960s have by now adapted to the artificial conditions of cell cultures (plates of agar filled with nutrients to feed them and antibiotics to prevent them from being infected with bacteria) and are no longer entirely normal. They are referred to as "immortalized cells" that might already be partially transformed. The scientists I interviewed acknowledged that cells from NIH 3T3 cell lines have already undergone the first step in transformation from normal cells to cancer cells. However, they continued to use these cells as the "normal" base from which to examine the processes of "transformation."
18. After these events had occurred, a *New York Times* science journalist spent six months observing Weinberg's laboratory in 1986. From interviews with Weinberg and his postdoctural and graduate students, Natalie Angier (1988) reconstructed the events of 1981 and 1982 when a few laboratories raced for priority in cloning transfected human cancer genes. Although Angier was obviously a sympathetic observer in that laboratory, she presents another useful perspective on those events.
19. The desire to avoid "messiness" recalls Scott Gilbert's discussion at the end of Chapter 3 of the removal of the messiness and randomness of the organism in molecular genetics.
20. Weinberg's claims have since been modified. Current views are that at least two events, and perhaps up to eight events, are necessary to transform "truly normal" cells into cancer cells.
21. These are specific changes in DNA sequences that some scientists claimed to be

the initial trigger causing a cell to begin reproducing itself with no control mechanism to end its growth. There are other "steps" in the progression of cancer to the point where it kills the organism. In addition, theories of metastasis argue that many cancers consist of many different kinds of cells; cancer cells in one tumor are not all descended from one transformed cell.

22. Interview with the director of a private, nonprofit research institute that focuses on breast cancer research, especially on early cell transformation properties. The director had been active in basic cell transformation research in the university sector for many years before founding the research institute. I use the pseudonym "Norma Oakdale" because the director chose to remain anonymous. The interview was conducted in California on September 5, 1984.
23. But see note 5 to this chapter.
24. See DeVita 1984 and Gallo 1986.
25. The VCP was not regarded as a failure by everyone. A major contribution attributed to the program was the discovery of the reverse transcriptase in the late 1960s and early 1970s. However, even this research was attributed by some of my respondents to earlier research done by virologists far before the VCP was instituted.
26. For accounts of the controversies surrounding the 1971 National Cancer Act, see Chubin and Studer 1978; DeVita 1984; Epstein 1978; Moss 1980, 1989; Rettig 1977; and Strickland 1972.
27. However, some critics have argued that oncogene research itself was an outcome of research done throughout the National Institutes of Health and not only through National Cancer Institute funding contracts (Brown 1991).
28. Interview on August 29, 1985, with Vincent DeVita, M.D., then director of the National Cancer Institute, Bethesda, Maryland.
29. Biologists such as Ruth Hubbard (1990, Hubbard and Wald 1993) and historians of biology such as Evelyn Fox Keller (1985, 1991) have criticized this view of the gene as the "Master Molecule" that governs all subsequent development and discuss its widespread acceptance in Western scientific and popular thought. See also Ayala 1994 and Lewontin 1992 on the contributions of genetic variation in relation to environmental variation for determining phenotypes.
30. Reported NIH funding data is from the CRISP system of references to United States Public Health Service grants and contracts and to NIH intramural projects and retrieved through their keywords. I am grateful to F. M. Biggs and Seu-Lain Chen of the Division of Research Grants and N. Sue Meadows of the Office of Grants Inquiries for their assistance in this data collection.
31. Varmus later established his own laboratory also at the University of California, San Francisco, Medical Center. In 1993 he began his tenure as director of the NIH.

5. Distributing Authority and Transforming Biology

1. It should be noted that oncogene research is not all of cancer research. Clinical and epidemiological lines of research with no ties to molecular biological meth-

ods continue. Molecular biological cancer research occupies only part of the vast cancer research world.

2. NCI commitments to oncogenes and other molecular biological cancer research studies meant cutbacks in other basic research lines, as a 1983 article (Shapley 1983:5) in the journal *Nature* indicates. If the NCI allowed spending on oncogene research to expand naturally, did this mean less prominence for important traditional fields such as chemotherapy? At the time, DeVita predicted some other work would have to go, given the unlikelihood of the NCI receiving any budget increases in the next few years. He noted that chemotherapy had been cut by about 30 percent in the past six years on scientific grounds: "some things we didn't need to do any more." As viewed by Alan Rabson, director of NCI's division of cancer biology and diagnosis, "if you understand oncogenes you may learn where to go in chemotherapy. It may open up whole new areas of chemotherapy." Although chemotherapy was the only area mentioned in the article, other areas of research were also neglected simply because Congress did not increase their budget while oncogene research expanded. The article also hints that chemotherapists would benefit by taking oncogenes as their starting point.
3. Interview on November 22, 1983, with the director, a professor of molecular biology and virology, at a major Californian university and a noted viral oncogene researcher, hereafter identified as "Yuzen." I also interviewed him on many other occasions. I chose to keep his name confidential and use the pseudonym "Yuzen," because I also conducted fieldwork in his laboratory full-time (morning through evening) during four months of 1984. I continued to visit the laboratory to keep up with the laboratory's research thereafter until 1986. Subsequent quotations cited as taken from "Yuzen field notes" refer to discussions with members of the laboratory during my fieldwork.
4. Interview on May 17, 1985, with a postdoctoral researcher in a viral oncogene research laboratory in a major Californian university, hereafter identified as "Jones." I use pseudonyms such as "Jones" for all graduate students, postdoctoral researchers, and laboratory technicians.
5. Since the period examined in this chapter, even the FBI has been working with molecular biological tools and skills, especially polymerase chain reaction. PCR is a process, partially housed literally in a black box, that quickly and reliably reproduces multiple copies of a sequence of DNA. The FBI has been exploring the PCR as a potential "DNA fingerprint" to identify assailants in criminal cases. However, there has been considerable debate about PCR's capabilities in making such identifications because most DNA samples available to law enforcement agencies are not as unique as an actual fingerprint. In other arenas, hospitals will soon also add PCR to their array of diagnostic technologies.
6. Interview on May 16, 1984, with a graduate student in a cellular immunology laboratory in a major Californian university, hereafter identified as "Medina" (a pseudonym).
7. Cetus has since been acquired by Roche and was formally dissolved (except for a group incorporated into Chiron, another biotechnology start-up company) as of December 12, 1991.

8. See Draper 1991 for a more recent argument along the same lines.
9. Somatic cell hybridization has two genealogical lines of development: cell clones and mutant cells. Culturing techniques were a critical part of these efforts. Shortly after World War II geneticists conducted cloning experiments involving single cell isolation and colony isolations. Cells sparsely distributed in culture created additional nutritional requirements of culture, for example, different concentrations of nutrients (Willmer 1965). The second line came from mutant cells created in the 1960s through a hybridization technique that fused a human cell with a mouse cell. The resulting cells were heterokaryons containing two genetically different nuclei. By altering the conditions of the cell culture, researchers changed the survival outcomes of the cells. For example, by adding hypoxanthine aminopterin thymidine to the media, they allowed particular mutants to survive and grow, and others to die (Littlefield 1964, cf. Weiss and Green 1967).
10. Their work became redundant after the development of hybridomas and monoclonal antibodies. Other somatic cell geneticists were interested in the mechanisms of gene regulation, for example, how a gene operated in kidney regulation. Their work sat in the interstices between genetics and developmental biology. This line of research also "failed," in Scott Gilbert's view, when the subsequent development of recombinant DNA techniques presented researchers with another tool for exploration.
11. See Mintz and Illmensee 1975, Papaioannou et al. 1975, and Stevens 1970.
12. See Bishchoff and Bryson 1964 and Karp et al. 1973; cf. Dunham 1972.
13. See also Hacking 1992, Pickering 1990, and Star and Griesemer 1989 for discussions of malleable or plastic concepts.
14. These comments are based on interviews. See also Boffey 1986 and Moss 1989 on Sloan-Kettering's more general shift in research from immunological to molecular biological approaches to understanding cancer.
15. Donna Haraway (1995) points to the OncoMouse™ as an example of the laboratory creation of the transgressive and monstrous in material form.
16. I do not imply that innovation occurs only at universities such as those named. In fact, innovation occurs everywhere, often not in universities.
17. Koch's postulates state that to prove that an infectious agent is the cause of an illness, it is necessary to establish that: (1) The parasite is present in every case of the disease under appropriate circumstances. (2) The parasite should occur in no other disease as a fortuitous and nonpathogenic parasite. (3) The agent must be isolated from those infected subjects, cultivated in vitro (in pure culture), and induce the disease when introduced into healthy subjects. See Fujimura and Chou 1994.

6. Problems and Work Practices

1. Compare this view with Lynch's (1985) and Garfinkel, Lynch, and Livingston's (1981) detailed analyses of the performance of scientific methods and the "or-

dinary discoveries" that constitute the actual daily activities of lab work.

2. This chapter's subtitle reflects my focus on concrete stories about scientific problem-solving. I do not take the "lines of research" perspective, told post hoc from the historian's or sociologist's vantage point as I did in Chapter 4. In this chapter I was just as "involved" as the participants I interviewed, and my stories try to tell their stories from their perspectives at the time.
3. Resources include knowledge, workers, skills, technologies, materials, funds, and infrastructural requirements such as building space, electricity, benches, running water, and so on.
4. See, for example, Clarke and Fujimura 1992, Gooding et al. 1989, Hacking 1992, Kuhn 1957, Latour and Woolgar 1986, Pickering 1990, Pinch 1985, and Traweek 1988 for discussions of the adjustments made between theory, experiment, instrument, and other elements of experimental work.
5. In contemporary research environments, university administrations often require researchers to obtain large grants and research has become so expensive that laboratories cannot exist without external funds. A research proposal is a formal statement of plans for organizing work. The format of a proposal is largely a requirement of the agency and is less a description of how a project is set up. In many cases, a proposal is actually an account of work already completed but framed in terms of procedures to be carried out in the future.
6. Computer scientists working in artificial intelligence face similar issues in their attempts to develop problem solving strategies in open systems (e.g., Hewitt 1985).
7. Despite the open-ended and evolutionary nature of scientific work, researchers have some commitments to and expectations of the sequence of future events. As Becker (1960, 1982), Strauss (1978a), and Strauss and associates (1985) have pointed out, they are based in the conventions of work and tacit, long-standing assumptions operative in an occupation, organization, and local work site at any point. See also Bourdieu's (1977, 1990) and Rosaldo's (1989) related discussions of improvisation in rituals.
8. Unexpected events can turn out to be quite felicitous. Serendipitous findings are the stuff of many success stories in science (cf. Barber and Fox 1958). In other situations, unexpected events can turn out to be deleterious, and scientists may drop research problems when faced with such events. Other events can at first appear to be negative but eventually be dealt with in ways that allow the researchers to go on with their work. See March and Simon 1958 on unintended consequences in organizational decision-making, Hanson 1958 on "backing into discovery," Merton 1973 on unanticipated consequences, and Zeldenrust 1989 for a study of unintended consequences in technological research. Becker 1994 discusses examples of coincidence that people use to explain major life and work decisions.
9. Note that even the recognition of an event as a contingency is an interactional and contingent achievement. In the examples that follow, I take the participants' view that an event is a contingency which affects their problem process.
10. See also Barnes 1983 and Galison 1987 on "how experiments end" in physics.

11. See Chapter 4 for more on these events, which are part of a bandwagon of proto-oncogene research.
12. See Chapter 3 for more discussion of criticisms of the original NIH guidelines for recombinant DNA experimentation. Congress and NIH officials relaxed these guidelines even further in 1982.
13. There has since been a range of claims regarding the nature of the trigger mechanisms, from one gene mutation as enough to cause cancer to more conservative estimates that several independent changes are necessary for carcinogenesis to occur. See Chapter 4.
14. Interview on January 10, 1985, with "Kensington," Ph.D. and vice-president of research and development at one of the first Californian biotechnology companies. "Kensington" is a pseudonym, used to protect the confidentiality of the members of the Xavier laboratory.
15. This is an example of March and Simon's (1958) garbage can model of organizational decision-making. For a treatment of technology research using their model, see Zeldenrust 1989.
16. Other ongoing experiments not directly related to these questions included an effort to transform mouse breast epithelial cells with two hypothetically interactive oncogenes; a study of metastasis in nude mice; and the subcloning of a gene that researchers think will prove to be an oncogene. Finally, the laboratory was also collaborating with another branch of the company in a long-term product development effort to manufacture monoclonal antibodies to the transforming proteins or antigens by using peptides related to the family of proteins produced by oncogene *ras*.
17. See Shapin 1994 for his view of trust in science.
18. See Latour 1987 and Rheinberger 1992 on the collapsing of these categories.
19. Wimsatt similarly views scientific theories as generative structures and argues that this model allows one to see the generative entrenchment, or resistance toward change, which results from such embedded structures. He describes (1986:1) a simple model of generative entrenchment: "If one proposition is used to generate a much larger part of the theoretical structure than another, and testing indicates that one or the other is to be given up, it seems reasonable that the proposition on which less depends should be given up or modified in preference to the more centrally entrenched one, on the grounds that changes in the latter will induce much more global changes, with a higher probability that some of these changes will be disadvantageous." Note, however, that this formulation preserves the logic of propositional reasoning, whereas my formulation delves into heterogeneous considerations that are not readily "weighted" in a unitary space.
20. Within each modular piece, the sequence in which scientists do their tasks might matter.
21. Kauffman (1971) discusses the integration of the results of subproblems as the articulation of parts. Wimsatt (1974) discusses the complexity that arises in trying to put the parts together again in building models.
22. I speak in relative terms here. Most graduate students I observed had greater

control over their problem choice than did most laboratory technicians. As noted earlier, graduate and postdoctoral students are also subject to their "boss's" discretion with respect to problem construction.

23. Kimmelman (1983, 1987) provides examples of commercial agriculture's impact on biological research in genetics in the early 1900s. Similar historical studies might also confirm fears that recent university-industry links have influenced problem choice and changed problem structures in some academic biological laboratories. We may see no differences between patterns of problem choices and structures in some academic and industrial laboratories. Zeldenrust's (1985) examples show that problem choice in academic polymer research labs changed in response to industrial demands.
24. Problem decomposition has been discussed by others in terms of, first, different ways of cutting up phenomena to structure solvable problems and, second, the complexity that arises in trying to put the parts together again in building models (Kauffman 1971; Lewontin 1992; Schaffner 1993; Simon 1969, 1973; and Wimsatt 1974). These discussions of decomposition are efforts to frame a problem and translate it into sets of technoscientific (in contrast to sociotechnical) tasks. They frame the problem in terms of available and required scientific and technical tools and not in terms of the daily and long-term work conditions.
25. This phrasing sounds very much like a manager's perspective of the laboratory. I do not intend to promote this managerial perspective, but I am presenting the views held by both lab directors and students of the process of assigning problems to students. Whether they like it or not, members recognize that lab directors more often than not define and assign problems in this manner. Many students call their lab directors "boss." However, there is room for independent work by students in some laboratories.
26. This is the basic thesis of Marxist and neo-Marxist discussion of capitalist modes of production and consumption and of the disciplining of people to fit these modes of production. Scientific work organization does not escape from its location in post–World War II capitalist society.
27. See Friedson 1970, Merton 1970, and Strauss 1959 on professionalization. Freidson's and Strauss's perspectives differ from Merton's on the professional habits people learn in the professionalization process.
28. The professional scientist is not a neutral entity. It is an entity historically constituted in a particular time and place, in the "Enlightenment" period in the Western world (Keller 1985, Schiebinger 1989, Shapin and Schaffer 1985, Strathern 1992). In the United States, academia itself has been designed to serve the needs of the industrial sector. The original curriculum of many universities was designed to produce graduates with a general education that would allow them to be shaped to the particular demands of industrial organizations. The training of technical professionals gradually became more specialized with the increasing complexity and diversity of technologies. Emily Martin (1994) has argued that this Fordist model is giving way to a post-Fordist model of being, in which people are trained to be flexible in order to best respond to rapidly changing demands in this postmodern era.

29. This example is taken from an interview in 1984 with a professor of virology and molecular biology at a major Californian university, whom I have given the pseudonym "Chaucer."
30. This example is taken from an interview in 1984 with a tumor virologist at a major Californian University, whom I have given the pseudonym "Grant."
31. Hacking (1983) discusses this kind of analytic style as representing by intervening. Molecular biological technologies allowed researchers not just to observe but to actually *change nature* in the laboratory. This is the same analytic style referred to by Bishop in his definition of "direct evidence" of the cellular oncogene's transforming capacity in Chapter 4.
32. This quotation is taken from a journalist's interview with Oakdale. The article appeared in a San Francisco Bay Area newspaper. I omit the citation because it would reveal the identity of the researcher and I have promised her confidentiality.
33. Oakdale's funding for the institute's work first came from private industrial sources. She later managed to convince NIH that epithelial cell cultures were worth funding. However, as she states, "It's been very difficult. I've hung by my fingernails . . . by gentle persuasion . . . I have drawers of grants that have been turned down. It's persistence. But NCI, in its wisdom, and the peer review people in their wisdom, realized the fundamental importance of this work and that somebody has to do it. And I think I have the reputation of doing it well" (Oakdale interview).

7. The Articulation of Doable Problems in Cancer Research

1. See Chapters 1 and 3, Blumer 1969, Callon 1986a,b, Callon and Latour 1992, Collins and Yearley 1992, Haraway 1985, and Latour 1986 on "nonhuman" actors.
2. See, e.g., Knorr-Cetina 1981, Latour and Woolgar 1986, Lynch 1985, and Traweek 1988. These last two works were carried out in the late 1970s and published later. In the traditional anthropological commitment to study the same community over a long period of time, Traweek's ethnography of several physics laboratories and communities continues today.
3. See Callon 1986a,b, Latour 1987, and Law 1986.
4. See Strauss 1978b and Clarke 1991 on social worlds and arena analysis.
5. See Bendifallah and Scacchi 1987; Fujimura 1987; Strauss 1988; and Strauss et al. 1985 on articulation.
6. Becker (1982) points to the many participants involved in the collective work in art production. These include canvas stretchers, state regulators, and audiences, as well as those given the title "artists."
7. Cf. Pickering 1990 and Hacking 1992, which discuss the tinkering and adjusting of instruments, experimental results, and theories to make them fit and work well together. See also Knorr-Cetina 1979 and 1981, which address "the occasioned character of research" as exemplified by tinkering and patching things together within the contingencies of the local situation.

8. In this case the lab director and head of research and development planned the problem and project together. This management-organized arrangement was due to the project's location in the biotechnology company. In an academic laboratory, I found that students and postdoctoral researchers were able to propose and plan their own research problems, if they could persuade the lab director of the merits of their plans.
9. For more detailed but still accessible descriptions of this process, see Bishop 1982, 1983; Hunter 1984; and Weinberg 1983.
10. *Ras* protein products have still not produced cancer therapeutic agents, but they form one of the largest subfields of contemporary oncogene research and are being explored as diagnostic agents (Marx 1984). See McCormick 1989 for a technical research summary of the field.
11. The degree to which research programs are shaped by profit hopes is not surprising to anyone familiar with the history of industrial research. What is interesting to note in this book is that the line between industrial and publicly funded research as well as the line between profit- and nonprofit-oriented research programs are not clear. I note elsewhere that biological research programs earlier in the century were tightly tied with industrial agricultural breeding programs. Clarke 1985 also points to the tight connections between academic reproductive biology programs and animal breeding programs in industry. The latest example of this private-public merging is the first laboratory to map and sequence a breast cancer gene called *BRCA1* on chromosome 17. Mark Skolnick's laboratory is at the University of Utah, but it is also part of a biotechnology company called Myriad Genetics, Inc., founded by Skolnick and Walter Gilbert in 1991.
12. Callon 1987 provides a similar example of decisions made on the basis of projected futures by engineers and sociologists at Electricité de France (EDF) in competition with engineers and sociologists at Renault. In his example, Renault wins the competition between projections, and the EDF engineers' plans to develop an electric car and to reduce Renault from a powerful automobile manufacturer to a chassis producer fail.
13. Note that constraints can also be opportunities in other situations and on other occasions. And constraints can change over time. I am looking specifically at one situation here.
14. As an aside, the project also allowed the lab director, who had come from an academic setting, to simultaneously satisfy sponsor demands and his desire to do innovative basic biological research. When asked what he missed about academic research, the lab director implied that he had missed doing basic research in his field of training, but then said that this was no longer a problem. "Actually now I don't [miss anything], because now the work I'm doing is much more in gear with . . . my sort of peers in university departments. I guess for the first couple of years I was here, I was doing more applied research towards product goals . . . So, I was a little out of the circle of people that I knew before . . . I'm not a major player back in it, but at least I'm going to the same meetings again, and I can contribute to the field that I really like most." This had the added benefit of keeping his career options open should he decide to reenter academia.

15. See Chapter 4 for another view of *ras* oncogene research.
16. "In vitro" here refers to biochemical tests of antibodies (or pieces of antibodies) against antigens in reagents rather than on living cells.
17. Public announcement is not the usual forum for publicizing scientific results. The researchers were pressured to do this in part because of the company's efforts at the time to raise the share prices. Luckily for this laboratory, they did not meet the same fate as Pons and Fleischman upon public announcement of their cold fusion experimental "success"!
18. Other experiments attempted to answer questions about the the structures and activities of the family of *ras* oncogenes. Three other experiments were not at all related to *ras* oncogenes.
19. However, see Collins 1985 and Pinch 1991 for discussions that make replication and testing problematic concepts.
20. Histories of science are peppered with stories of recalcitrant research materials that resist researchers' efforts to control them. See, for example, Clarke 1987 and Clarke and Fujimura 1992.
21. See Keating, Cambrosio, and MacKenzie 1992 for a detailed historical analysis of specificity as it relates to concepts of affinitiy and avidity.
22. "Serum" here refers to the blood fluid containing all the antibodies and other bodies in the immune system. In contrast, monoclonal antibodies are single antibodies produced by synthetic means.
23. See Riemer 1979 for a study of mistakes at work—some hilarious, many hazardous—in the construction industry.
24. On the use of standardized procedures to simplify teaching, see Latour and Woolgar 1986 and Rip 1982. See Jordan and Lynch 1992 for a contrasting discussion of the difficulties involved in reproducing even a standardized technique.
25. Cambrosio and Keating (1988) argue that tacit knowledge or "art" need not always be "black magic" and can be passed on from one researcher to another, even if it cannot be standardized. Their example is taken from hybridoma technology. Coincidentally, hybridoma technology has "revolutionized" protein work in immunology by making it possible to quickly produce huge quantities of the same antibody (protein).

8. Conclusion

1. See Hughes 1971, Shalin 1986, and Thomas and Znaniecki 1918 on indeterminacy and the definition of the situation.
2. See Crane 1972 on the "invisible college."
3. See Strauss 1978b on arenas.
4. For molecular biologists, recall that "homology" means identity or at least similarity. See Chapter 3, Fujimura 1996, and Hall 1994.
5. Another large part of the viral research funded by the Virus Cancer Program focused on DNA tumor viruses.
6. Temin (1971) had earlier suggested a similar mechanism by which "protovi-

ruses" were formed as an intermediate stage in the process of the cell's normal genetic activities. See note 10 to Chapter 4.

7. I do not regard the theory-methods package as constituting a necessary connection. The coupling of the oncogene theory and recombinant DNA with other molecular biology technologies is constructed and not born in nature. The theory may in the future continue to exist as an entity separate from these techniques or coupled to another set of techniques. Similarly, the technologies are coupled with quite different theories in other lines of biological research.
8. My use of the term "crafting science" is inspired by anthropologist Dorinne Kondo's *Crafting Selves* (1990) and Latour and Woolgar's (1986) use of "craftwork." Ravetz (1971) also discusses science as craftwork, but he takes a more realist view of the process and focuses more on technical craftwork and less on the activities of coherence construction. See also Polanyi 1958 for a view of craftwork framed in terms of skills.
9. While it is not a laboratory study per se, Suchman's book (1987) also belongs to this category of studies of situated local practice. For historical studies of instruments in experimentation, see Galison 1987; Gooding, Pinch, and Schaffer 1989; Holmes 1992; Lenoir 1988; Lenoir and Elkana 1988; Pickering 1984; and Shapin and Schaffer 1985.
10. Philosopher William Whewell (1847) sought a method for achieving truth via a "consilience of inductions." Larry Laudan (1990) has since taken up the banner.
11. Although oncogene research is pursued in many other countries, this book focuses on research in the United States.
12. Two recent volumes deal specifically with the construction and uses of research tools and inscriptions. See Clarke and Fujimura 1992 and Ruse and Taylor 1991. A third volume was published as a special issue of the journal *History of Biology*. See especially Kohler 1993 and Clause 1993. For recent work on instruments and entities, see Dodier 1991, Mol 1991, and Cambrosio and Keating 1992.
13. In Wilson's diagram, genetic transmission is shown as a continuous causal history moving from germ cell to germ cell, with no role of somatic cells and development of particular individuals. In contrast to this continuity of germ cells, Weismann's diagram "shows only a single organism's partial development (in the idealized sense of a sequence of cell divisions). Moreover, it shows that the first germ-cells arise, as products of somatic differentiation, in cell generation 9 . . . By contrast [then], Weismann's diagram shows that the germ-plasm is continuous in its passage from zygote to germ-cells through somatic intermediates while the somato-plasm is a discontinuous product of germinal plasm" (Griesemer 1991:3–4).
14. "In the early to mid-19th century, acceptance of the inheritance of acquired characteristics allowed many social hereditarians to also be social reformers: individual moral conduct would have a bearing on the heritage that parents pass to their offspring . . . With the 'hardening' of the concept of heredity, the prospects for social reform through individual moral conduct were lessened because individuals were no longer thought to have precise control over the structure of their hereditary material, but only over its fate in choice of mates. The need for popu-

lational 'control' by genetics experts to achieve eugenic social ends became part of the rhetoric of the newly professionalizing biological specialties" (Griesemer 1991:3).

15. See Cambrosio, Jacobi, and Keating 1993 for a related case.
16. For biotechnologies that have commercial or legal ramifications (e.g., monoclonal antibodies and DNA forensics), other controls such as patents and trademarks play a large role in this determination.
17. See, for example, Allen 1978, Fogle 1990, Jacob 1982, Keller 1983, Kitcher 1982, and Lloyd 1988.
18. See especially Ackerman 1985, Latour and Woolgar 1986, and Pickering 1984, 1990.
19. See Callon and Latour 1981, 1992; Callon 1986a; Latour 1988b.
20. In a 1994 article announcing the designated twentieth-anniversary celebration of oncogene research to be held in January 1995, the science writer Jean Marx (1994:1944) wrote that "one sobering note, almost all in the field say, is that despite the remarkable progress made in understanding the genesis of cancer, the work so far has had little impact in the clinic." She quotes Steve Martin saying that "the disappointment has been the lag between the understanding of basic mechanisms and applications." She also quotes Frank McCormick of Onyx Pharmaceuticals in Richmond, California, as saying, "It's been a real blast over the past ten years, pulling out all the genes and figuring out what they do. But the next challenge will be to put the information to work." McCormick's research has been on the *ras* oncogenes and their protein products.
21. This remaking of the world and the subsequent demonstrations out in the world of the microbe's existence and activity, in Hacking's framework, do not "prove" Pasteur's theory—that microbes caused fermentation, anthrax, and many human illnesses—to be true in any metaphysical sense. The theory also does not "explain" the effects. Hacking (1992:60) considers it "a metaphysical mistake [to think] that truth or the world explains anything . . . [as a realist would explain it:] 'If the treatment works, then the world or the truth about the world makes it work, and that is what we found out in the laboratory and then applied to the world.' "
22. See, for example, Duesberg 1983, 1985, 1987a,b; Habeshaw 1983; Rubin 1983, 1985, Shubik 1983; and Teitelman 1985 for early arguments against these claims. See Temin 1983 for an early review of claims and counterclaims.
23. See also Finch 1990:446 for an example of other scientists' use of Duesberg's critique.
24. See also Littlefield 1982a,b, Pollack 1982, Rubin 1983.
25. Recall that I use their term "bandwagon" descriptively and not pejoratively. Some of my respondents used it derisively, some used in a matter-of-fact manner. As I shall discuss, bandwagons are as common in science as in other kinds of work and play.
26. See also Weinberg comments later in this chapter and Duesberg and Schwartz 1992 for a critical view of these adjustments.
27. Interview on June 15, 1991, with Robert Weinberg, professor of biology and

member of the Whitehead Institute for Biomedical Research at the Massachusetts Institute of Technology.

28. Duesberg and Schwartz (1992) argue similarly that the oncogene research program keeps shoring itself up by making continual adjustments to deal with difficulties that cannot be explained by the original theoretical formulation. However, Duesberg and Schwartz view this shoring up process as a sign of error in the theory rather than as a sign of progress as does Lakatos.
29. Lakatos would disclaim any connections between bandwagons and his model of research programs. He considers a bandwagon to be an example of "mob psychology" (p. 178) à la Kuhn. Kuhn (1957:77), however, uses the term "bandwagon" quite differently than I do.
30. There are many disagreements with this reading of Kuhn. However, my point here is not to analyze Kuhn's framework but instead to explore its uses for presenting the different "faces" of proto-oncogene research.
31. See Fujimura 1996 and Fujimura and Fortun 1996.
32. Whether the theories buoyed by bandwagons are truths or facts is not my concern here as it is Rothman's (1989) concern. Rothman's *Science a la Mode* discusses several bandwagons (as he calls them)in science (e.g., the big bang theory of the origins of the universe in astrophysics) and suggests that alternative explanations are possible for each phenomenon.

Appendix

1. For recent reviews of the literature in the social and cultural studies of science, see Clarke and Fujimura 1992, Fischer 1991, Haraway 1994, Hess and Layne 1992, Lynch 1993, Pickering 1992, Sismondo 1993, Traweek 1993.
2. Merton and his students examined "the social system of science" or, more precisely, the system of norms that are supposed to govern scientists. These writers proposed that scientists operate according to the norms universalism, communism, disinterestedness, and organized skepticism. They studied the behaviors of scientists, not the content of scientific knowledge or the processes of knowledge production. Indeed, their claim was that the validity of scientific knowledge escapes sociological analysis. Their implicit assumption was that scientific knowledge per se is not subject to social influence and therefore not open to sociological inquiry. See Mendelsohn 1989 for a review of this work.
3. See Lynch 1992 and Sismondo 1993 for careful treatments of the similarities and differences among various constructivist lines. In comparison to Haraway's (1993) and Traweek's (1993) discussions of the "field" of social, cultural, and feminist studies of science, these histories focus on a narrower circle of writers.
4. The Strong Program includes Barnes 1977; Bloor 1976; MacKenzie 1981; Pickering 1984; Shapin 1979, 1982; Shapin and Schaffer 1985; and Wallis 1979. By now, many Strong Programmers are no longer Strong Programmers in their most recent writings. Compare Pickering 1993, for example, with Pickering 1984. The Relativist Program's early works include Collins and Pinch 1982, Collins

1985, and Pinch 1986. See the Conclusion for more discussion of these lines of research.

5. In the 1960s a generation of writers in the United States and the United Kingdom (some of whom were scientists, others of whom were sociologists and historians) studied science as a social problem. This group used Marxist-influenced terms to study science. These included sociologists Ruth Cowan Schwartz and Dorothy Nelkin and leftist scientists Ruth Hubbard, Richard Lewontin, Richard Levins, and Jonathan Beckwith. Also in the 1960s, Hilary Rose, Steven Rose, and their colleagues and students at the Open University in England began to examine science from Marxist perspectives.
6. "Like the sociologists of scientific knowledge, ethnomethodologists try to transform the traditional themes in epistemology into topics for 'empirical' research. But instead of advocating a 'sociological turn' in which philosophy's problems are given sociological explanations, ethnomethodologists initiate a 'praxiological turn' through which they turn the sociological aim to explain social facts into a situated phenomenon to be described. Sociology's loss becomes society's accomplishment" (Lynch 1993:162). Lynch further argues with the SSKers' representation of the entire content of science as knowledge. Instead, "much of what goes under the heading of 'knowledge' in science studies can be decomposed into embodied practice of handling instruments, making experiments work, and presenting arguments in texts or demonstrations . . . Knowledge become[s] more tangible—and less monolithic—when translated into various practical activities and textual productions (Lynch 1993:310–311).
7. See, for example, Bijker and Law 1992; Callon 1986a,b, 1987; Callon and Latour 1981, 1992; Callon, Law, and Rip 1986; Latour 1983, 1987, 1988a, 1989, 1990c, 1993; and Law 1986.
8. Latour (1993) has summarized his framework and methodology for talking about science, society, and history. Using the sociology of associations or actor-networks, he extends his frame beyond science to the historical development of our "society(ies)."
9. For discussions and debates on nonhumans, see Collins and Yearley 1992 and Callon and Latour 1992. (See also Chapters 2, 4, and 7.) For different treatments of nonhuman "objects," see Blumer 1969, Haraway 1985, and Star and Griesemer 1989.
10. This is Latour's twist on Greimas's work on "actants" in his semiotic system. In a conference presentation ("Rediscovering Skills in Science, Technology, and Medicine," Bath, England, August 1990), Latour changed his use of "actor" to "actant" in concession to Collins's critique. See Lynch 1993 for discussion of this interpretation of the actant notion.
11. Historians of science Steve Shapin and Simon Schaffer argue in the *Leviathan and the Air-Pump* (1985) that science was offered by particular historical actors such as Robert Boyle and Thomas Hobbes as an answer to the seventeenth-century problem of order. For these natural philosophers, the enduring and vexing question was how to create and maintain order in a world in which the war of each against all was the natural state of humankind left to its own devices.

(This view of human nature was the antithesis of Marx's view.) Before Boyle and Hobbes, the church decided all matters of truth and fact. After Boyle and Hobbes, "scientists" of the Royal Society decided matters of natural facts, and sovereign leaders decided on matters of the polity, of human social interaction and organization. In either case, power became vested in the hands of a few.

More important for my story, Shapin and Schaffer's history contributes a new twist on this division of authority. They carefully demonstrate that Boyle engaged in political work through his organization of the practices and conventions of working and interpreting the results of his air pump. Shapin and Schaffer describe the social ("trustworthy" witnesses), material (the air pump) and literary ("virtual witnesses") technologies that Boyle constructed, advocated, and instituted as the means to determine truth about natural matters. However, although his scientific efforts were preserved, his politics disappeared from the chronicles. Through this historical erasure, Boyle's efforts, as well as the efforts of others who came after him, helped to create a world in which scientists became the authority over interpretations of nature, in which experiments became the focus of attention, and in which politics and theology appeared to disappear from this new kind of authority. Meanwhile, Hobbes's design of a political science was not preserved, while his political philosophy of an absolutist state for the maintenance of order survived the seventeenth century. To counterbalance this authority, Shapin and Schaffer argue that this version of "authentic" knowledge is by now also a powerful constituent of the political actions of the state. They propose that the knowledge that is wielded by the most astute politicians—those who are most savvy at "insinuating themselves into the activities of other institutions and other interest groups" (Shapin and Schaffer 1985:342)—is the knowledge that becomes authenticated.

12. See also Ashmore 1989, Fujimura 1991a, Haraway 1988, Myers 1990, Pickering 1992, Star 1992, and Woolgar 1988 for discussions of reflexivity in social and cultural studies of science, meaning how one should write stories about science given our understandings of how the scientific stories are constructed and our concerns about positivist statements.
13. See Anzaldua 1987; Butler 1990, 1993; Haraway 1989, 1992; Harding 1986, 1994; Kondo 1990; Mani 1992; Minh-ha 1989; Rao 1991; Rosaldo 1989; Steedly 1994; Traweek 1992; Tsing 1993.
14. See, e.g., Clifford and Marcus 1986, Gupta and Ferguson 1992, Marcus and Fischer 1986, Pratt 1986, and Rosaldo 1989.

References

Abir-Am, P. 1982. "The Discourse of Physical Power and Biological Knowledge in the 1930s: A Reappraisal of the Rockfeller Foundation's 'Policy' in Molecular Biology." *Social Studies of Science* 12: 341–382.

———. 1991. "Schools of Thought and Research Traditions in Molecular Biology." Paper presented at the colloquium "The Philosophy and History of Molecular Biology," Boston University, April 15–16.

Ackerman, Robert S. 1985. *Data, Instruments, and Theory: A Dialectical Approach to Understanding Science.* Princeton, N.J.: Princeton University Press.

Adams, M. B., ed. 1990. *The Well-Born Science: Eugenics in Germany, France, Brazil, and Russia.* New York: Oxford University Press.

Alberts, Bruce, Dennis Bray, Julian Lewis, Martin Raff, Keith Roberts, and James D. Watson. 1983. *Molecular Biology of the Cell.* New York: Garland.

Allen, G. 1975. *Life Science in the Twentieth Century.* Cambridge: Cambridge University Press.

———. 1978. *Thomas Hunt Morgan: The Man and His Science.* Princeton, N.J.: Princeton University Press.

"A Machine That Decodes DNA." 1986. *San Francisco Chronicle,* June 12.

Amann, K., and K. Knorr-Cetina. 1990. "The Fixation of (Visual) Evidence." In M. Lynch and S. Woolgar, eds., *Representation in Scientific Practice.* Cambridge, Mass.: MIT Press.

Ames, B. N., W. E. Durston, E. Yamasaki, and F. D. Lee. 1973. "Carcinogens Are Mutagens: A Simple Test System Combining Liver Homogenates for Activation and Bacteria for Detection." *Proceedings of the National Academy of Sciences (USA)* 70: 2281–2285.

Angier, N. 1988. *Natural Obsessions: The Search for the Oncogene.* Boston: Houghton Mifflin.

Anzaldua, B. 1987. *Borderlands/La Frontera: The New Mestiza.* San Francisco: Aunt Lute Book.

Ashmore, M. 1989. *The Reflexive Thesis: Wrighting Sociology of Scientific Knowledge.* Chicago: University of Chicago Press.

Ayala, Francisco. 1994. "The Nature/Nurture Controversy, or, When Both Sides of an Issue Are Wrong." Paper presented to the Program on the Ethical, Legal, and Social Implications of the Human Genome Project, Interdisciplinary Colloquy on Genetic Predisposition, Washington, D.C., October.

Baltimore, D. 1970. "Viral RNA-Dependent DNA Polymerase." *Nature* 226 (June 27): 1209–1211.

———. 1974–75. "The Strategy of RNA Viruses." *Harvey Lectures,* series 46: 57–74.

———. 1976. "Viruses, Polymerases, and Cancer." *Science* 192: 632–636.

Baltimore, D., A. S. Huang, and M. Stampfer. 1970. "Ribonucleic Acid Synthesis of Vesicular Stomatitis Virus, II: An RNA Polymerase in the Virion." *Proceedings of the National Academy of Sciences (USA)* 66: 572–576.

Barber, Bernard, and Renee Fox. 1958. "The Case of the Floppy-Eared Rabbits: An Instance of Serendipidity Gained and Serendipidity Lost." *American Journal of Sociology* 64: 128–136.

Barnes, B. 1977. *Interests and the Growth of Knowledge.* London: Routledge and Kegan Paul.

———. 1983. "On the Conventional Character of Knowledge and Cognition." In K. Knorr-Cetina and M. Mulkay, eds., *Science Observed.* Beverly Hills, Calif.: Sage.

Barnes, B., and J. Law. 1976. "Whatever Should Be Done with Indexical Expression?" *Theory and Society* 3: 223–237.

Bartels, D. 1981. "Genes and Proteins: Two Historical Case Studies Employing the Methodology of Scientific Research Programmes." Ph.D. diss., Department of History and Philosophy of Science, University of New South Wales, Australia.

Bartels, D., and R. Jonston. 1984. "The Sociology of Goal-Directed Science: Recombinant DNA Research." *Metascience* 1: 37–45.

Becker, H. S. 1960. "Notes on the Concept of Commitment." *American Journal of Sociology* 66: 32–40.

———. 1967. "Whose Side Are We On?" *Social Problems* 14: 239–248.

———. 1974. "Art as Collective Action." *American Sociological Review* 39: 767–776.

———. 1978. "Arts and Crafts." *American Journal of Sociology* 83: 862–889.

———. 1982. *Art Worlds.* Berkeley: University of California Press.

———. 1986. "Telling about Society." In *Doing Things Together: Selected Papers.* Evanston, Ill.: Northwestern University Press.

———. 1994. " 'Foi por Acaso': Conceptualizing Coincidence." *Sociological Quarterly* 35 (May): 183.

Becker, H. S., and M. McCall, eds. 1990. *Symbolic Interaction and Cultural Studies.* Chicago: University of Chicago Press.

Bendifallah, S., and W. Scacchi. 1987. "Understanding Software Maintenance Work." *IEEE Transactions on Software Engineering,* SE-13, January.

Bennett, J. A. 1989. "A Viol of Water or a Wedge of Glass." In D. Gooding, T. Pinch, and S. Schaffer, eds., *The Uses of Experiment.* Cambridge: Cambridge University Press.

Benson, S. A., and R. K. Taylor. 1984. "A Rapid Small-Scale Procedure for Isolation of Phage [Lambda] DNA." *BioTechniques* 2: 126–127.

Bentley, A. F. 1926. *Relativity in Man and Society.* New York: Octagon Books.

———. 1954. *Inquiry into Inquiries: Essays in Social Theory.* Westport, Conn.: Greenwood Press.

Benz, C. C., and E. T. Liu, eds. 1993. *Oncogenes and Tumor Suppressor Genes in Human Malignancies.* Boston: Kluwer Academic Publishers.

Berenblum, I. 1952. *Man against Cancer: The Story of Cancer Research.* Baltimore: Johns Hopkins University Press.

Berenblum, I., and P. Shubik. 1947. "The Role of Croton Oil Applications, Associated with a Single Painting of a Carcinogen, in Tumour Induction of the Mouse's Skin." *British Journal of Cancer* 1: 379–382.

———. 1949. "The Persistence of Latent Tumour Cells Induced in the Mouse's Skin by a Single Application of 9:10-Dimethyl-1:2-Benzanthracene." *British Journal of Cancer* 3: 384–386.

Berg, P., D. Baltimore, H. W. Boyer, S. N. Cohen, R. W. Davis, D. S. Hogness, D. Nathans, R. Robin, J. D. Watson, S. Weissman, and N. D. Zinder. 1974. "Potential Biohazards of Recombinant DNA Molecules." *Science* 185: 303.

Bijker, W., and J. Law, eds. 1992. *Shaping Technology/Building Society.* Cambridge, Mass.: MIT Press.

"BioFeedback." 1994. *Biotechniques,* May/June, p. 126.

Bischoff, F., and G. Bryson. 1964. "Carcinogenesis through Solid State Surfaces." *Progress in Experimental Tumor Research* 5: 85–133.

Bishop, J. M. 1982. "Oncogenes." *Scientific American* 246: 80–92.

———. 1983. "Cellular Oncogenes and Retroviruses." *Annual Review of Biochemistry* 52: 301–354.

———. 1989. "Retroviruses and Oncogenes, II." Nobel Lecture, Stockholm, Sweden, December 8.

Bishop, J. M., and H. E. Varmus. 1982. "Functions and Origins of Retroviral Transforming Genes." In R. Weiss et al., eds., *Molecular Biology of Tumor Viruses,* 2d ed., *RNA Tumor Viruses,* pp. 999–1108. Cold Spring Harbor, N.Y.: Cold Spring Harbor Laboratory.

Bittner, J. J. 1936. "Some Possible Effects of Nursing on the Mammary Gland Tumor Incidence in Mice." *Science* 84: 162.

———. 1942. "Possible Relationship of the Estrogenic Hormones, Genetic Susceptibility, and Milk Influence in the Production of Mammary Cancer in Mice." *Cancer Research* 2: 710–721.

Bloom, A. D. 1972. "Induced Chromosomal Abberations in Man." In H. Harris and K. Hirschhorn, eds., *Advances in Human Genetics,* vol. 3, pp. 99–172. New York: Plenum Press.

Bloor, D. 1976. *Knowledge and Social Imagery.* London: Routledge and Kegan Paul.

Blumer, H. 1969. *Symbolic Interactionism: Perspective and Method.* Englewood Cliffs, N.J.: Prentice-Hall.

Boffey, P. M. 1986. "Dr. Marks' Crusade: Shaking Up Sloan-Kettering for a New Assault on Cancer." *New York Times Magazine,* April 26, pp. 25–31, 60–67.

Borofsky R. 1987. *Making History: Pukapukan and Anthropological Constructions of Knowledge.* New York: Cambridge University Press.

Bourdieu, P. 1977. *Outline of a Theory of Practice.* Cambridge: Cambridge University Press.

———. 1990. *The Logic of Practice.* Stanford, Calif.: Stanford University Press. (Originally published as *Le sens pratique,* Les Editions de Minuit, 1980.]

Boveri, T. 1929. *The Origin of Malignant Tumors.* Baltimore: Williams and Wilkins.

Brandon, R. N. 1990. *Adaptation and Environment.* Princeton, N.J.: Princeton University Press.

Brandon, R. N., and R. M. Burian, eds. 1984. *Genes, Organisms, Populations: Controversies over the Units of Selection.* Cambridge, Mass.: MIT Press.

Braun, A. C. 1969. *The Cancer Problem: A Critical Analysis and Modern Synthesis.* New York: Columbia University Press.

Britten, R. J. 1994. "Evidence That Most Human *Alu* Sequences Were Inserted in a Process That Ceased about *30* Million Years Ago." *Proceedings of the National Academy of Sciences (USA)* 91(13): 6148–6150.

Britten, R., D. Stout, and E. Davidson. 1989. "The Current Source of Human *Alu* Retroposons Is a Conserved Gene Shared with Old World Monkey." *Proceedings of the National Academy of Sciences (USA)* 86: 3718–3722.

Brown, D. D. 1991. "Views of the Genome Project" (letter to the editor). *Science* 251 (February 22): 854–855.

Brutlag, D. 1994. Presentation to the National Research Council on GenBank.

Bucher, R. 1962. "Pathology: A Study of Social Movements within a Profession." *Social Problems* 10: 40–51.

———. 1988. "On the Natural History of Health Care Occupations." *Work and Occupations* 15: 131–147.

Bucher, R., and A. L. Strauss. 1961. "Professions in Process." *American Journal of Sociology* 66: 325–334.

Bud, R. F. 1978. "Strategy in American Cancer Research after World War II: A Case Study." *Social Studies of Science* 8: 425–459.

Budiansky, S. 1985. "Cetus in Revolt over Gene-Splicing Royalties." *San Francisco Chronicle,* April 16.

Burian, Richard M. 1991. "Underappreciated Pathways toward Molecular Genetics." Paper presented at the colloquium "The Philosophy and History of Molecular Biology," Boston University, April 15–16.

Burke P., ed. 1991. *New Perspectives on Historical Writing.* 1986. Reprint, University Park: Pennsylvania State University Press.

Butler, J. 1990. *Gender Trouble: Feminism and the Subversion of Identity.* New York and London: Routledge.

———. 1993. *Bodies That Matter: On the Discursive Limits of "Sex."* New York and London: Routledge.

Cairns, J. 1978. *Cancer: Science and Society.* San Francisco: W. H. Freeman.

Cairns, J., G. S. Stent, and J. D. Watson, eds. 1966. *Phage and the Origins of Molecular Biology.* Cold Spring Harbor, N.Y.: Cold Spring Harbor Laboratory.

Callon, M. 1986a. "The Sociology of an Actor-Network." In M. Callon, J. Law, and A. Rip, eds., *Mapping the Dynamics of Science and Technology.* London: Macmillan.

———. 1986b. "Some Elements of a Sociology of Translation: Domestication of the Scallops and the Fishermen of St. Brieuc Bay." In J. Law, ed., *Power, Action, and Belief: Sociological Review Monograph.* London: Routledge and Kegan Paul.

———. 1987. "Society in the Making: The Study of Technology as a Tool for Sociological Analysis." In W. Bijker, T. Pinch, and T. Hughes, eds., *New Directions in the Sociology and History of Technology.* Cambridge, Mass.: MIT Press.

Callon, M., and B. Latour. 1981. "Unscrewing the Big Leviathan: How Actors Macro-Structure Reality and How Sociologists Help Them to Do So." In K. Knorr-Cetina and A. V. Cicourel, eds., *Advances in Social Theory and Methodology: Toward an Integration of Micro- and Macro-Sociologies.* London: Routledge and Kegan Paul.

———. 1992. "Don't Throw the Baby Out with the Bath School!" In A. Pickering, ed., *Science as Practice and Culture.* Chicago: University of Chicago Press.

Callon, M., J. Law, and A. Rip, eds. 1986. *Mapping the Dynamics of Science and Technology.* London: Macmillan.

Cambrosio, A., and P. Keating. 1983. "The Disciplinary Stake: The Case of Chronobiology." *Social Studies of Science* 13: 323–353.

———. 1988. "Going Monoclonal: Art, Science, and Magic in the Day-to-Day Use of Hybridoma Technology." *Social Problems* 35: 244–260.

———. 1992. "A Matter of FACS: Constituting Novel Entities in Immunology." *Medical Anthropology Quarterly* 6 (December): 362–384.

Cambrosio, A., D. Jacobi, and P. Keating. 1993. "Ehrlich's 'Beautiful Pictures' and the Controversial Beginnings of Immunological Imagery." *ISIS* 84: 662–699.

Casalino, L. 1991. "Decoding the Human Genome Project: An Interview with Evelyn Fox Keller." *Socialist Review* 21: 111–128.

Cherfas, J. 1982. *Man-Made Life: An Overview of the Science, Technology, and Commerce of Genetic Engineering.* New York: Pantheon Books.

Chubin, D. E., and T. Connolly. 1982. "Research Trails and Science Policies: Local and Extra-Local Negotiation of Scientific Work." In N. Elias, H. Martins, and R. Whitley, eds., *Scientific Establishments and Hierarchies: Sociology of the Sciences,* vol. 6, pp. 293–311. Dordrecht: Reidel.

Chubin, D. E., and K. E. Studer. 1978. "The Politics of Cancer." *Theory and Society.* 6: 55–74.

Cicourel, A. 1964. *Method and Measurement in Sociology.* New York: Free Press.

———. 1981. "Notes on the Integration of Micro- and Macro-Levels of Analysis." In K. Knorr-Cetina and A. V. Cicourel, eds., *Advances in Social Theory and Methodology: Toward an Integration of Micro- and Macro-Sociologies.* London: Routledge and Kegan Paul.

Ciurczak, E. 1986. "Oakland Research Institute Works against Breast Cancer." *Daily Californian,* January 15.

Clark, M. 1983. "Spotting the Cancer Genes." *Newsweek,* May 3, p. 84.

Clark, M., and D. Witherspoon. 1984. "Cancer: The Enemy Within." *Newsweek,* March 5, pp. 66–67.

Clark, M., M. Gosnell, D. Shapiro, and M. Hager. 1984. "Medicine: A Brave New World." *Newsweek,* March 5, pp. 64–70.

Clarke, A. E. 1985. "Emergence of the Reproductive Research Enterprise: A Sociology of Biological, Medical, and Agricultural Science in the United States, 1910–1940." Ph.D. diss., Department of Sociology, University of California, San Francisco.

———. 1987. "Research Materials in Reproductive Physiology in the United States, 1910–1940." In G. L. Geison, ed., *Physiology in the American Context.* Bethesda, Md.: American Physiological Society.

———. 1990a. "A Social Worlds Research Adventure: The Case of Reproductive Science." In S. Cozzens and T. Gieryn, eds., *Theories of Science in Society.* Bloomington: Indiana University Press.

———. 1990b. "Controversy and the Development of Reproductive Science." *Social Problems* 36: 18–37.

———. 1991. "Social Worlds Theory as Organization Theory." In D. Maines, ed., *Social Organization and Social Process: Festschrift in Honor of Anselm L. Strauss.* Hawthorne, N.Y.: Aldine de Gruyter.

Clarke, A., and J. H. Fujimura, eds. 1992. *The Right Tools for the Job: At Work in Twentieth-Century Life Sciences.* Princeton, N.J.: Princeton University Press.

Clause, B. 1993. "The Wistar Rat as a Right Choice: Establishing Mammalian Standards and the Ideal of Standardized Mammal." *Journal of the History of Biology* 26: 233–267.

Clifford, J., and G. E. Marcus, eds. 1986. *Writing Culture: The Poetics and Politics of Ethnography.* Berkeley: University of California Press.

CMSHG. 1988. *Report of the Committee on Mapping and Sequencing the Human Genome, Board on Basic Biology, Commission on Life Sciences, National Research Council.* Washington, D.C.: National Academy Press.

Collins, H. M. 1975. "The Seven Sexes: A Study in the Sociology of a Phenomenon, or the Replication of Experiments in Physics." *Sociology* 9: 205–224.

———. 1981. "Contributions to the Sociology of Scientific Knowledge." Ph.D. diss., Bath University, England.

———. 1985. *Changing Order: Replication and Induction in Scientific Practice.* Beverly Hills, Calif.: Sage.

———. 1992. *Artificial Experts: Social Knowledge and Intelligent Machines.* Cambridge, Mass.: MIT Press.

Collins, H. M., and T. Pinch. 1982. *Frames of Meaning: The Social Construction of Extraordinary Science.* London: Routledge and Kegan Paul.

Collins, H. M., and S. Yearley. 1992. "Epistemological Chicken." In A. Pickering, ed., *Science as Practice and Culture.* Chicago: University of Chicago Press.

Colwell, R. R., ed. 1989. *Biomolecular Data: A Resource in Transition.* Oxford: Oxford University Press.

Cooper, G. M. 1982. "Cellular Transforming Genes." *Science* 217: 801–806.

———. 1995. *Oncogenes,* 2d ed. Boston: Jones and Bartlett.

Crane, D. 1969. "Fashion in Science: Does It Exist?" *Social Problems* 16: 433–441.

———. 1972. *Invisible Colleges: Diffusion of Knowledge in Scientific Communities.* Chicago: University of Chicago Press.

Crick, F. H. C. 1958. "On Protein Synthesi." *Symposium of the Society of Experimental Biology* 12: 138–163.

Culliton, B. J. 1974. "HeLa Cells: Contaminating Cultures around the World." *Science,* June 7, p. 1059.

Darden, L. 1991. *Theory Change in Science: Stategies from Mendelian Genetics.* New York: Oxford University Press.

Darnell, James, Harvey Lodish, and David Baltimore. 1990. *Molecular Cell Biology,* 2d ed. New York: Scientific American Books.

Darnovsky, M. 1991. "Overhauling the Meaning Machines: An Interview with Donna Haraway." *Socialist Review* 21: 65–84.

Davis, J. 1991. *Mapping the Code: The Human Genome Project and the Choices of Modern Science.* New York: Wiley.

De Certeau, M. 1988. *The Writing of History.* New York: Columbia University Press.

Dening G. 1988. *History's Anthropology: The Death of William Gooch* (ASAO special publications, no. 2). Lanham, Md.: University Press of America.

Der, C. J., T. G. Krontiris, and G. M. Cooper. 1982. "Transforming Genes of Human Bladder and Lung Carcinoma Cell Lines Are Homologous to the Ras Genes of Harvey and Kirsten Sarcoma Viruses." *Proceedings of the National Academy of Sciences (USA)* 79: 3637–3640.

Dermer, G. B. 1983. Letter to the editor. *Science* 221: 318.

Derrida, J. 1978. *Writing and Difference,* trans. A. Bass. Chicago: University of Chicago Press.

De Saussure, F. 1983. *Course in General Linguistics,* trans. Roy Harris. London: Duckworth.

DeVita, V. T. 1984. "The Governance of Science at the National Cancer Institute: A Perspective on Misperceptions." In *Management Operations of the National Cancer Institute That Influence the Governance of Science* (National Cancer Institute Monograph 64). Bethesda, Md.: U.S. Department of Health and Human Services (NIH Publication no. 84–2651).

Dewey, J. 1920. *Reconstruction in Philosophy.* New York: Henry Holt.

———. 1925. *Experience and Nature.* LaSalle, Ill.: Open Court Publishing Company.

———. 1929. *The Quest for Certainty: A Study of the Relation of Knowledge and Action.* New York: Minton, Balch.

———. 1938. *Logic: The Theory of Inquiry.* New York: Henry Holt.

Dixon, M. 1970. "The History of Enzymes and of Biological Oxidations." In J. Needham, ed., *The Chemistry of Life: Lectures on the History of Biochemistry.* Cambridge: Cambridge University Press.

Dodier, N. 1990. "How Are Medical Judgements Transcribed? Occupational Physicians and Administrative Protocols." Paper prepared for the workshop "The Social Construction of Health," Bielefeld, Germany, September.

———. 1991. "Agir Dans Plusieurs Mondes." *Critique* 529–530: 427–458.

Doolittle, R. F., M. W. Hunkapiller, L. E. Hood, S. G. DeVare, K. C. Robbins, S. A. Aaronson, and H. N. Antoniades. 1983. "Simian Sarcoma Virus *Onc* Gene, V-*sis,* Is Derived from the Gene (or Genes) Encoding a Platelet-Derived Growth Factor." *Science* 221: 275–276.

Draper, E. 1991. *Risky Business: Genetic Testing and Exclusionary Practices in the Workplace.* New York: Cambridge University Press.

Duesberg, P. H. 1968. "Physical Properties of Rous Sarcoma Virus RNA." *Proceedings of the National Academy of Sciences (USA)* 94: 1511–1518.
———. 1983. "Retroviral Transforming Genes in Normal Cells?" *Nature* 304: 219–225.
———. 1985. "Activated Proto-Onc Genes: Sufficient or Necessary for Cancer?" *Science* 228: 669–677.
———. 1987a. "Cancer Genes: Rare Recombinants instead of Activated Oncogenes" (a review). *Proceedings of the National Academy of Sciences (USA)* 94: 2117–2124.
———. 1987b. "Retroviruses as Carcinogens and Pathogens: Expectations and Reality." *Cancer Research* 47: 1199–220.
Duesberg, P. H., and J. Schwartz. 1992. "Latent Viruses and Mutated Oncogenes: No Evidence for Pathogenicity." *Progress in Nucleic Acid Research and Molecular Biology* 43: 135–204.
Duesberg, P. H., and P. K. Vogt. 1970. "Differences between the Ribonucleic Acids of Transforming and Nontransforming Avian Tumor Viruses." *Proceedings of the National Academy of Sciences (USA)* 67: 1673–1680.
Duesberg, P. H., G. S. Martin, and P. K. Vogt. 1970. "Glycoprotein Components of Avian and Murine RNA Tumor Virues." *Virology* 41 (March 30): 631–646.
Dulbecco, R. 1976. "From the Molecular Biology of Oncogenic DNA Viruses to Cancer." *Science* 192: 436–440.
———. 1986. "A Turning Point in Cancer Research: Sequencing the Human Genome." *Science* 231: 1055–1056.
Dulbecco, R., and M. G. Stoker. 1970. "Conditions Determining Initiation of DNA Synthesis in 3T3 Cells." *Proceedings of the National Academy of Sciences (USA)* 66(1): 204–210.
Dulbecco, R., and M. Vogt. 1954. "Plaque Formation and Isolation of Pure Lines with Poliomyelitis Viruses." *Journal of Experimental Medicine* 99: 176–177.
Dunham, L. J. 1972. "Cancer in Man at a Site of Prior Benign Lesion of Skin or Mucous Membrane: A Review." *Cancer Research* 32: 1359–1374.
Dupré, John. 1993. *The Disorder of Things: Metaphysical Foundations of the Disunity of Science.* Cambridge, Mass.: Harvard University Press.
Duster, T. 1981. "Intermediate Steps between Micro- and Macro-Integration: The Case of Screening for Inherited Disorders." In K. Knorr-Cetina and A. V. Cicourel, eds., *Advances in Social Theory and Methodology: Toward an Integration of Micro- and Macro-Sociologies.* London: Routledge and Kegan Paul.
———. 1990. *Eugenics through the Back Door.* Berkeley: University of California Press.
Emerson, R. 1983. *Contemporary Field Research.* Boston: Little Brown.
Epstein, S. S. 1978. *The Politics of Cancer.* Garden City, N.Y.: Anchor Press.
Farmer, P., and J. Walker. 1985. *The Molecular Basis of Cancer.* New York: Wiley.
Fausto-Sterling, A. 1985. *Myths of Gender: Biological Theories about Women and Men.* New York: Basic Books.
Feyerabend, P. 1975. *Against Method.* London: New Left Books.
———. 1987. *Farewell to Reason.* London: Verso.

Finch, C. E. 1990. *Longevity, Senescence, and the Genome.* Chicago: University of Chicago Press.
Fischer, M. 1991. "Anthropology as Cultural Critique: Inserts for the 1990s Cultural Studies of Science, Visual-Virtual Realities, and Post-Trauma Polities." *Cultural Anthropology* 6: 525–537.
Fleck, L. 1979. *Genesis and Development of a Scientific Fact.* 1935. Reprint, Chicago: University of Chicago Press.
Fleming, D. 1962. "Emigre Physicists and the Biological Revolution." In D. Fleming and B. Bailyn, eds., *The Intellectual Migration: Europe and America, 1930–1960,* pp. 152–189. Cambridge, Mass.: Harvard University Press.
Fogle, Thomas. 1990. "Are Genes Units of Inheritance?" *Biology and Philosophy* 5: 349–371.
Foreman, J. 1990. *Boston Globe.*
Fortun, Michael A. 1993. "Mapping and Making Genes and Histories: The Genomics Project in the United States, 1980–1990." Ph.D. diss., Harvard University, Cambridge, Mass.
Foucault, M. 1970. *The Order of Things: An Archaeology of the Human Sciences.* (Translation of *Les mots et les choses.*) New York: Random House.
———. 1975. *The Birth of the Clinic: An Archaeology of Medical Perception.* 1963. Reprint, New York: Vintage Books.
———. 1978. *The History of Sexuality.* Vol. 1. New York: Pantheon.
———. 1980a. *Power/Knowledge; Selected Interviews and Other Writings, 1972–77,* ed. C. Gordon. New York: Pantheon.
———. 1980b. *Language, Counter-Memory, Practice,* ed. D. F. Bouchard. Ithaca, N.Y.: Cornell University Press.
Franks, F. 1981. *Polywater.* Cambridge, Mass.: MIT Press.
Freidson, E. 1970. *Profession of Medicine.* New York: Dodd, Mead.
Friedewald, W. F., and P. Rous. 1944. "The Initiating and Promoting Elements in Tumor Production." *Journal of Experimental Medicine* 80: 101–126.
Friedland, P., and L. H. Kedes. 1985. "Discovering the Secrets of DNA." *Communications of the ACM* 28: 1164–1186.
Fuerst, J. A. 1982. "The Role of Reductionism in the Development of Molecular Biology: Peripheral or Central?" *Social Studies of Science* 12: 241–278.
Fujimura, J. H. 1987. "Constructing Doable Problems in Cancer Research: Articulating Alignment." *Social Studies of Science* 17: 257–293.
———. 1988. "The Molecular Biological Bandwagon in Cancer Research: Where Social Worlds Meet." *Social Problems* 35: 261–283.
———. 1989. "Denaturing DNA and Deleting Society: Strategies for Constructing DNA as a Tool for Rationalized Problem-Solving." Paper presented to the International Society for the History, Philosophy, and Social Studies of Biology, London, Ontario.
———. 1991a. "On Methods, Ontologies, and Representation in the Sociology of Science: Where Do We Stand?" In D. Maines, ed., *Social Organization and Social Process: Festschrift in Honor of Anselm L. Strauss.* Hawthorne, N.Y.: Aldine de Gruyter.

———. 1991b. "Reconstructing Nature in Bytes: The Molecular Genetic Sequence Databases." Paper presented at the conference "Genes R Us, But Who Is That?" at the University of California Humanities Research Institute, Irvine, California, May 2–3.

———. 1991c. "Representing Science and Representing Genes: On Methods and Theory in the Sociology of Science and Medicine." Paper presented to the Department of Social and Behavioral Sciences, University of California, San Francisco, February 6.

———. 1992. "Crafting Science: Standardized Packages, Boundary Objects, and 'Translations.' " In A. Pickering, ed., *Science as Practice and Culture.* Chicago: University of Chicago Press.

———. 1994. "Culture(s) in Bioinformatics." Paper presented at the workshop "Biology, Computers, and Society: At the Intersection of the 'Real' and the 'Virtual,' " Stanford, Calif., June 2–4.

———. 1995. "Ecologies of Action: Recombining Genes, Molecularizing Cancer, and Transforming Biology." In S. Leigh Star, ed., *Ecologies of Knowledge: New Directions in the Sociology of Science and Technology.* Albany: State University of New York Press.

———. 1996. "The Practices and Politics of Producing Meaning in the Human Genome Project." *Sociology of Science Yearbook,* vol. 19.

Fujimura, J. H., and D. Chou. 1994. "Dissent in Science: Styles of Scientific Practice and the Controversy over the Cause of AIDS." *Social Science and Medicine* 38 (April): 1017–1036.

Fujimura, J. H., and M. Fortun. 1996. "Constructing Knowledge across Social Worlds: The Case of DNA Sequence Databases in Molecular Biology." In Laura Nader, ed., *Naked Science: Anthropological Inquiry into Boundaries, Power, and Knowledge.* New York and London: Routledge.

Fujimura, J. H., S. L. Star, and E. M. Gerson. 1987. "Methodes de recherche en sociologie des sciences: Travail, pragmatisme et interactionnisme symbolique" (Research methods in the sociology of science and technology: Work, pragmatism, and symbolic interactionism). *Cahiers de Recherche Sociologique* 5: 65–85.

Fuller, S. 1992. "Social Epistemology and the Research Agenda of Science Studies." In A. Pickering, ed., *Science as Practice and Culture.* Chicago: University of Chicago Press.

Furmanski, P., J. C. Hager, and M. A. Rich, eds. 1985. *RNA Tumor Viruses, Oncogenes, Human Cancer, and Aids—On the Frontiers of Understanding: Proceedings of the International Conference on RNA Tumor Viruses in Human Cancer* (Denver, Colorado, June 10–14, 1984). Boston: Martinus Nijhoff.

Galison, P. 1987. *How Experiments End.* Chicago: University of Chicago Press.

Gallo, R. C. 1986. "The First Human Retrovirus." *Scientific American* 255: 88–98.

Garfinkel, H. 1967. *Studies in Ethnomethodology.* Englewood Cliffs, N.J.: Prentice-Hall.

———. 1982. *Ethnomethodological Studies of Work in the Discovering Sciences.* London: Routledge and Kegan Paul.

Garfinkel, H., M. Lynch, and E. Livingston. 1981. "The Work of Discovering Science Construed with Materials from the Optically Discovered Pulsar." *Philosophy of the Social Sciences* 11: 131–158.

Gasser, L. 1984. "The Social Dynamics of Routine Computer Use in Complex Organizations." Ph.D. diss., Department of Information and Computer Science, University of California, Irvine.

Gerson, E. M. 1983. "Scientific Work and Social Worlds." *Knowledge* 4: 357–377.

Gieryn, T. F. 1983. "Boundary Work and the Demarcation of Science from Non-Science: Strains and Interests in Professional Ideologies of Scientists." *American Sociological Review* 48: 781–785.

Gieryn, T. F., and A. Figert. 1990. "Ingredients for a Theory of Science in Society: O-rings, Ice Water, C-clamps, Richard Feynman, and the Press." In S. Cozzens and T. Gieryn, eds., *Theories of Science in Society.* Bloomington: Indiana University Press.

Gilbert, G. 1981. "DNA Sequencing and Gene Structure." *Science* 214: 1305–1312.

Gilbert, N., and M. Mulkay. 1984. *Opening Pandora's Box: A Sociological Analysis of Scientists' Discourse.* Cambridge: Cambridge University Press.

Gilbert, S. F. 1982. "Intellectual Traditions in the Life Sciences: Molecular Biology and Biochemistry." *Perspectives in Biology and Medicine* 26 (Autumn): 151–162.

———. 1991. "Enzymatic Adaptation and the Entrance of Molecular Biology into Embryology." Paper presented at the colloquium "The Philosophy and History of Molecular Biology," Boston University, April 15–16.

Gilbert, Walter. 1991. "Towards a Paradigm Shift in Biology." *Nature* 349 (January 10): 99.

Gilroy, Paul. 1991. *"There Ain't No Black in the Union Jack": The Cultural Politics of Race and Nation.* Chicago: University of Chicago Press.

———. 1993. *The Black Atlantic: Modernity and Double Consciousness.* Cambridge, Mass.: Harvard University Press.

Goodfield, J. 1975. *The Siege of Cancer.* New York: Random House.

Gooding, D. 1990. "Mapping Experiment as a Learning Process: How the First Electromagnetic Motor Was Invented." *Science, Technology, and Human Values* 15: 165–201.

Gooding, D., T. Pinch, and S. Schaffer. 1989. *The Uses of Experiment: Studies in the Natural Sciences.* New York: Cambridge University Press.

Gold, M. 1986. *A Conspiracy of Cells: One Woman's Immortal Legacy and the Medical Scandal It Caused.* Albany: State University of New York Press.

Goldfarb, M., K. Shimizu, M. Perucho, and M. Wigler. 1982. "Isolation and Preliminary Characterization of a Human Transforming Gene from T24 Bladder Carcinoma Cells." *Nature* 296: 404–409.

Gramsci, A. 1985. *Selections from Cultural Writings,* ed. D. Forgacs and G. Nowell-Smith, trans. W. Boelhower. Cambridge, Mass.: Harvard University Press.

Gregory, K. L. 1984. "Signing-up: The Culture and Careers of Silicon Valley Computer People." Ph.D. diss., Department of Anthropology, Northwestern University, Evanston, Ill.

Griesemer, J. R. 1991. "Weismannism and the Interpretation of Genetic Information." Paper presented at the conference "Genes-R-Us, But Who Is That?" University of California Humanities Research Institute, Irvine, May 2–3.

———. 1992. "The Role of Instruments in the Generative Analysis of Science." In A. E. Clarke and J. H. Fujimura, eds., *The Right Tools for the Job: At Work in Twentieth-Century Life Sciences.* Princeton, N.J.: Princeton University Press.

———. 1994. "Visual Tools for Talking: Genetic Determination and the Social Hybridization of DNA." Paper presented at the Henry R. Luce Colloquium Series on Biotechnology and Society, Stanford University, Stanford, Calif., May 26.

Griesemer, J., and W. C. Wimsatt. 1989. "Picturing Weismannism: A Case Study of Conceptual Evolution." In M. Ruse, ed., *What the Philosophy of Biology Is,* pp. 75–137. Dordrecht: Kluwer Academic Publishers.

Gross, L. 1957. "Development and Serial Cell-Free Passage of a Highly Potent Strain of Mouse Leukemia Virus." *Proceedings of the Society of Experimental Biology and Medicine* 94: 767–771.

———. 1961. "Induction of Leukemia in Rats with Mouse Leukemia (Passage A) Virus." *Proceedings of the Society of Experimental Biology and Medicine* 106: 890–983.

———. 1970. *Oncogenic Viruses.* 1961. Reprint, New York: Pergamon Press.

———. 1974. "The Role of Viruses in the Etiology of Cancer and Leukemia." *Journal of the American Medicial Association* 230: 1029–1032.

Gupta, A., and J. Ferguson. 1992. "Beyond 'Culture': Space, Identity, and the Politics of Difference." *Cultural Anthropology* 7: 6–23.

Gussak, L. S. 1986. "Biosynthetic Human Insulin Four Years Later." *GeneWATCH* 3: 6–8.

Habeshaw, J. A. 1983. Letter to the editor. *Nature* 301: 652.

Hacking, I. 1983. *Representing and Intervening.* Cambridge: Cambridge University Press.

———. 1988. "On the Stability of the Laboratory Sciences." *Journal of Philosophy* 85(10): 507–514.

———. 1992. "The Self-Vindication of the Laboratory Sciences." In A. Pickering, ed., *Science as Practice and Culture.* Chicago: University of Chicago Press.

Hall, B. K. 1994. *Homology: The Hierarchial Basis of Comparative Biology.* San Diego, Calif.: Academic Press.

Haller, M. H. 1963. *Eugenics: Hereditarian Attitudes in American Thought.* New Brunswick, N.J.: Rutgers University Press.

Hanafusa, H., and T. Hanafusa. 1971. "Noninfectious RSV Deficient in DNA Polymerase." *Virology* 43: 313–316.

Hanson, N. R. 1958. *Patterns of Discovery.* Cambridge: Cambridge University Press.

Haraway, D. 1981–82. "The High Cost of Information in Post–World War II Evolutionary Biology." *Philosophical Forum* 13: 244–278.

———. 1985. "A Manifesto for Cyborgs: Science, Technology, and Socialist Feminism in the 1980s." *Social Review* 15.

———. 1988. "Situated Knowledges: The Science Question in Feminism and the Privilege of Partial Perspective." *Feminist Studies* 14: 575–599.

———. 1989. *Primate Visions: Gender, Race, and Nature in the World of Modern Science.* New York and London: Routledge.

———. 1991. *Simians, Cyborgs, and Women: The Reinvention of Nature.* New York and London: Routledge.

———. 1992. "When Man Is on the Menu." In J. Crary and S. Kwinter, eds., *Incorporations.* New York: Zone Press.

———. 1993. "The Promises of Monsters: A Regenerative Politics for Inappropriate/d Others." In P. Treichler, C. Nelson, and L. Grossberg, eds., *Cultural Studies Now and in the Future.* New York and London: Routledge.

———. 1994. "A Game of Cat's Cradle: Science Studies, Feminist Theory, Cultural Studies." *Configurations* 1: 59–71.

———. 1995. "Universal Donors in a Vampire Culture—It's All in the Family: Biological Kinship Categories in the Twentieth-Century United States." In W. Cronon, ed., *Uncommon Ground: Toward Reinventing Nature.* New York: Norton.

Harding, S. 1986. *The Science Question in Feminism.* Ithaca, N.Y.: Cornell University Press.

———. 1993. *The "Racial" Economy of Science: Toward a Democratic Future.* Bloomington: Indiana University Press.

———. 1994. "Is Science Multicultural? Challenges, Resources, Opportunities, Uncertainties." *Configurations* 2: 301–330.

Harris, M. 1964. *Cell Culture and Somatic Variation.* New York: Holt, Rinehart and Winston.

Harrison, R. 1907. "Observations on the Living Developing Nerve Fibre." *Proceedings of the Society of Experimental Biology (N.Y.)* 4: 140.

Hayles, N. Katherine. 1984. *The Cosmic Web: Scientific Field Models and Literary Strategies in the Twentieth Century.* Ithaca, N.Y.: Cornell University Press.

———. 1990. "Designs on the Body: Norbert Wiener, Cybernetics, and the Play of Metaphor." *History of the Human Sciences* 3: 211–228.

Hess, D., and L. Layne, eds. 1992. *Knowledge and Society: The Anthropology of Science and Technology.* Greenwich, Conn.: JAI Press.

Hewitt, C. 1985. "The Challenge of Open Systems." *Byte,* pp. 223–242.

Hine, C. 1994. "Scientific Instruments and Social Boundaries: The Use of IT in UK Genome Research." Paper presented at the workshop entitled "Biology, Computers, and Society: At the Intersection of the 'Real' and the 'Virtual,' " Stanford, Calif., June 2–4.

Hirschauer, S. 1991. "The Manufacture of Bodies in Surgery." *Social Studies of Science* 21: 279–320.

Holmes, F. L. 1985. *Lavoisier and the Chemistry of Life: An Exploration of Scientific Creativity.* Madison: University of Wisconsin Press.

———. 1992. "Manometers, Tissue Slices, and Intermediary Metabolism." In A. E. Clarke and J. H. Fujimura, eds., *The Right Tools for the Job: At Work in Twentieth-Century Life Sciences.* Princeton, N.J.: Princeton University Press.

Holtzman, N. 1989. *Proceed with Caution: Predicting Genetic Risks in the Recombinant DNA Era.* Baltimore, Md.: Johns Hopkins University Press.

Hornstein, G., and S. Star. 1990. "University Biases: How Theories about Human Nature Succeed." *Philosophy of the Social Sciences* 20: 421–436.

Hubbard, R. 1990. *The Politics of Women's Biology.* New Brunswick, N.J.: Rutgers University Press.

Hubbard, R., and E. Wald. 1993. *Exploding the Gene Myth.* Boston: Beacon Press.

Huebner, R. J., and G. J. Todaro. 1969. "Oncogenes of RNA Tumor Viruses as Determinants of Cancer." *Proceedings of the National Academy of Sciences (USA)* 64: 1087–1094.

Huggins, C. B. 1979. *Experimental Leukemia and Mammary Cancer: Induction, Prevention, Cure.* Chicago: University of Chicago Press.

Hughes, E. C. 1971. *The Sociological Eye.* Chicago: Aldine.

Hughes, T. P. 1986. "The Seamless Web: Technology, Science, Etcetera, Etcetera." *Social Studies of Science* 16: 281–292.

Hunter, T. 1984. "The Proteins of Oncogenes." *Scientific American* 251: 70–79.

Itakura, K., T. Hirose, R. Crea, A. D. Riggs, H. L. Heyneker, F. F. Bolivar, and H. W. Boyer. 1977. "Expression in *Escherichia Coli* of a Chemically Synthesized Gene for the Hormone Somatostatin." *Science* 198: 1056–1063.

Jackson, D. A., and S. P. Stich, eds. 1979. *The Recombinant DNA Debate.* Englewood Cliffs, N.J.: Prentice-Hall.

Jacob, F. 1982. *The Logic of Life: A History of Heredity.* New York: Pantheon.

James, W. 1907. *Pragmatism: A New Name for Some Old Ways of Thinking.* New York: Longmans, Green.

———. 1914. *The Meaning of Truth.* New York: Longmans, Green.

Johnson, R. S. 1984. "Oncor, Oncogene Diagnostics Venture, Is 'Encore' for BRL Cofounder Turner." *Genetic Engineering News* 4.

Jordan, K., and M. Lynch. 1992. "The Sociology of a Genetic Engineering Technique: Ritual and Rationality in the Performance of the Plasmid Prep." In A. E. Clarke and J. H. Fujimura, eds., *The Right Tools for the Job: At Work in Twentieth-Century Life Sciences.* Princeton, N.J.: Princeton University Press.

———. 1993. "The Mainstreaming of a Molecular Biological Tool: A Case Study of a New Technique." In G. Button, ed., *Technology in Working Order: Studies in Work, Interaction and Technology,* pp. 160–180. London: Routledge and Kegan Paul.

Judson, H. F. 1979. *The Eighth Day of Creation.* New York: Simon and Schuster.

———. 1992. "A History of the Science and Technology behind Gene Mapping and Sequencing." In D. J. Kevles and L. Hood, eds., *The Code of Codes: Scientific and Social Issues in the Human Genome Project,* pp. 37–80. Cambridge, Mass.: Harvard University Press.

Juma, C. 1989. *The Gene Hunters: Biotechnology and the Scramble for Seeds.* Princeton, N.J.: Princeton University Press.

Karp, R. D., K. H. Johnson, L. C. Bouen, H. K. Ghobrial, I. Brand, and K. G. Brand. 1973. "Tumorigenesis by Millipore Filters in Mice: Histology and Ultrastructure of Tissue Reactions as Related to Pore Size." *Journal of the National Cancer Institute* 51: 1275–1279.

Kauffman, S. A. 1971. "Articulation of Parts Explanation in Biology and the Rational Search for Them." In R. Buck and R. Cohen, eds., *PSA 1970,* pp. 257–272. Dordrecht: Reidel.

Kay, L. 1991. "Representing, Intervening, and Molecularizing." Paper presented at the colloquium "The Philosophy and History of Molecular Biology," Boston University, April 15–16.

———. 1993. *The Molecular Vision of Life: Caltech, the Rockefeller Foundation, and the Rise of the New Biology.* New York: Oxford University Press.

Keating, P., A. Cambrosio, and M. MacKenzie. 1992. "Tools of the Discipline? Standards, Models, and Measures in the Affinity-Avidity Controversy in Immunology." In A. E. Clarke and J. H. Fujimura, eds., *The Right Tools for the Job: At Work in Twentieth-Century Life Sciences.* Princeton, N.J.: Princeton University Press.

Keller, E. F. 1983. *A Feeling for the Organism.* San Francisco: W. H. Freeman.

———. 1985. *Reflections on Gender and Science.* New Haven: Yale University Press.

———. 1991. "Genetics, Reductionism, and the Normative Uses of Biological Information." Paper presented at the conference "Genes-R-Us, But Who Is That?" University of California Humanities Research Institute, Irvine, May 2–3.

———. 1992a. "Nature, Nurture, and the Human Genome Project." In D. J. Kevles and L. Hood, eds., *The Code of Codes: Scientific and Social Issues in the Human Genome Project.* Cambridge, Mass.: Harvard University Press.

———. 1992b. *Secrets of Life, Secrets of Death: Essays on Language, Gender, and Science.* New York and London: Routledge.

———. 1995. *Refiguring Life: Changing Metaphors of Twentieth-Century Biology.* New York: Columbia University Press.

Keller, E. F., and E. A. Lloyd, eds. 1992. *Keywords in Evolutionary Biology.* Cambridge, Mass.: Harvard University Press.

Kenney, M. 1986. *Biotechnology: The University-Industry Complex.* New Haven: Yale University Press.

———. 1987. "The Impact of the International Political Economy on National Biotechnology Programs." Paper presented at the National Symposium on the Role of Biotechnology in Crop Protection, Kalyani, West Bengal, India, January.

Kevles, D. J. 1985. *In the Name of Eugenics: Genetics and the Uses of Human Heredity.* New York: Alfred A. Knopf.

———. 1992. "Renato Dulbecco and the New Animal Virology: Medicine, Methods, and Molecules." Paper presented at the Conference "Building Molecular Biology: Comparative Studies of Ideas, Institutions, and Practices," Massachusetts Institute of Technology, Cambridge, April 3–4.

Kevles, D. J., and L. Hood, eds. 1992. *The Code of Codes: Scientific and Social Issues in the Human Genome Project.* Cambridge, Mass.: Harvard University Press.

Kimmelman, B. 1983. "The American Breeders' Association: Genetics and Eugenics in an Agricultural Context." *Social Studies of Science* 13: 163–204.

———. 1987. "The Progressive Era Discipline: Genetics at American Agricultural Colleges and Experiment Stations, 1890–1920." Ph.D. diss., Department of History of Science, University of Pennsylvania.

———. 1992. "Organisms and Interests in Scientific Research: R. A. Emerson's Claims for the Unique Contributions of Agricultural Genetics." In A. Clarke and J. H. Fujimura, eds., *The Right Tools for the Job: At Work in Twentieth-Century Life Sciences.* Princeton, N.J.: Princeton University Press.

King, M.-C., et al. 1990. "Linkage of Early-Onset Familial Breast Cancer to Chromosome 17q21." *Science* 250: 1684–1689.

Kitcher, P. 1982. "Genes." *British Journal of Philosophy of Science* 33: 337–359.

———. 1992. "Gene: Current Usages." In E. F. Keller and E. A. Lloyd, eds., *Keywords in Evolutionary Biology,* pp. 128–131. Cambridge, Mass.: Harvard University Press.

Knorr-Cetina, K. 1979. "Tinkering toward Success: Prelude to a Theory of Scientific Practice." *Theory and Society* 8: 347–376.

———. 1981. *The Manufacture of Knowledge.* Oxford: Pergamon Press.

———. 1982. "Scientific Communities or Transepistemic Arenas of Research: A Critique of Quasi-Economic Models of Science." *Social Studies of Science* 12: 101–130.

Kodama, F., and C. Nishigata. 1991. "Structural Changes in the Japanese Supply/Employment Systems of Engineers: Are We Losing or Gaining?" in D. S. Zinberg, ed., *The Changing University,* pp. 101–128. Dordrecht: Kluwer Academic Publishers.

Koenig, R. 1985. "Technology: Product Payoffs Prove Elusive after a Cancer Research Gain." *Wall Street Journal,* June 28, p. 25.

Kohler, R. E. 1993. "Drosophila: A Life in the Laboratory." Special issue of the *Journal of the History of Biology* 26: 233–367.

Kondo, D. K. 1990. *Crafting Selves: Power, Gender, and Discourses of Identity in a Japanese Workplace.* Chicago: University of Chicago Press.

Kong, D. 1989. *Boston Globe.* October 10.

Koop, B., and L. Hood. 1994. "Striking Sequence Similarity over Almost 100 Kilobases of Human and Mouse T-cell Receptor DNA." *Nature Genetics* 7: 48–53.

Kranakis, E. 1989. "Social Determinants of Engineering Practice: A Comparative View of France and America in the Nineteenth Century." *Social Studies of Science* 19: 5–70.

———. 1992. "Hybrid Careers and the Interaction of Science and Technology." In P. Kroes and M. Bakker, eds., *Technological Development and Science in the Industrial Age.* Dordrecht: Kluwer Academic Publishers.

Krimsky, S. 1982. *Genetic Alchemy: The Social History of the Recombinant DNA Controversy.* Cambridge, Mass.: MIT Press.

Krimsky, S., J. Ennis, and R. Weissman. 1991. "Academic-Corporate Ties in Biotechnology: A Quantitative Study." *Science, Technology, and Human Values.* 16: 275–287.

Kuhn, T. S. 1957. *The Copernican Revolution: Planetary Astronomy in the Development of Western Thought.* Cambridge, Mass.: Harvard University Press.

———. 1970. *The Structure of Scientific Revolutions.* 1962. Reprint, Chicago: University of Chicago Press.

Kuklick, B. 1977. *The Rise of American Philosophy: Cambridge, Massachusetts, 1860–1930.* New Haven: Yale University Press.

Kumar, A., ed. 1984. *Eukaryotic Gene Expression.* New York: Plenum Press.

Lai, M. M. C., and P. H. Duesberg. 1972a. "Adenylic Acid-Rich Sequence in RNAs of Rous Sarcoma Virus and Rauscher Mouse Leukemia Virus." *Nature* 135 (February 18): 383–386.

———. 1972b. "Differences between the Envelope Glycoproteins and Glycopeptides of Avian Tumor Viruses Released from Transformed and from Nontransformed Cells." *Virology* 50: 359–372.

Lai, M. M. C., P. H. Duesberg, J. Horst, and P. K. Vogt. 1973. "Avian Tumor Virus RNA: A Comparison of Three Sarcoma Viruses and Their Transformation-Defective Derivatives by Oligonucleotide Fingerprinting and DNA-RNA Hybridization." *Proceedings of the National Academy of Sciences (USA)* 70(8): 2266–7220.

Lakatos, I. 1978. *The Methodology of Scientific Research Programmes: Philosophical Papers,* vol. 1, ed. J. Worrall and G. Currie. Cambridge: Cambridge University Press.

Lakatos, I., and A. Musgrave, eds. 1970. *Criticism and the Growth of Knowledge.* New York: Cambridge University Press.

Land, H., L. F. Parada, and R. A. Weinberg. 1983. "Cellular Oncogenes and Multistep Carcinogenesis." *Science* 222: 771–778.

———. 1984. "Cellular Oncogenes and Multistep Carcinogenesis." In P. A. Abelson, ed., *Biotechnology and Biological Frontiers.* Washington, D.C.: American Association for the Advancement of Science.

Lappe, M. 1984. *Broken Code: The Exploitation of DNA.* San Francisco: Sierra Club Books.

Laquer, T. 1990. *Making Sex: Body and Gender from the Greeks to Freud.* Cambridge, Mass.: Harvard University Press.

Latour, B. 1983. "Give Me a Laboratory and I Will Raise the World." In K. Knorr-Cetina and M. Mulkay, eds., *Science Observed.* Beverly Hills, Calif.: Sage.

———. 1986. "A Relativistic Approach to Einstein's Relativity." Paper presented at the Eleventh Annual Meeting of the Social Studies of Science, Pittsburgh, Penn., October.

———. 1987. *Science in Action: How to Follow Scientists and Engineers through Society.* Cambridge, Mass.: Harvard University Press.

———. 1988a. *The Pasteurization of France.* Cambridge, Mass.: Harvard University Press. (Translation of *Les microbes: Guerre et paix suivi de irreductions,* A. M. Metailie, 1984.)

———. 1988b. "The Politics of Explanation: An Alternative." In S. Woolgar, ed., *Knowledge and Reflexivity: New Frontiers in the Sociology of Knowledge.* London: Sage.

———. 1989. "Do We Really Need the Notion of Ideology? A Case to Get Rid of the Notion by Using Pasteur's Historiography." Paper presented at the Conference on Ideology in the Life Sciences, Harvard University, April.

———. 1990a. "Drawing Things Together." In M. Lynch and S. Woolgar, eds., *Representation in Scientific Practice.* Cambridge, Mass.: MIT Press.

———. 1990b. "Are We Talking about Skills or about the Redistribution of Skills?" Paper presented at the conference "Rediscovering Skill in Science, Technology, and Medicine," Bath, England, September.

———. 1990c. "Postmodern? No, Simply Amodern! Steps towards an Anthropology of Science." *Studies in the History and Philosophy of Science* 21: 145–171.

———. 1993. *We Have Never Been Modern.* Cambridge, Mass.: Harvard University Press.

Latour, B., and S. Woolgar. 1986. *Laboratory Life: The Social Construction of Scientific Facts.* 1979. Reprint, Beverly Hills, Calif.: Sage.

Laudan, Larry. 1990. "Demystifying Underdetermination." In C. Wade Savage, ed., *Scientific Theories.* Minneapolis: University of Minnesota Press.

Lave, J. 1988. *Cognitition in Practice: Mind, Mathematics, and Culture in Everyday Life.* Cambridge: Cambridge University Press.

Law, J. 1973. "The Development of Specialties in Science: The Case of X-ray Crystallography." *Science Studies* 3: 275–303.

———. 1986. "On the Methods of Long-Distance Control: Vessels, Navigation, and the Portuguese Route to India." In J. Law, ed., *Power, Action, and Belief.* London: Routledge and Kegan Paul.

———. 1994. *Organizing Modernity.* Oxford: Blackwell Publishers.

Lenoir, T., and Y. Elkana, eds. 1988. "Practice, Context, and the Dialogue between Theory and Experiment." *Science in Context* 2: 3–232.

Lenoir, T., and C. Lécuyer. 1995. "Instrument Makers and Discipline Builders: The Case of Nuclear Magnetic Resonance." *Perspectives on Science* 3 (3): 276–345.

Lewin, R. 1986a. "DNA Sequencing Goes Automatic." *Science* 232: 24.

———. 1986b. "DNA Databases Are Swamped." *Science* 232: 1599.

———. 1986c. "Proposal to Sequence the Human Genome Stirs Debate." *Science* 232: 1598–1600.

Lewontin, R. C. 1992. *Biology as Ideology: The Doctrine of DNA.* New York: HarperPerennial.

Lippman, A. 1988. "Prenatal Genetic Screening and Testing: Constructing Needs and Reinforcing Inequities." *American Journal of Law and Medicine* 17: 15–50.

Little, C. C. 1924. "The Genetics of Tissue Transplantation in Mammals." *Journal of Cancer Research* 8: 75–95.

Little, C. C., and E. E. Tyzer. 1916. "Further Experimental Studies of the Susceptibility of Transplantable Tumor Carcinoma of the Japanese Waltzing Mouse." *Journal of Medical Research* 3: 393–427.

Littlefield, J. W. 1964. "Selection of Hybrids from Matings of Fibroblasts in Vitro and Their Presumed Recombinants." *Science* 145: 709–710.

———. 1982a. Letter to the editor. *Nature,* 301: 369.

———. 1982b. Letter to the editor. *Science,* 218: 214–216.

Lloyd, E. A. 1988. *The Structure and Confirmation of Evolutionary Theory.* Westport, Conn.: Greenwood Press.

———. 1991. "What Is Disease?" Paper presented at the conference "Genes-R-Us,

But Who Is That?" University of California Humanities Research Institute, Irvine, May 2–3.

Longino, H. E. 1990. *Science as Social Knowledge: Values and Objectivity in Scientific Inquiry.* Princeton, N.J.: Princeton University Press.

Löwy, I. 1990. "Variances in Meaning in Discovery Accounts: The Case of Contemporary Biology." *Historical Studies of the Physical and Biological Sciences* 21: 87–121.

———. 1993. "Experimental Systems and Clinical Practices: Tumor Immunology and Cancer Immunotherapy 1895–1980." Paper presented at the colloquium "Conceptual Issues in Immunology," Boston University, May 6.

Ludmerer, K. M. 1972. *Genetics and American Society: A Historical Appraisal.* Baltimore, Md.: Johns Hopkins University Press.

Luria, S. E. 1984. *A Slot Machine, a Broken Test Tube: An Autobiography.* New York: Harper and Row.

Lynch, M. 1985. *Art and Artefact in Laboratory Science.* London: Routledge and Kegan Paul.

———. 1991a. "Ordinary and Scientific Measurement as Ethnomethodological Phenomena." In G. Button, ed., *Ethnomethodology and the Human Sciences: A Foundational Reconstruction.* Cambridge: Cambridge University Press.

———. 1991b. "Science in the Age of Mechanical Reproduction: Moral and Epistemic Relations between Diagrams and Photographs." *Biology and Philosophy* 6 (2): 205–226.

———. 1992. "Extending Wittgenstein: The Piviotal Move from Epistemology to the Sociology of Science." In A. Pickering, ed., *Science as Practice and Culture.* Chicago: University of Chicago Press.

———. 1993. *Scientific Practice and Ordinary Action: Ethnomethodology and Social Studies of Science.* New York: Cambridge University Press.

Lynch, M., and S. Woolgar, eds. 1990. *Representation in Scientific Practice.* 1988. Reprint, Cambridge, Mass.: MIT Press.

Lyotard, J.-F. 1984. *The Postmodern Condition: A Report on Knowledge.* Minneapolis: University of Minnesota Press.

MacKenzie, D. 1981. *Statistics in Britain, 1865–1930.* Edinburgh: University of Edinburgh.

Macpherson, I., and L. Montagnier. 1964. "Agar Suspension Culture for the Selective Assay of Cells Transformed by Polyoma Virus." *Virology* 23: 291–294.

Maienschein, J. 1992. "Gene: Historical Perspectives." In E. F. Keller and E. A. Lloyd, eds., *Keywords in Evolutionary Biology,* pp. 122–127. Cambridge, Mass.: Harvard University Press.

Mani, L. 1992. "Multiple Mediations: Feminist Scholarship in the Age of Multinational Reception." In H. Crowley and S. Himmelweit, eds., *Knowing Women: Feminism and Knowledge.* Cambridge: Open University.

Maniatis, T., E. F. Fritsch, and J. Sambrook. 1982. *Molecular Cloning: A Laboratory Manual.* Cold Spring Harbor, N.Y.: Cold Spring Harbor Laboratory.

March, J., and H. Simon. 1958. *Organizations.* New York: Wiley.

Marcus, G. E., and M. J. Fischer. 1986. *Anthropology as Cultural Critique: An*

Experimental Moment in the Human Sciences. Chicago: University of Chicago Press.

Martin, E. 1987. *The Woman in the Body: A Cultural Analysis of Reproduction*. Boston: Beacon Press.

———. 1994. *Flexible Bodies: Tracking Immunity in American Culture from the Days of Polio to the Age of AIDS*. Boston: Beacon Press.

Martin, G. S. 1970. "Rous Sarcoma Virus: A Function Required for the Maintenance of the Transformed State." *Nature* 227: 1021–1023.

Martin, G. S., and P. H. Duesberg. 1972. "The Alpha Subunit in the RNA of Transforming Avian Tumor Viruses, I: Occurrence in the Different Strains," and "II: Spontaneous Loss Resulting in Nontransforming Variants." *Virology* 47: 494–497.

Marx, J. L. 1984. "Oncogene Overview: What Do Oncogenes Do?" *Science* 223: 673–676.

———. 1994. "Oncogenes Reach a Milestone (Research News)." *Science* 226: 1942–1944.

Maxam, A. M., and W. Gilbert. 1977. "A New Method for Sequencing DNA." *Procedures of the National Academy of Sciences (USA)* 74: 560–564.

McCann, J., E. Choi, E. Yamasaki, and B. N. Ames. 1975. "Detection of Carcinogens as Mutagens in the Salmonella Microsome Test: Assay of 300 Chemicals." *Proceedings of the National Academy of Sciences (USA)* 72: 5135–5139.

McCarty, M. 1985. *The Transforming Principle: Discovering that Genes Are Made of DNA*. New York: W. W. Norton.

McCormick, F. 1989. "*Ras* Oncogenes." In R. A. Weinberg, ed., *Oncogenes and the Molecular Origins of Cancer.* Cold Spring Harbor, N.Y.: Cold Spring Harbor Laboratory.

McLuhan, M. 1964. *Understanding Media: The Extensions of Man*. New York: McGraw Hill.

Mead, G. H. 1932. "The Objective Reality of Perspectives." In *Philosophy of the Present.* LaSalle, Ill.: Open Court Publishing.

———. 1934. *Mind, Self, and Society.* Chicago: University of Chicago Press.

———. 1938. *The Philosophy of the Act*. Chicago: University of Chicago Press.

Medawar, P. B. 1982. *Pluto's Republic*. Oxford: Oxford University Press.

Mendelsohn, E. 1989. "Robert K. Merton: The Celebration and Defense of Science." *Science in Context* 3: 3–302.

Merton, R. K. 1970. *Science, Technology, and Society in Seventeenth Century England.* 1938. Reprint, New York: Harper and Row.

———. 1973. *The Sociology of Science: Theoretical and Empirical Investigations*. Chicago: University of Chicago Press.

Miller, J. A. 1970. "Carcinogensis by Chemicals: An Overview." *Cancer Research* 30 (March): 559–576.

Miller, J. A., and E. C. Miller. 1969. "Metabolic Activation of Carcinogenic Aromatic Amines and Amides via N-hydroxylation and N-hydroxy Esterification and Its Relationship to Ultimate Carcinogens as Electrophilic Reactants." In E. Bergmann and E. Pullman, eds., *The Jerusalem Symposia on Quantum Chemistry*

and Biochemistry, vol. 1, *Physiochemical Mechanisms of Carcinogenesis*. Jerusalem: Israel Academy of Sciences and Humanities.
Miller, E. C., J. A. Miller, R. R. Brown, and J. C. MacDonald. 1958. "On the Protective Action of Certain Polycyclic Aromatic Hydrocarbons against Carcinogenesis by Aminoazo Dyes and 2-Acetylaminofluorene." *Cancer Research* 18: 469–477.
Mills, C. W. 1966. *Sociology and Pragmatism: The Higher Learning in America*. 1964. Reprint, New York: Oxford University Press.
Minh-ha, T. 1989. *Woman Native Other*. Bloomington: Indiana University Press.
Mintz, B., and K. Illmensee. 1975. "Normal Genetically Mosaic Mice Produced from Malignant Teratocarcinoma Cells." *Proceedings of the National Academy of Sciences (USA)* 72: 3585–3589.
Mittman, G., and A. Fausto-Sterling. 1992. "Whatever Happened to Planaria? C. M. Child and the Physiology of Inheritance." In A. E. Clarke and J. H. Fujimura, eds., *The Right Tools for the Job: At Work in Twentieth-Century Life Sciences*. Princeton, N.J.: Princeton University Press.
Mizutani, S., D. Boettiger, and H. M. Temin. 1970. "A DNA-Dependent DNA Polymerase and a DNA Endonuclease in Virions of Rous Sarcoma Virus." *Nature* 118 (October 31): 424–427.
Mol, A. 1991. "Where Is the Category? The Example of Arteriosclerosis." Paper presented at the Conference of the International Society for the History, Philosophy, and Social Studies of Biology, Evanston, Ill., July.
Morange, M. 1993. "The Discovery of Cellular Oncogenes." *History and Philosophy of the Life Sciences* 15: 45–58.
Morris, N. 1976. *The Cancer Blackout: A History of Denied and Suppressed Remedies, 1762–1976*. Los Angeles: Regent House.
Morrow, J. F., S. N. Cohen, A. C. Y. Chang, H. W. Boyer, H. M. Goodman, and R. B. Helling. 1974. "Replication and Transcription of Eucaryotic DNA in *Escherichia Coli*." *Proceedings of the National Academy of Sciences (USA)* 71: 1743–1747.
Moss, R. W. 1980. *The Cancer Syndrome*. New York: Grove Press.
———. 1989. *The Cancer Industry: Unraveling the Politics*. New York: Paragon House.
Mulkay, M., and G. Gilbert. 1982. "Accounting for Error: How Scientists Construct Their Social World When They Account for Correct and Incorrect Belief." *Sociology: Journal of the British Sociological Association* 16: 165–183.
Muller, H. J. 1937. "Physics in the Attack of the Fundamental Problems of Genetics." *Scientific Monthly* 44: 210–214.
Mulvagh, S. L., L. Michael, M. B. Perryman, R. Roberts, and M. Schneider. 1987a. "Cardiac Expression of C-*myc* and Muscle Creatine Kinase after a Hemodynamic Load in-vivo Corresponds to the Effects of Mitogens in-vitro." *Journal of Cell Biology* 105: 107A.
———. 1987b. "A Hemodynamic Load in-vivo Can Alter Cardiac Muscle Gene Expression That Is Responsive to Mitogens in-vitro." *Journal of Molecular and Cellular Cardiology* 19: S25.

Mulvagh, S. L., R. Roberts, and M. D. Schneider. 1988. "Cellular Oncogenes in Cardiovascular Disease." *Journal of Molecular and Cellular Cardiology* 20: 657–662.
Myers, G. 1990. *Writing Biology: Texts in the Social Construction of Scientific Knowledge.* Madison: University of Wisconsin Press.
Nagley, P., A. W. Linnane, W. J. Peacock, and J. A. Pateman, eds. 1983. *Manipulation and Expression of Genes in Eukaryotes.* Sydney, Australia: Academic Press.
Nandi, S. 1984. Lecture, Department of Zoology, University of California, Berkeley, spring.
Narayan, U. 1989. "The Project of Feminist Epistemology: Perspectives from a Non-Western Feminist." In A. M. Jaggar and S. R. Bordo, eds., *Gender/Body/Knowledge: Feminist Reconstructions of Being and Knowledge.* London: Routledge and Kegan Paul.
National Center for Human Genome Research. 1990. *Annual Report I—FY 1990.* Bethesda, Md.: Department of Health and Human Services, Public Health Service, National Institutes of Health.
Nelkin, D., and S. Lindee. 1995. *The DNA Mystique: The Gene as a Cultural Icon.* New York: Freeman.
Nelkin, D., and L. Tancredi. 1989. *Dangerous Diagnostics: The Social Power of Biological Information.* New York: Basic Books.
Nelson-Rees, W. A., R. R. Flandermeyer, and P. K. Hawthorne. 1974. "Banded Marker Chromosomes as Indicators of Intraspecies Cellular Contamination." *Science* (June 7): 1093.
New England Biolabs 1985–86 Catalogue.
Newmark, P. 1983. "Oncogenic Intelligence: The *Ras*matazz of Cancer Genes." *Nature* 305: 470–471.
Nietzsche, F. 1966. *Beyond Good and Evil.* New York: Vintage.
Office of Technology Assessment. 1984. *Commercial Biotechnology: An International Analysis.* Washington, D.C.: U.S. Congress, OTA-BA-218.
———. 1988. *Mapping Our Genes—Federal Genome Projects: How Vast, How Fast.* Contractor Reports, vol. 1. Washington, D.C.: U.S. Congress.
Olby, R. C. 1974. *The Path to the Double Helix.* London: Macmillan.
———. 1984. "The Sheriff and the Cowboys, or, Weaver's Support of Astbury and Pauling." *Social Studies of Science* 14: 244–247.
———. 1991. "The Development and Spread of Molecular Genetics in the United Kingdom." Paper presented at the colloquium "The Philosophy and History of Molecular Biology," Boston University, April 15–16.
Oudshoorn, N. 1990. "On the Making of Sex Hormones: Research Materials and the Production of Knowledge." *Social Studies of Science* 20: 5–34.
Oyama, S. 1985. *The Ontogeny of Information: Developmental Systems and Evolution.* Cambridge: Cambridge University Press.
Palmiter, R. D., H. Y. Chen, and R. L. Brinster. 1982. "Differential Regulation of Metallothionein-Thymidine Kinase Fusion Genes in Transgenic Mice and Their Offspring." *Cell* 29: 701–710.
Palmiter, R. D., G. Norstedt, R. E. Gelinas, R. E. Hammer, and R. L. Brinster. 1983.

"Metallothionein-Human GH Fusion Genes Stimulate Growth of Mice." *Science* 222: 809–814.

Panem, S. 1984. *The Interferon Crusade: Public Policy and Biomedical Dreams.* Washington, D.C.: Brookings Institute.

Papaioannou, V. E., M. W. McBurney, R. L. Gardner, and M. J. Evans. 1975. "Fate of Teratocarcinoma Cells Injected into Early Mouse Embryos." *Nature* 258: 70–73.

Parada, L. F., C. J. Tabin, C. Shih, and R. A. Weinberg. 1982. "Human EJ Bladder Carcinoma Oncogene Is Homologue of Harvey Sarcoma Virus *Ras* Gene." *Nature* 297: 474–479.

Patient, R. 1984. "DNA Hybridization—Beware [News]." *Nature* 308 (5954): 15–16.

Patterson, J. 1987. *The Dread Disease.* Cambridge, Mass.: Harvard University Press.

Paul, D., and B. Kimmelman. 1988. "Mendel in America: Theory and Practice." In R. Rainger, K. R. Benson, and J. Maienschein, eds., *The American Development of Biology.* Philadelphia: University of Pennsylvania Press.

Peirce, C. S. 1923. *Chance, Love, and Logic,* ed. M. R. Cohen. New York: Harcourt, Brace.

Penley, C., and A. Ross. 1990. "Cyborgs at Large: Interview with Donna Haraway." *Social Text* 25/26: 8–23.

Perbal, B. 1984. *Practical Guide to Molecular Cloning.* New York: Wiley.

Perrow, C. 1984. *Normal Accidents: Living with High-Risk Technologies.* New York: Basic Books.

Pickering, A. 1984. *Constructing Quarks: A Sociological History of Particle Physics.* Chicago: University of Chicago Press.

———. 1990. "Knowledge, Practice, and Mere Construction." *Social Studies of Science* 20: 682–729.

———. 1992. *Science as Practice and Culture.* Chicago: University of Chicago Press.

———. 1993. "The Mangle of Practice." *American Journal of Sociology* 99 (3): 559–589.

Piller, C., and K. Yamamoto. 1988. *Gene Wars: Military Control over the New Genetic Technologies.* New York: Beech Tree Books.

Pinch, T. 1985. "Towards an Analysis of Scientific Observation: The Externality and Evidential Significance of Observation Reports in Physics." *Social Studies of Science* 15.

———. 1986. *Confronting Nature: The Sociology of Solarneutrino Detection.* Dordrecht: Reidel.

———. 1991. "Opening Black Boxes: Science, Technology, and Society." Mullins Lecture, Virginia Polytechnic Institute, April 19.

Polanyi, M. 1958. *Personal Knowledge: Towards a Post Critical Philosophy.* Chicago: University of Chicago Press.

Pollack, R. E. 1981. *Readings in Mammalian Cell Culture,* 2d ed. Cold Spring Harbor, N.Y.: Cold Spring Harbor Laboratory.

———. 1982. Letter to the editor. *Science* 218: 1069–1070.

Portugal, F. H., and J. S. Cohen. 1977. *A Century of DNA: A History of the Discovery of the Structure and Function of the Genetic Substance.* Cambridge, Mass.: MIT Press.

Pratt, M. L. 1986. "Fieldwork in Common Places." In J. Clifford and G. Marcus, eds., *Writing Culture: The Poetics and Politics of Ethnography.* Berkeley: University of California Press.

———. 1992. *Imperial Eyes: Travel Writing and Transculturation.* New York and London: Routledge.

Prehn, R. T., and J. M. Main. 1957. "Immunity to Methylcholanthrene-Induced Sarcomas." *Journal of the National Cancer Institute* 18: 769–778.

———. 1975. "Relationship of Tumor Immunogenicity to Concentration of the Oncogen." *Journal of the National Cancer Institute* 55: 189–190.

Proctor, R. N. 1988. *Racial Hygiene: Medicine under the Nazis.* Cambridge, Mass.: Harvard University Press.

———. 1995. *Cancer Wars.* New York: Basic Books.

Rabinow, P. 1986. "Representations Are Social Facts: Modernity and Post-Modernity in Anthropology." In J. Clifford and G. E. Marcus, eds., *Writing Culture: The Poetics and Politics of Ethnography.* Berkeley: University of California Press.

———. 1992. "Artificiality and Enlightenment: From Sociobiology to Biosociality." In J. Crary and S. Kwinter, eds., *Incorporations.* New York: Urzone.

———. 1996. *Making PCR: A Story of Biotechnology.* Chicago: University of Chicago Press.

Rader, K. A. 1995. "Making Mice: C. C. Little, the Jackson Laboratory, and the Standardization of Mus Musculus for Research." Ph.D. diss., Indiana University at Bloomington.

Rainger, R., K. Benson, and J. Maienschein, eds. 1988. *The American Development of Biology.* Philadelphia: University of Pennsylvania Press.

Rao, B. 1991. "Dominant Constructions of Women and Nature in Social Science Literature." CES/CNS Pamphlet 2.

Rather, L. J. 1978. *The Genesis of Cancer: A Study in the History of Ideas.* Baltimore: Johns Hopkins University Press.

Ravetz, J. R. 1971. *Scientific Knowledge and Its Social Problems.* Oxford: Clarendon Press.

"Repository of Human DNA Probes and Libraries." 1986. *Science* 232: 170.

Research and Educational Association. 1982. *Genetic Engineering.* New York: Research and Educational Association.

Restivo, S. 1983. *The Social Relations of Physics, Mysticism, and Mathematics.* Dordrecht: Reidel.

Rettig, R. A. 1977. *Cancer Crusade: The Story of the National Cancer Act of 1971.* Princeton, N.J.: Princeton University Press.

Rheinberger, H.-J. 1992. "The Laboratory Production of Transfer RNA." *Studies in History and Philosophy of Science* 23: 389–422.

Richards, V. 1978. *The Wayward Cell Cancer: Its Origins, Nature, and Treatment,* 2d ed. Berkeley: University of California Press.

Ricoeur, P. 1970. *Freud and Philosophy.* New Haven: Yale University Press.

———. 1984. *Time and Narrative,* vol. 1. Chicago: University of Chicago Press.

Riemer, J. W. 1979. *Hard Hats: The Work World of Construction Workers.* Beverly Hills, Calif.: Sage.

Rip, A. 1982. "The Development of Restrictedness in the Sciences." In N. Elias, H. Martins, and R. Whitley, eds., *Scientific Establishments and Hierarchies,* vol. 6 of *Sociology of the Sciences Yearbook.* Dordrecht: Reidel.

Robertson, M. 1983. "Clues to the Genetic Basis of Cancer." *New Scientist,* June 9, pp. 688–691.

Rolls-Hansen, N. 1980. "Eugenics before World War II: The Case of Norway." *History and Philosophy of the Life Sciences* 2: 269–298.

———. 1988. "The Progress of Eugenics: Growth of Knowledge and Change in Ideology." *History of Science* 26: 295–331.

Rorty, R. 1982. *Consequences of Pragmatism.* Minneapolis: University of Minnesota Press.

———. 1989. *Contingency, Irony, and Solidarity.* Cambridge: Cambridge University Press.

Rosaldo, R. 1989. *Culture and Truth: The Remaking of Social Analysis.* Boston: Beacon Press.

Rothman, T. 1989. *Science a la Mode: Physical Fashions and Fictions.* Princeton, N.J.: Princeton University Press.

Rous, P. 1911a. "Transmission of a Malignant New Growth by Means of a Cell-Free Filtrate." *Journal of the American Medical Association* 56: 198.

———. 1911b. "A Sarcoma of Fowl Transmissible by an Agent Separable from Tumor Cells." *Journal of Experimental Medicine* 13: 397–411.

Rous, P., and F. S. Jones. 1916. "A Method for Obtaining Suspensions of Living Cells from the Fix Tissues, and for the Plating Out of Individual Cells." *Journal of Experimental Medicine* 23: 549–555.

Rubenstein, I., R. L. Phillips, C. E. Green, and R. J. Desnick, eds. 1977. *Molecular Genetic Modification of Eucaryotes.* New York: Academic Press.

Rubin, H. 1964. "A Defective Cancer Virus." *Scientific American* 210: 46–52.

———. 1983. Letter to the editor. *Science* 219: 1170–1171.

———. 1985. "Cancer as a Dynamic Developmental Disorder." *Cancer Research* 45: 2935–2942.

Ruse, M., and P. Taylor, eds. 1991. "Special Issue on Pictorial Representation in Biology." *Biology and Philosophy* 5.

Russell, E. S., ed. 1981. *Mammalian Genetics and Cancer: The Jackson Laboratory Fiftieth Anniversity Symposium.* New York: Alan R. Liss.

Saltus, R. 1989. *Boston Globe,* October 10.

San Francisco Chronicle. June 12, 1986.

Sanger, F., and A. R. Coulson. 1975. "A Rapid Method for Determining Sequences in DNA by Primed Synthesis with DNA Polymerase." *Journal of Molecular Biology* 94: 441–448.

Santos, E., S. Tronick, S. Aaronson, S. Pulciani, and M. Baracid. 1982. "T24 Human Bladder Carcinoma Oncogene Is an Activated Form of Normal Human Homologue of BALAB-Harvey MSV Transforming Genes." *Nature* 298: 343–347.

Sapp, Jan. 1987. *Beyond the Gene: Cytoplasmic Inheritance and the Struggle for Authority in Genetics.* New York: Oxford University Press.

Sayre, A. 1975. *Rosalind Franklin and DNA.* New York: W. W. Norton.

Schaffer, S. 1989. "Glass Works: Newton's Prisms and the Uses of Experiment." In D. Gooding, T. Pinch, and S. Schaffer, eds., *The Uses of Experiment: Studies in the Natural Sciences.* New York: Cambridge University Press.

Schaffner, K. 1993. *Discovery and Explanation in Biology and Medicine.* Chicago: University of Chicago Press.

Schiebinger, L. 1989. *The Mind Has No Sex? Women in the Origins of Modern Science.* Cambridge, Mass.: Harvard University Press.

Schmeck, H. 1987. "Young Science of Cancer Genes Begins to Yield Practical Applications." *New York Times,* October 6.

Schneider, W. H. 1990. *Quality and Quantity: The Quest for Biological Renegeration in Twentieth-Century France.* Cambridge: Cambridge University Press.

Schwartz, D. E., R. Tizard, and W. Gilbert. 1983. "Nucleotide Sequence of Rous Sarcoma Virus." *Cell* 32: 853–869.

Searle, G. R. 1976. *Eugenics and Politics in Britain, 1900–1914.* Leyden: Noordhoff International Publishing.

Shalin, D. 1986. "Pragmatism and Social Interactionism." *American Sociological Review* 51: 9–29.

Shapin, S. 1979. "The Politics of Observation: Cerebral Anatomy and Social Interests in the Edinburgh Phrenology Disputes." In R. Wallis, ed., *On the Margins of Science: The Social Construction of Rejected Knowledge.* London: Routledge and Kegan Paul.

———. 1982. "History of Science and Its Sociological Reconstructions." *History of Science* 20: 157–211.

———. 1989. "The Invisible Technican." *American Scientist* 77: 554–562.

———. 1994. *A Social History of Truth: Civility and Science in Seventeenth-Century England.* Chicago: University of Chicago Press.

Shapin, S., and S. Schaffer. 1985. *Leviathan and the Air-Pump: Hobbes, Boyle, and the Experimental Life.* Princeton, N.J.: Princeton University Press.

Shapley, D. 1983. "U.S. Cancer Research: Oncogenes Cause Cancer Institute to Change Tack." *Nature* 301: 5.

Shibutani, T. 1955. "Reference Groups as Perspectives." *American Journal of Sociology* 60: 562–569.

———. 1962. "Reference Groups and Social Control." In A. Rose, ed., *Human Behavior and Social Processes.* Boston: Houghton Mifflin.

———. 1978. *The Derelicts of Company K: A Sociological Study of Demoralization.* Berkeley: University of California Press.

———. 1986. *Social Processes: An Introduction to Sociology.* Berkeley: University of California Press.

Shih, C., B. Z. Shilo, M. P. Goldfarb, A. Dannenberg, and R. A. Weinberg. 1979. "Passage of Phenotypes of Chemically Transformed Cells via Transfection of DNA and Chromantic." *Proceedings of the National Academy of Sciences (USA)* 76: 5714–5718.

Shimizu, K., M. Goldfarb, Y. Suard, M. Perucho, Y. Li, T. Kamata, J. Feramiso, E. Starnezer, J. Fogh, and M. Wigler. 1983. "Three Human Transforming Genes

Are Related to the Viral Ras Oncogenes." *Proceedings of the National Academy of Sciences (USA)* 80: 2112–2116.

Shimkin, M. B. 1977. *Contrary to Nature: Being an Illustrated Commentary on Some Persons and Events of Historical Importance in the Development of Knowledge Concerning Cancer.* Washington, D.C.: Department of Health, Education, and Welfare.

Shubik, P. 1983. Letter to the editor. *Science* 220: 1226–1228.

Simon, H. 1969. *The Sciences of the Artificial.* Cambridge, Mass.: MIT Press.

———. 1973. "The Structure of Ill-Structured Problems." *Artificial Intelligence* 4: 181–201.

———. 1977. "How Complex Are Complex Systems?" In F. Suppe and P. D. Asquith, eds., *PSA 1976,* vol. 2, pp. 507–522. East Lansing: Philosophy of Science Association.

Sismondo, S. 1993. "Some Social Constructions." *Social Studies of Science* 23: 515–553.

Skolnick, M. H., et al. 1994. "A Strong Candidate for the Breast and Ovarian Cancer Susceptibility Gene BRCA1." *Science* 266: 66–71.

Smith, L. M. 1986. "The Synthesis and Sequence Analysis of DNA." *Science* 232: G63.

Soprano, K. J. 1984. "Use of Cloned SV40 DNA Fragments to Study Signals for Cell Proliferation." In G. S. Stein and J. L. Stein, eds., *Recombinant DNA and Cell Proliferation,* pp. 3–24. Orlando, Fla.: Academic Press.

Southern, Edward M. 1975. "Detection of Specific Sequences among DNA Fragments Separated by Gel Electrophoresis." *Journal of Molecular Biology* 98(3): 503–517.

Spector, D. H., H. E. Varmus, and J. M. Bishop. 1978. "Nucleotide Sequences Related to the Transforming Gene of Avian Sarcoma Virus Are Present in the DNA of Uninfected Vertebrates." *Proceedings of the National Academy of Sciences (USA)* 75: 5023–5027.

Sperber, I. 1990. *Fashions in Science: Opinion Leaders and Collective Behavior in the Social Sciences.* Minneapolis: University of Minnesota Press.

Stanbridge, E. J. 1990. "Human Tumor Suppressor Genes." *Annual Review of Genetics* 24: 615–657.

Stanley, L. A. 1995. "Molecular Aspects of Chemical Carcinogenesis: The Roles of Oncogenes and Tumour Suppressor Genes." *Toxicology* 96: 173–194.

Star, S. L. 1983. "Simplification in Scientific Work: An Example from Neuroscience Research." *Social Studies of Science* 13: 205–228.

———. 1986. "Triangulating Clinical and Basic Research: British Localizationists, 1870–1906." *History of Science* 24: 29–48.

———. 1989. *Regions of the Mind: Brain Research and the Quest for Scientific Certainty.* Stanford, Calif.: Stanford University Press.

———. 1992. "How the Right Tool Becomes the Wrong One: Taxidermy and Natural History, 1880–1925." In A. Clarke and J. H. Fujimura, eds., *The Right Tools for the Job: At Work in Twentieth-Century Life Sciences.* Princeton, N.J.: Princeton University Press.

Star, S. L., and J. R. Griesemer. 1989. "Institutional Ecology, 'Translations,' and Boundary Objects: Amateurs and Professionals in Berkeley's Museum of Vertebrate Zoology, 1907–39." *Social Studies of Science* 13: 205–228.

Steedly, M. M. 1993. *Hanging without a Rope: Narrative Experience in Colonial and Postcolonial Karoland.* Princeton, N.J.: Princeton University Press.

Stehelin, D., H. E. Varmus, J. M. Bishop, and P. K. Vogt. 1976. "DNA Related to the Transforming Gene(s) of Avian Sarcoma Viruses Is Present in Normal Avian DNA." *Nature* 260: 170–173.

Stein, G. S., and J. L. Stein, eds. 1984. *Recombinant DNA and Cell Proliferation.* Orlando, Fla.: Academic Press.

Stent, G. S. 1968. "That Was the Molecular Biology That Was." *Science* 160: 390–395.

Stevens, L. C. 1970. "The Development of Transplantable Teratocarcinomas from Intratesticular Grafts of Pre- and Post-implantation Mouse Embryos." *Developmental Biology* 21: 364–382.

Stokes, T. D. 1982. "The Double Helix and the Warped Zipper—An Exemplary Tale." *Social Studies of Science* 12: 207–240.

———. 1985. "The Role of Molecular Biology in an Immunological Institute." Paper presented at the International Congress of History of Science, University of California, Berkeley, July 31–August 8.

Strathern, M. 1992. *After Nature: English Kinship in the Late Twentieth Century.* Cambridge: Cambridge University Press.

Strauss, A. L. 1959. *Mirrors and Masks: The Search for Identity.* Chicago: Free Press. Reprint, San Francisco: Sociology Press, 1969.

———. 1978a. *Negotiations: Varieties, Contexts, Processes, and Social Order.* San Francisco: Jossey-Bass.

———. 1978b. "A Social World Perspective." *Studies in Symbolic Interaction* 1: 119–28.

———. 1985. "Work and the Division of Labor." *Sociological Quarterly* 26: 1–19.

———. 1988. "The Articulation of Project Work: An Organizational Process." *Sociological Quarterly* 29: 163–178.

———. 1993. *Continual Permutations of Action.* Hawthorne, N.Y.: Aldine de Gruyter.

Strauss, A. L., S. Fagerhaugh, B. Suczek, and C. Weiner. 1985. *The Organization of Medical Work.* Chicago: University of Chicago Press.

Strauss, A. L., L. Schatzman, R. Bucher, D. Erlich, and M. Sabshin. 1963. "The Hospital and Its Negotiated Order." In E. Freidson, ed., *The Hospital in Modern Society.* New York: Free Press.

———. 1964. *Psychiatric Ideologies and Institutions.* New York: Free Press.

Strickland, S. P. 1972. *Politics, Science, and Dread Disease: A Short History of United States Medical Research Policy.* Cambridge, Mass.: Harvard University Press.

Studer, K. E., and D. E. Chubin. 1980. *The Cancer Mission: Social Contexts of Biomedical Research.* London: Sage.

Suchman, L. 1987. *Plans and Situated Action: The Problem of Human-Machine Communication.* Cambridge: Cambridge University Press.

Suzuki, D. T., and P. Knudtson. 1989. *Genethics: The Clash between the New Genetics and Human Values.* Cambridge, Mass.: Harvard University Press.

Suzuki, D. T., A. J. F. Griffiths, J. H. Miller, and R. C. Lewontin. 1989. *An Introduction to Genetic Analysis,* 4th ed. New York: W. H. Freeman.

Tabin, C. J., S. M. Bradley, C. I. Bargmann, R. A. Weinberg, A. G. Papageorge, E. M. Scolnick, R. Dhar, D. R. Lowy, and E. H. Chang. 1982. "Mechanism of Activation of a Human Oncogene." *Nature* 300: 143–149.

Taylor, P. 1992. "Re/constructing Socio-Ecologies: System Dynamics Modeling of Nomadic Pastoralists in Sub-Saharan Africa." In A. Clarke and J. H. Fujimura, eds., *The Right Tools for the Job: At Work in Twentieth-Century Life Sciences.* Princeton, N.J.: Princeton University Press.

Teitelman, R. 1985. "The Baffling Standoff in Cancer Research." *Forbes,* July 15, pp. 110–114.

———. 1989. *Gene Dreams: Wall Street, Academia, and the Rise of Biotechnology.* New York: Basic Books.

Temin, H. M. 1964a. "Nature of the Provirus of the Rous Sarcoma." *National Cancer Institute Monograph* 17 (March 31): 557–570.

———. 1964b. "The Participation of DNA in Rous Sarcoma Virus Production." *Virology* 23 (April 13): 486–494.

———. 1964c. "Homology between RNA from Rous Sarcoma Virus and DNA from Rous Sarcoma Virus-Infected Cells." *Proceedings of the National Academy of Sciences (USA)* 52: 323–329.

———. 1971. "The Protovirus Hypothesis: Speculations on the Significance of RNA Directed DNA Synthesis for Normal Development and Carcinogenesis." *Journal of the National Cancer Institute* 46: 3–7.

———. 1980. "Origin of Retroviruses of Cellular Genetic Moveable Elements." *Cell* 21: 599–600.

———. 1983. "We Still Don't Understand Cancer." *Nature* 302: 656.

Temin, H. M., and D. Baltimore. 1972. "RNA-directed DNA Synthesis and RNA Tumor Viruses." *Advances in Virus Research* 17: 129–186.

Temin, H. M., and S. Mizutani. 1970. "RNA-dependent DNA Polymerase in Virions of Rous Sarcoma Virus." *Nature* 226 (June 27): 1211–1213.

Temin, H. M., and H. Rubin. 1958. "Characteristics of an Assay for Rous Sarcoma Virus and Rous Sarcoma Cells in Tissue Culture." *Virology* 6: 669–688.

Thomas, W. I., and D. S. Thomas. 1928. *The Child in America: Behavior Problems and Programs.* New York: Alfred A. Knopf.

Thomas, W. I., and F. Znaniecki. 1918. *The Polish Peasant in Poland and America.* New York: Alfred A. Knopf.

Todaro, G. J., and R. J. Huebner. 1972. "The Viral Oncogene Hypothesis: New Evidence." *Proceedings of the National Academy of Sciences (USA)* 69: 1009–1015.

Traweek, S. 1988. *Beamtimes and Lifetimes: The World of High Energy Physicists.* Cambridge, Mass.: Harvard University Press.

———. 1992. "Border Crossings: Narrative Strategies in Science Studies and among

Physicists in Tsukuba Science City, Japan." In A. Pickering, ed., *Science as Practice and Culture.* Chicago: University of Chicago Press.

———. 1993. "An Introduction to Cultural and Social Studies of Sciences and Technologies." *Culture, Medicine, and Psychiatry* 17: 3–25.

———. 1995. "Bodies of Evidence: Law and Order, Sexy Machines, and the Erotics of Fieldwork among Physicists." In S. Forster, ed., *Choreographing History.* Indiana: Indiana University Press.

Tsing, A. 1993. *In the Realm of the Diamond Queen.* Princeton, N.J.: Princeton University Press.

Tucker, J. B. 1984. "Gene Machines: The Second Wave." *High Technology* 4: 50–59.

Varmus, H. T. 1989a. "Retroviruses and Oncogenes, I." Nobel Lecture, December 8, Stockholm, Sweden, published in *Bioscience Reports* 10 (1990): 413–430.

———. 1989b. "An Historical Overview of Oncogenes." In R. Weinberg, ed., *Oncogenes and the Molecular Origins of Cancer,* pp. 3–44. Cold Spring Harbor, N.Y.: Cold Spring Harbor Laboratory.

Vecchio, G. 1993. "Oncogenes of DNA and RNA Tumor Viruses and the Origin of Cellular Oncogenes." *History and Philosophy of the Life Sciences* 15: 59–74.

Vogelstein, B., and K. W. Kinsler. 1993. "The Multistep Nature of Cancer." *Trends in Genetics* 9: 138–141.

Vogt, M., and R. Dulbecco. 1960. "Virus-Cell Interaction with a Tumor-Producing Virus." *Proceedings of the National Academy of Sciences (USA)* 46: 369.

Volberg, R. A. 1983. "Constraints and Commitments in the Development of American Botany, 1880–1920." Ph.D. diss., Department of Sociology, University of California, San Francisco.

Waddington, C. H. 1969. "Some European Contributions to the Prehistory of Molecular Biology." *Nature* 221: 318–321.

Wade, N. 1980. "Three New Entrants in Gene Splicing Derby." *Science* 208: 690.

Walker, J. M., and W. Gaastra, eds. 1983. *Techniques in Molecular Biology.* New York: Croom Helm/Macmillan.

Wallis, R., ed. 1979. *On the Margins of Science: The Social Construction of Rejected Knowledge.* Staffordshire: J. H. Brookes.

Warburg, O. 1930. *The Metabolism of Tumours.* London: Constable.

Waterfield, M. D., G. T. Scrace, N. Whittle, P. Stroobant, A. Johnsson, A. Wasteson, B. Westermark, C. H. Heldin, J. S. Huang, and T. F. Deuel. 1983. "Platelet-Derived Growth Factor Is Structurally Related to the Putative Transforming Protein p28 *sis* of Simian Sarcoma Virus." *Nature* 304: 35–39.

Watson, J. D. 1968. *The Double Helix.* New York: Mentor Books.

———. 1970. *Molecular Biology of the Gene,* 2d ed. New York: W. A. Benjamin.

———. 1990. "The Human Genome Project: Past, Present, and Future." *Science* 249 (April 6): 44–49.

Watson, J. D., and J. Tooze. 1981. *The DNA Story: A Documenary History of Gene Cloning.* San Francisco: W. H. Freeman.

Watson, J. D., J. Tooze, and D. T. Kurtz. 1983. *Recombinant DNA: A Short Course.* New York: W. H. Freeman.

Watson, J. D, N. H. Hopkins, J. W. Roberts, J. A. Steitz, and A. M. Weiner. 1987.

Molecular Biology of the Gene, vol. 2. 1965. Reprint, Menlo Park: W. A. Benjamin.
Weber, B. 1992. "An Eye for Danger." *New York Times Magazine,* January 19, pp. 18–24.
Weinberg, R. A. 1982. "Review: Oncogenes of Human Tumor Cells." In S. Prentis, ed., *Trends in Biochemical Sciences,* vol. 7. Amsterdam: Elsevier Biomedical Press.
———. 1983. "A Molecular Basis of Cancer." *Scientific American* 249: 126–143.
———. 1991. "Tumor Suppressor Genes." *Science* 254: 1138–1146.
———., ed. 1989. *Oncogenes and the Molecular Origins of Cancer.* Cold Spring Harbor, N.Y.: Cold Spring Harbor Laboratory.
Weindling, P. 1990. *Health, Race, and German Politics between National Unification and Nazism, 1870–1945.* Cambridge: Cambridge University Press.
Weiss, M. C., and H. Green. 1967. "Human-Mouse Hybrid Cell Lines Containing Partial Complements of Human Chromosomes and Functioning Human Genes." *Proceedings of the National Academy of Sciences (USA)* 58: 1104–1111.
Weiss, S. F. 1987. *Race Hygiene and National Efficiency: The Eugenics of Wilhelm Schallmayer.* Berkeley: University of California Press.
West, C. 1989. *The American Evasion of Philosophy: A Genealogy of Pragmatism.* Madison: University of Wisconsin Press.
Willis, R. A. 1960. "The Experimental Production of Tumours." In *Pathology of Tumours.* London: Butterworth.
Willmer, E. N., ed. 1965. *Cells and Tissues in Culture: Methods, Biology, and Physiology,* vols. 1, 2. New York: Academic Press.
Wimsatt, W. C. 1974. "Complexity and Organization." In K. Schaffner and R. S. Cohen, eds., *PSA 1972,* pp. 67–86. Dordrecht: Reidel.
———. 1980. "Reductionist Research Strategies and Their Biases in the Units of Selection Controversy." In T. Nickles, ed., *Scientific Discovery,* vol. 2, *Case Studies.* Dordrecht: Reidel.
———. 1981. "Robustness, Reliability, and Overdetermination." In M. B. Brewer and B. E. Collins, eds., *Scientific Inquiry and the Social Sciences.* San Francisco: Jossey-Bass.
———. 1986. "Developmental Constraints, Generative Entrenchment, and the Innate-Acquired Distinction." In William Bechtel, ed., *Integrating Scientific Disciplines.* Dordrecht: Nijhoff.
Wise, G. 1985. "Science and Technology." *Osiris* 2d ser., 1: 229–246.
Witkowski, J. A. 1979. "Alexis Carrel and the Mysticism of Tissue Culture." *Medical History* 23: 279–296.
Wittgenstein, L. 1970. *Philosophical Investigations.* New York: Macmillan.
Wofsy, L. 1986. "Biotechnology and the University." *Journal of Higher Education* 57 (5): 477–492.
Wolgom, W. 1913. *The Study of Experimental Cancer: A Review.* New York: Columbia University Press.

Woolgar, S. W. 1976. "Writing an Intellectual History of Scientific Development: The Use of Discovery Accounts." *Social Studies of Science* 6: 395–422.

———. 1981. "Critique and Criticism: Two Readings of Ethnomethodology." *Social Studies of Science* 11: 504–514.

———. 1988. *Knowledge and Reflexivity: New Frontiers in the Sociology of Knowledge.* London: Sage.

Woolgar, Steve, and Dorothy Pawluch. 1985. "Ontological Gerrymandering: The Anatomy of Social Problems Explanations." *Social Problems* 32: 214–227.

Wright, S. 1986. "Recombinant DNA Technology and Its Social Transformation, 1972–1982." *Osiris* 2d ser., 2: 303–360.

Yanagisako, S., and C. Delaney. 1995. *Naturalizing Power: Essays in Feminist Cultural Analysis.* New York and London: Routledge.

Yoxen, E. J. 1983. *The Gene Business: Who Should Control Biotechnology?* New York: Oxford University Press.

Zallen, Doris. 1991. "Molecular Biology in France: Historical Case Studies." Paper presented at the colloquium "The Philosophy and History of Molecular Biology," Boston University, April 15–16.

Zeldenrust, S. 1985. "Strategic Action in the Laboratory: (Inter)organizational Resources and Constraints in Industrial and University Research." Paper presented at the Tenth Annual Meeting of the Society for Social Studies of Science, Rensselaer Polytechnic Institute, Troy, N.Y., October.

———. 1989. "Ambiguity, Choice, and Control in Research." Ph.D. diss., University of Amsterdam, Amsterdam, Netherlands.

Zimmerman, B. K. 1984. *Biofuture: Confronting the Genetic Era.* New York: Plenum Press.

Index

Abbot Laboratory, 148
Accountability, 157, 158, 172
Action: ecologies of, 12, 16–18, 21, 73; consequences of, 18–19; in problem-solving, 170; social, 184; commitments and, 226. *See also* Collective action; Conventions of action; Ecologies of action; Situated action
Actor-network theory, 15–17, 78–79, 112, 113, 206–207, 239–242; analysis of, 220
Agency/agents, 214, 258n19
AIDS, 52, 53, 252nn32,39
Alternate strategies, 72, 166, 174–181
Ambiguity, 11, 34, 152, 208, 265n17
American Cancer Society, 24, 76, 145
American Type Culture Collection (ATCC), 6, 99–100, 141, 142
Amersham, 86
Ames, Bruce/Ames test, 61–62, 148–149
Angier, Natalie, 125, 265n18
Animals, experimental, 25, 27, 28–30, 41, 42, 66, 70; inbred, 136; in viral studies, 167–168. *See also* Inbred mouse colonies; Non-human actors
Anomalies, 170, 218, 230
Anthropology, 17, 19, 220, 239, 242
Anti-oncogenes, 20, 65, 227
Articulation, 11, 16, 77, 152, 186–187; of research programs, 188–194; standardization and, 200–202; organization of work and, 203, 208, 216, 233. *See also* Organization of scientific work
Artifacts, 7, 13, 16, 78, 107, 206; creation/production of, 17, 71, 172, 212, 238; laboratory, 31, 51, 52, 219, 239
Art to science, 39–41, 203
Audiences, 4, 9, 10, 15, 32, 69, 75, 87, 133–134, 141, 191, 199; demands of, 208, 211
Authority, 72, 114, 153, 207; discretionary power, 172–173; centers of, 206. *See also* Distributed authority

Bacterial molecular genetics, 63, 74, 81, 82
Baltimore, David, 49, 52
Bandwagon(s), 2–3, 9, 110, 155, 181, 230–234, 276n25; of proto-oncogene research, 22, 137–141, 159, 162, 224, 228, 229, 231; of molecular biology, 225–227
Barbacid, Mariano, 63, 124, 160
Barnes, Barry, 238
Basic vs. applied science, 76–77, 233
Becker, Howard S., 12, 16, 110, 145, 269nn7,8, 273n6, 274n14
Becton-Dickinson, 98
Berenblum, I., 56–57, 58, 59, 60
Bethesda Research Laboratories (BRL), 86, 91, 98–99, 114
Biogen, 75
Biohazard, 256n11
Biological information/bioinformatics, 248n16; molecular information databases, 7, 101–104, 112. *See also specific companies*
Biology, 105; experimental, 29, 53, 115; research, 68–69, 76, 80; evolutionary, 117–121, 135, 141, 262n43; population, 216. *See also* Developmental biology; Transplantation biology
Biomedicine, 52
Biotechnology companies, 2, 9, 21, 69, 75–77, 113, 145, 151, 162, 194. *See also specific companies*
Bishop, J. Michael, 1, 6, 9, 17, 53, 118–127, 130, 132, 135, 136, 141, 142, 160, 179, 180, 206, 208, 210, 222, 224–226, 229
Bittner, J. J., 35, 37, 50
Black box technologies, 33, 71, 78, 81, 111, 112, 141, 212, 213, 215, 216
Bloor, David, 238
Blumer, Herbert, 13–14
Boehringer Mannheim, 86
Boundaries, 24, 31
Bourdieu, Picrre, 145, 246n13
Boveri, Theodore, 32, 55
Boyer, Herbert, 73–74, 83
Boyle, Robert, 241, 242

Braun, A. C., 59
Breast cancer/mammary tumors: in mice, 4, 26, 34–37, 45, 48, 50, 65, 153–154; in humans, 177–179, 180, 227
Burian, Richard, 232, 249n18

Cairns, John, 47, 51, 52, 55–56, 61–62, 63
California Institute of Technology, 46, 99, 100
Callon, Michael, 78–79, 206, 217, 220, 239, 240, 242, 258n19
Cambrosio, Alberto, 216, 239
Cancer: definitions of, 1, 3, 23, 64–65, 153, 207; genetic theories of, 1, 2, 5, 16, 25–27, 30, 34, 36, 51, 53, 62, 119, 123–124, 134, 205, 207, 215, 222; biology, 2, 70, 124–125; causation, 4, 9, 11, 16, 23, 30, 54, 55, 59, 62, 64, 117, 118, 127, 129, 134, 149, 169, 222; initiation and promotion theory of, 4, 56, 58–60, 62; types of, 4, 23; viral theories of, 4, 5, 9, 16, 25, 30, 36, 47, 52, 57–58, 66, 117, 125, 127, 130–131, 153, 210, 219, 225; representations of, 14, 63, 112, 153, 205; molecular basis of, 18, 69; sarcoma, 25, 27, 48, 49, 50, 125, 166; multistage process of, 60–61, 62, 253n47; environmental theories of, 70, 148, 149, 256n11, 259n21; potential cures, 76, 127, 151, 161, 225; as a genetic disease, 119, 154, 205; prevention, 149; nongenetic, 150; reconstruction of, 153–154. *See also* Common genetic pathway; Proto-oncogene(s)
Carcinogenesis, 6, 16, 55, 135, 150; chemical, 3, 24, 27–30, 37, 47, 51, 54, 56–61, 63, 117, 123, 126, 127, 131, 148; radiation, 3, 28–29, 30, 37, 47, 54–57, 59, 63, 117, 127; hormonal, 28, 59, 63, 117; testing for, 42, 61–62; research, 53, 55, 117; latency period, 56, 58; ultimate, 60–61; natural vs. experimental, 120, 141
Careers, 140, 143, 145, 156, 167, 173–174, 177, 273n14
Carrel, Alexis, 38
Castle, William, 33
Cell: lines, 3, 5, 7, 41–44, 79, 136; biology, 4, 6, 9, 25, 38, 68, 121, 142, 146, 218, 232; cultures, 28, 37–53, 54, 63, 65, 66, 130, 179; transformation, 41–44, 63–65, 69, 117, 119, 123–124, 126, 130, 142, 168–169, 176–178, 223; fibroblast, 42, 43, 44; types, 43, 142, 178, 250n23; somatic, 75; fusion, 149; donors, 178, 179; epithelial, 178, 179, 180; production, 179–180; germ, 214; immortalized, 265n17
Cellular oncogenes, 34, 50, 122, 126, 127, 141, 145, 160–161, 167, 222, 226; *abl*, 7; *fos*, 7; *myc*, 7, 50, 71, 145, 168, 226; *ras*, 7, 50, 124, 125, 146, 162, 163, 166, 168, 171, 188, 191, 192, 195; *int-1*, 36, 50, 154; *src*, 49, 50, 51, 118, 119, 125, 166, 264n10; *crk*, 53; *jun*, 53; *c-ras*, 162, 191; *sis*, 121
Centocor Oncogene Research Partners, 148
Central Dogma, 49, 214
Cetus, 75, 148, 254n3, 260n28
Chemistry, 105, 113, 145
Chemotherapy, 54, 178, 253n44, 267n2
Chiron, 267n7
Chromosome(s): theory of heredity, 33, 35; theory of cancer, 55, 57, 64; X and Y, 261n37
Clarke, Adele, 239, 246n12, 248n12, 273nn11,20
Classical genetics, 24–36, 65, 66, 71, 117
Classification of cancers, 23
Clones/cloning, 45; molecular, 51, 71, 74, 75, 113; kits, 69, 91; gene, 75, 78, 81; manuals ("cookbooks"), 78, 84–86, 91, 105, 111, 113, 201, 209; procedures, 81–86; of human cancer genes, 124
Co-construction (co-production), 9–11, 16–18, 24, 109, 138, 152–153, 206–208, 217, 224, 229
Code, genetic, 75, 150
Co-definition, 152
Cohen, Stanley, 73–74
Coherence, 209, 211, 212, 216–220
Cold Spring Harbor Laboratory, 63, 78, 84, 85, 98, 99, 101, 105, 124, 132, 162, 201, 261n34
Collective action, 3–5, 11–16, 31–32, 71, 72, 112, 113, 157, 207, 212, 216, 221–222; in scientific work, 14–15; standardization and, 109–111; proto-oncogene theory as, 135–136, 152; problem-solving as, 160, 175; vs. fashion, 227–230
Collins, Harry, 212, 215, 238
Commitment, 2, 9–10, 12, 13, 31–32, 65, 72, 104, 113, 114, 156, 207, 216, 243; to procedures, 14; to recombinant DNA technology, 86–100, 112; to oncogene and proto-oncogene research, 136–138, 150, 210, 224, 226; to molecular biological cancer research, 139–140, 225–227; of established researchers, 143, 145–146, 147,

151, 206; of new investigators, 143, 144–146, 148, 151, 206; of students, 143–144, 145, 147, 173; by private industry, 145, 147–148, 162; by research institutes, 145, 147; to careers, 173–174; continuity of, 222
Common genetic pathway, 59, 63, 65, 119, 127–128, 134, 135
Comparative cultural studies of science, 242–243
Competition, 21, 110, 124–125, 157, 163–164, 165, 191, 228–229, 230–231
Complexity, 34, 111–112, 151, 224
Conflict, 13, 14, 21
Consensus, 2, 14
Constructivism, 220, 237
Contingency, 10, 11, 14, 34, 66, 152, 238; problem-solving and, 156–157, 159–160, 170, 173, 182, 199; continuity and, 210–222, 225; local, 211
Continuity, 11, 14–16, 22, 64, 66, 135, 136, 151, 176, 231; traces of, 16, 136; historical, 17, 209–210, 225; in practices and concepts, 209–210, 216; across situations, 210–222
Controversy, 67, 120, 131–132, 222–224
Conventions of action, 4, 12, 13, 14–15, 81, 110
Cooper, Geoffrey, 126
Coulson, A. R., 107
Crafting, 5, 11, 135, 136, 140, 148, 182, 207, 208, 209, 211
Credibility, 144; hierarchies of, 145
Crick, Francis, 49, 84, 214
Criticism of oncogene research, 2, 11, 120, 210, 222–224, 252n39, 262n51
Cross-breeding, 29, 30, 35, 36, 70, 73
Cultural anthropology, 258n20
Cultural views of disease, 15
Culture and science, 12, 206, 208, 237, 242–243
Customization, 71, 91, 111, 112
Cyborgs, 52–53, 207

Daily life in the laboratory, 1, 10, 11, 21, 77, 155, 175, 182, 184, 202–204, 208
Data bases, 101–104, 112
Delaney, Carol, 258n20
DeOme, Kenneth, 25–29, 32, 36–37, 45, 56, 70, 131, 180–181
Dermer, G. B., 223–224
Developmental biology, 6, 25, 74, 121–123, 135
DeVita, Vincent T., Jr., 132–133, 134–135
Dewey, John, 245n9
Diagnostic technologies, 2, 69, 143, 162, 267n5; kits, 171, 190, 194–195; detection of cancer, 178, 223; reagents, 189–190; research, 218
Dialectical reproduction, 114, 221
Difference, 64, 74, 213–216
Disciplinary cultures, 14, 18, 24, 206, 216, 239
Discontinuity, 16, 64
Discourses, 7, 211, 212
Dissenters, 170
Distributed authority, 16, 17, 114, 153, 181, 206
Distributed coordination, 103
DNA, 3, 11, 33, 49, 53, 61–62, 65, 210; experimentation, 34, 69; production of, 49, 73, 108–109; transfection, 54, 63–64, 68, 75, 117, 123; function of, 59, 74, 75, 180; structure of, 59, 74, 75; transfer, 68, 73, 123; manipulation, 71, 73, 74, 80, 81, 113, 152, 160; cloning, 78, 80; probes, 79, 83, 97, 99, 100, 108–109, 113; synthesizers, 98, 100, 108–109, 113; sequenators, 108, 118, 210; moratorium on techniques, 160
DNA Data Base of Japan, 101
DNA tumor viruses, 47–48, 130
Doable problems, 10–11, 16, 104, 140, 141, 143, 150, 151, 208; in oncogene research, 184–185; articulation and, 187; construction of, 187–200, 228, 230; daily detail of, 202–204; definition of, 208
Doolittle, Russell, 121
Duesberg, Peter, 49, 222–223, 264n11
Dulbecco, Renato, 18, 46–47, 48, 49, 219, 251n28
Du Pont Corporation, 7, 151
Duster, Troy, 19, 45, 246n20, 255n5

Ecologies of action, 12, 16–18, 21, 73
"Efficiency," 103, 104–112, 113
Enrollment, 71, 81, 141, 151, 170, 175, 206, 211, 220, 229, 239–240
Enzyme(s), 9, 49, 91, 106. *See also* Restriction enzymes
Epidemiology, 4, 23, 63, 129, 141
Epigenetics, 150
Epistemology, 213
Epstein, Sydney, 149
Equivalence, 213–216
Ethnographies of science and practice, 4, 5, 17, 19, 159, 182–185, 203, 209, 211, 225, 238–239, 242

Ethnomethodology, 17, 212, 215
Eugenics, 24, 214, 257n11
Eukaryotes, 53, 64, 69, 70, 73–76, 84, 113, 120, 152, 227
European Molecular Biology Laboratory (EMBL), 101, 104
Evolutionary theory, 5, 53, 119, 120, 121, 122, 218, 262n43
Experimental production of similarity, 6–7, 25, 26–28
Experimental systems, 3, 16, 41, 43, 51, 52, 64, 65, 66, 202, 206; as symbotic systems, 66–67
Experimenter's regress, 211, 213

Failure, 78, 80, 81, 109, 189, 195, 268n10
Fashion, 125, 227–230
Feminist studies of science, 239, 242, 243
Feyerabend, Paul, 237
Fleck, Ludwig, 19, 237, 248n18
Flexibility, 112, 170, 174, 216, 221–222, 227, 242
Fortun, Michael, 260n31
Foucault, Michel, 137, 246n13, 252n35, 254n53
Frederick Cancer Research Facility, 135
Funding, 9, 10, 11, 15, 16, 21, 32, 74, 113, 135, 136, 145, 152, 156; government, 36, 71, 75, 76, 86, 87, 102, 130–132, 151, 210; commercial/industrial, 86–87, 100, 179; for recombinant DNA studies, 86–100, 109; for molecular biology, 134; private, 145, 179, 181; policies, 157, 203; continuation of, 170; cutbacks, 267n2

GenBank, 101–102, 104, 257n14, 261n40
Gender, 175, 242. *See also* Feminist studies of science
Gene(s): sequencing, 7, 71, 74, 75, 78, 80, 81, 100–104, 107, 118, 121, 136, 176, 210, 233; transfer, 7, 71, 75, 123, 126; concepts and definitions of, 9, 32–34, 64, 65, 66, 71, 125; identity, 27–28, 72, 119; transformation, 45–53; kinds, 49–50; expression, 53, 74, 75, 210, 251n3; manipulation, 66, 70, 176; therapy, 74; mapping, 74, 75, 77, 81, 83; function, 75, 80, 113, 161; structure, 75, 80, 100, 113, 161; libraries, 82–83, 99–100; conservation of, 117, 120, 121, 122; transfection, 123; cellular, 160; viral, 160; damage, 227
Genealogies: historical, 23–24, 64–65, 116–117, 154; tumor genealogies, 36; in plant and animal breeding, 73
Genentech, 75, 148
Genetic engineering, 11, 26, 27–28, 74, 75–76, 80, 81, 87, 100
Genetics, 4, 5, 24, 33, 55–56, 98, 225; definition of, 30, 32, 65, 207; cancer, 36, 70; clinical, 127–129; epidemiological, 127–129; medical, 141. *See also* Classical genetics
Genetic screening, 145
Genotype-phenotype differences, 26, 34, 35, 134
Geography of forces, 114
Gilbert, Scott, 115, 149, 261n36
Gilbert, Walter, 18, 107, 108, 115, 233, 265n13, 273n11
Glycosis theory, 25
Goldschmidt, Richard, 33
Griesemer, James, 214, 215, 249n18
Gross, Ludwig, 47
Growth factor, 119, 121

Hacking, Ian, 217–221, 268n15, 272n31, 276n21
Hanafusa, Hidesaburo, 49
Haraway, Donna, 31, 243, 249n18, 268n15
Harrison, Ross, 38, 41
Harvard University, 99
Hegemony, 72, 114, 153, 222
HeLa cells, 250n25
Heredity/heritability, 216; of tumors/cancer, 24–27, 30, 36, 55, 61, 120, 127; of characteristics, 32, 33, 34, 65; chromosome theory of, 33, 35, 55; genetic theory of, 62, 80–81; screening for, 145; social, 214; molecular theory of, 258n21
Heterogeneity: of practices, 65, 66, 193, 220, 221; of cancer genes, 126; of cancer cells, 129; of actors, 162, 182, 185, 240; of situations, 221. *See also* Homogeneity
Heuristics, 230, 231
Histories: of cancer research, 4, 17–18, 23–64; of molecular biology, 9, 34, 115, 227; of science, 19, 185, 206, 217, 225, 240, 243; of genetics, 34, 216; of proto-oncogenes, 35, 205, 219; of recombinant technologies, 70–73, 120; of problem-solving, 155–183
HIV virus, 52, 53, 252nn32,39
Hobbes, Thomas, 72, 241, 242
Hoffman-La Roche, Inc., 148
Homogeneity: of inbred animals, 26, 29–30, 44; of practices and representations, 207, 221. *See also* Heterogeneity
Homolog/homologies, 120, 124, 125, 132,

157; sequence, 103, 136, 210, 233; cellular, 119; in theory/methods packages, 208; definitions of, 261n43, 274n4
Hood, Leroy, 100, 108
Hormones, 28, 59, 63, 75, 117, 178, 253n50
Hubbard, Ruth, 266n29, 278n5
Huebner, R. J., 264n10
Hughes, Everett, 12, 14, 140, 148, 205–206
Human genome project, 18–19, 34, 75, 77, 80, 215, 233, 242–243, 258n16, 259n21; Human Genome Initiative, 257n11, 260n30
Hybridization, molecular, 210, 264n6, 268n9

Immunology, 4, 25, 27, 28, 68, 76, 145
Improvisation, 194–200
Inbred animal studies, 66, 70, 72, 73, 207
Inbred mouse colonies, 3, 5, 24–29, 44, 52, 65–66, 153–154, 180; standard, 29–32; transplantation research with, 35–36, 64; "nude" mice, 73
Innovation, 176
Inscription, 213, 215
Inside/outside: actions, 149–150, 156, 182; worlds, 192, 197, 204, 215, 219, 239
Instabilities, 215
Institutes, research, 9, 88, 140, 145, 147, 151, 178–179, 182
Institutions, 4, 14–16, 69, 135, 225, 232–233
Instruments/instrumentation, 7, 72, 100–104, 108–109, 111, 113, 213; gel electrophoresis, 186–187, 201
Intelligenetics, 104
Interaction, 3–4, 13, 21. *See also* Symbolic interaction
Interface, 152, 222
Interferon, 76, 83, 100, 161
Intervening, 6, 72, 79, 120, 134, 149, 176, 216, 240
"Invariability," 30, 31, 207
Invisible work, 11, 169, 185, 206, 215

Jackson Laboratory, 26, 29, 32, 154
Jacob, François, 33
James, William, 245n9
Johannsen, Wilhelm, 33
Jordan, K., 215, 260n28
Judson, Horace, 74

Keating, P. 216, 239
Keller, Evelyn Fox, 32, 115, 266n29
Kimmelman, Barbara, 248n8, 257n13
Kleep, Robert, 149
Knorr-Cetina, Karin, 19, 272n2, 273n7
Koch, Robert, 154, 268n17
Kondo, Dorinne, 258n20, 275n8
Krim, Mathilde, 76
Kuhn, Thomas, 16, 152, 231–232, 237, 277nn29,30

Laboratories, 73; organization of, 1, 12, 30, 31, 69, 71, 141, 169–173, 175–176, 233–234. *See also* Daily life in the laboratory; Organization of scientific work
Lakatos, Imre, 230, 231, 233
Language games, 7, 207
Latour, Bruno, 78–79, 144, 206, 213–214, 215, 217, 219, 220, 239, 240–241, 248n18, 258n19
Law, John, 206, 229, 239, 240
Leukemia, 23
Levins, Richard, 233, 278n5
Lewontin, Richard, 278n5
Little, Clarence C., 26–27, 29, 32, 36, 70
Lobbying. *See* Politics: of cancer
"Logic"/"logical steps" in procedures and research, 109–111, 113, 120, 144, 176, 182, 262n49
Lynch, Michael, 78, 211, 215, 239

Mackenzie, M., 216
Magic bullet, 5, 64, 117
Mammalian genetics. *See* Inbred mouse colonies
Maniatis, Tom, 83, 84, 85, 87
Marks, Paul, 147
Martin, Emily, 239, 258n20, 271n28
Martin, G. Steven, 46, 49, 106–107, 276n20
Marx, Jean, 276n20
Marx, Karl, 205, 271n26, 278n5
Materialization, 214
Material technologies/production, 7–8, 45, 66, 171
Maxam, Allan, 107
McKusick, Victor, 99
Mead, George Herbert, 13, 14
Medawar, Peter, 203
MedLine, 4, 171
Memorial Sloan-Kettering Cancer Center, 147
Mendel, Gregor, 32
Merck and Co., 148
Merton, Robert K., 237
Metaphor, 7
Metaphysics, 213, 218, 220

Metastasis, 129, 150, 266n21
Microbiology, 39, 240–241
Mistakes, 159, 170, 171, 199, 202, 209
MIT Whitehead Institute, 63, 99, 123, 124, 261n34
Model organisms, 77, 135
Model(s): explanatory, 115, 230; building, 152, 163–164, 165–166; of laboratory science, 217
Molecular biology, 2, 4–6, 9, 12–13, 34, 36, 64, 68–69, 71, 73, 75, 76, 138, 145, 146, 151, 232; molecular cloning, 7, 51, 71, 74, 75, 113; enzyme research, 79–82, 86; recombinant DNA research and, 88–91, 103–104; growth factor research, 119, 121; bandwagon, 225–227
Molecular genetic technologies, 5, 6, 7, 32, 59, 63–64, 71–73, 88, 175, 206, 221, 225; methods of, 136, 138; proto-oncogene theory and, 150–153; standardized, 152; analysis, 176; probes, 210; polymerase chain reaction (PCR), 260n28, 267n5. *See also* Gene(s)
Mouse Mammary Tumor Virus (MMTV), 35, 36, 154. *See also* Inbred mouse colonies
Mutation, 27, 51, 55; agents of, 56–57; theory, 58, 59–62, 63; of human cancer cells, 125; single point, 125, 227; of DNA, 148–149. *See also* Pathological growth and differentiation; Somatic mutation theory
Myriad Genetics, Inc., 273n11

National Academy of Sciences, 166
National Cancer Act/War on Cancer, 15, 16, 131–132, 133, 149
National Cancer Advisory Board, 132
National Cancer Institute (NCI), 9, 11, 16, 24, 37, 48, 102, 124, 130, 147, 183, 219; proto-oncogene research, 63, 136, 137; Biochemistry Laboratory, 80; Viral Cancer Program (VCP), 130–135, 151, 224, 274n5; funding to, 145
National Cancer Program (NCP), 130, 133
National Institutes of Health (NIH), 4, 24, 130, 134, 179, 245n4, 256n11; recombinant DNA guidelines and funding, 88, 160; viral cancer funding to, 132, 145; molecular genetics funding, 134–135. *See also* National Cancer Institute (NCI)
National Laboratory Gene Library Project, 99
National Science Foundation, 88, 101, 145, 261n38
Nature, 72, 77; reinvented, 66, 67, 72–73, 146, 207, 272n31; natural world as routine technology, 78–86; representation of, 114, 115, 182, 208, 209, 211, 237. *See also* Artifacts: laboratory
Neuroblastoma, 23, 226, 227
New England Biolabs, 86–87, 91
Nobel Prize, 1, 48, 49, 50, 52, 107, 119, 124, 128, 225, 261n34
Nongenetic cancers, 150
Non-human actors, 77, 79, 240, 241, 243, 257n16, 278n9
Normal growth and differentiation, 5, 105, 106, 116, 121, 122, 127, 135, 136, 206, 250n25, 252n38; oncogene research and, 141, 151, 161, 218, 219. *See also* Pathological growth and differentiation
Normalization, 252n35, 254n53; constructing the normal, 41–45; normal action, 277n2
Normal science, 151, 230, 233
Novelty, 9, 18, 44, 71, 73, 82, 110, 135, 140, 146, 149, 151, 187; of research programs, 188, 230; standardization and, 208
Nucleotides, 7, 71, 80, 100–101

Objects. *See* Technoscientific objects
Oncogen, 98, 148
Oncogene(s): research, 7, 9, 17–18, 32, 34, 36, 47, 50, 53, 97–98, 106, 123–127, 133, 137–154, 218–219, 224–225; theory, 116, 137–138, 141–148, 167; probes, 141–142; -antibodies, 187–200; definition of, 262n1. *See also* Anti-oncogenes; Cellular oncogenes; Proto-oncogenes; Viral oncogenes
Oncogene Science, Inc., 148
OncoMouse, 7, 151, 220, 268n15
Oncor, 97, 98
Ontology, 216
Open systems, 156, 181
Organization of scientific work, 3, 11–12, 15–16, 66, 79, 172, 173, 224, 233. *See also* Work organization

Package of theory and methods. *See* Theory-methods package
Parada, Luis, 124, 125, 157
Paradigms, 16, 125, 153, 228, 230–234, 237
Pasteur, Louis, 219, 220, 240–241

Patents, 156, 163, 164, 165, 173, 193, 276n16
Pathological growth and differentiation, 17, 119, 122, 136, 156, 213, 223. *See also* Mutation; Normal growth and differentiation; Somatic mutation theory
Pharmaceutical companies, 145, 147–148
Phenomenology, 156, 212
Phenotype. *See* Genotype-phenotype differences
Pickering, Andrew, 212, 277n4
Pinch, Trevor, 238
Politics: of cancer, 9, 15, 37, 49; of science, 83, 240–241, 279n11; of research, 109
Popper, Karl, 230
Portability, 215
Positioned rationality, 243
Practice(s), 4, 6, 7, 9–10, 16–17, 206–209, 226, 233–234; standardization of, 4, 30, 71, 77–78, 200–202, 221, 227; science as, 12; scientific, 14–15; changes and transformation in, 30, 153, 155; co-construction of, 207–208; continuity in, 209–210; ethnographies of, 238–239
Pragmatism, 12, 13–14, 16, 21
Priority, 157–159, 163–165, 171–172
Probes, 100, 141–142, 210; definition of, 118. *See also* DNA: probes
Problem-solving, 10–11, 69, 77, 109, 112, 140, 155; as process, 156–159, 182, 208, 211; histories, 159–170; problem structures as contexts of, 170–173, 184; decomposition of, 173–174; problem structures as response to situations, 173–174; alternate strategies, 174–181; as sociotechnical achievement, 181–183; through work organization, 185–187, 200; through standardization, 200–202, 207–208
Proctor, Robert, 149
Production: of knowledge, 17, 21, 65–67, 175, 206, 218, 220; of similarity, 26–28; situated, 34; material contexts of, 45, 65, 214; sites of, 45, 66, 151, 152; methods of, 64–67, 214; mutual, 150
Professionalization, 144, 174–175, 276n14
Prokayotes, 73–74, 76, 83–84
Propensity to develop tumors, 26, 27, 35. *See also* Heredity/heritability
Protein(s), 9, 53, 64, 65, 69, 71, 74, 75, 79, 106; synthesizers, 108; "p21" production and structure, 163–166, 170–171, 176–177, 191, 198; phosphorylation, 176, 177; transforming, 176–177
Protocols, 105, 138; experimental, 144; oncogene, 232. *See also* Standardization: of protocols
Proto-oncogene(s), 9, 50, 123–127; research, 1–3, 6–7, 9, 10, 11, 17, 20, 23–24, 63, 69, 116, 136, 161–163, 206, 219, 221; normal, 121, 167; definition of, 127; cellular, 167, 168
Proto-oncogene theory, 1, 2, 5, 7, 9, 45, 66, 69, 116–117, 121, 127, 128, 133–134, 167, 205, 210, 221, 225; as collective work, 135–136; commitments to, 136, 137; molecular genetic technolgies and, 150–153
Protovirus theory, 51, 210
Provirus theory, 136, 210
Publishing, 69, 91, 98, 143, 146, 156, 157–159, 163–164, 169, 173, 181, 199
Purification work, 79, 86, 91, 198, 215, 258n20

Rabinow, Paul, 19, 260n28
Radiation, 3, 28, 147. *See also* Carcinogenesis: radiation; Chemotherapy
Rauscher, Frank, 131, 143
Realism-constructivism debate, 19, 22, 216–220, 233, 237
Realism-relativism debate, 13–14
Realization, 151. *See also* Materialization
Recombinant DNA Advisory Committee (RAC), 15
Recombinant DNA technologies, 5, 6, 69–73, 80–82, 91–115, 120, 133, 150–151, 176, 205; research, 7, 68–69, 78–80, 97, 101, 105, 146; history of, 68, 69, 73–77, 116, 219; design of, 76, 78; public and private commitment to, 86–100; standardized, 140
Reductionism, 150
Reflexivity, 67, 279n12
Reification, 19, 214, 239
Relativist program, 238
Replication, 29, 30, 43, 44, 211, 212, 213, 215
Representation, 1–4, 7, 12–14, 15–16, 22, 65, 215, 225, 233–234, 237, 239; multiple, 4, 13, 208; of cancer, 14, 63, 112, 151, 210; consequences of, 18–19; standardization of, 30, 110; continuities of, 213–216
Reproducibility of experiments, 29, 31, 207, 213
Research: programs, 1, 51, 188–194, 230–234; design, 10, 15, 16, 17, 76; time

Research *(continued)*
frames, 146, 172, 174, 181, 195, 196–197, 199, 218, 224–225
Research and development companies, 147–148, 151, 203–204
Research tools, 4, 6, 9, 15–16, 18, 19–22, 24, 25, 27, 29–31, 34–36, 47, 52, 61, 64–66, 71, 72, 78, 79, 184, 187, 190–191, 195, 200–201, 212; standardization of, 22, 77–78, 108, 112–115, 141, 152, 215, 227; continuities in, 136; state-of-the-art, 152, 221; cost of, 187. *See also* Right tool for the job
Restriction enzymes, 79–80, 81–87, 101, 105, 112–113
Retinoblastoma, 227, 228
Retooling, 10
Retrovirology/retroviruses, 7, 48–53, 106, 117, 119, 124–125, 128, 130, 154, 160, 246n15; as "pets," 52, 53
Reverse transcription, 49, 53, 79, 263n3
Rheinberger, Hans-Jorg, 248n18
Rhetorical strategy, 74, 76, 78, 114
Ricouer, Paul, 19
Right tool for the job, 27, 29–30, 32, 38, 70
Rip, Arie, 239
RNA, 49, 53, 65, 209; manipulation, 71, 73; messenger, 83, 103, 118, 168; sequences, 101, 103; probes, 118
RNA tumor viruses. *See* Retrovirology/retroviruses
Roberts, Richard, 87, 101, 261n34
Roche Pharmaceuticals, 254n3
Rosaldo, Renato, 19
Rous, Peyton, 48, 49, 50, 58, 59, 60, 166
Routines, 15, 77–78
Rubin, Harry, 48, 49, 112, 222
Ruddle, Frank, 149

Sanger, Frederick, 101, 107
Schaffer, Simon, 72, 212, 241
Science: -society divide, 3, 225, 238, 241–242, 243; -technology divide, 31, 112, 238, 257n13; applied vs. basic, 76–77, 233; play of power in, 153; "good" standards of, 199; state-of-the-art, 233; as culture, 238. *See also* Social studies of science; Comparative cultural studies of science
Science (journal), 62, 98, 99, 138, 255n10, 264n11
Self-vindication, 217–218, 219, 220
Semiotics, 12, 212, 243, 278n10
Shapin, Steve, 72, 241
Sharp, Philip, 261n34
Shibutani, Tamotsu, 12
Shih, Chiaho, 124
Similarity/identity, 26–28, 64, 103–104, 213–216, 261n43
Simulations, 230
Singer, Maxine, 80, 81
Situated action/practices, 14–15, 34, 65, 158, 206, 209, 225; from contingency to continuity, 210–222
Situated knowledge, 22, 109, 243
Situated production, 34
Skolnick, Mark, 99, 263n1, 273n11
SmithKline Beecham Corporation, 148
Social studies of science, 17, 19–20, 182, 209, 217, 225, 237
Social worlds, 12–13, 152, 153, 186, 207, 208, 211, 221, 227
Socio-cultural theory, 2, 31, 238, 254n52
Sociology, 13–14, 17, 78, 175, 204, 213, 220, 238; of scientific knowledge, 19, 131, 206, 212, 237, 238, 278n6; of error, 202, 224
Socio-technical achievements, 181–183
Somatic mutation theory, 25, 55–58, 60, 65, 149, 275n13
Southern, Edward/Southern blotting technique, 83
Specificity/nonspecificity, 195–198, 199, 203–204, 216
Spectrometers, 144, 145
Speed, 77–78, 103, 104, 108, 109, 112–113, 114; in research, 111; in publishing, 169. *See also* Research: time frames
Stability/stabilization, 11–12, 152, 153, 206, 211, 212, 214, 220, 221; destabilization, 215; of oncogene theory, 226
Standardization, 13, 14, 18, 104–112, 114, 216, 233; concept of, 5, 78, 213–216, 221; of concepts, 5–6, 7, 9; of techniques, 5–6, 7, 9, 29–30, 39, 72, 77–79, 105, 113, 151, 152, 196, 200–202; of laboratories, 6, 7, 9, 97; of protocols, 7, 11, 71, 79, 111, 152, 200, 202, 215; of experimental systems, 22, 24, 30–32, 44–45, 64, 66, 67, 70, 207; of products, 91; as hegemony, 114; articulation and, 200–202
Standardized package. *See* Theory-methods package
Stanford University, 99, 115
Star, S. Leigh, 212, 239, 246nn18,21, 248n16, 251n27, 261n33
Stehelin, Dominique, 264n11
Strauss, Anselm, 12, 145, 269n7, 271n26, 273nn4,5, 275n3

Strong Program, 238
Studer, Kenneth, 23
Suchman, Lucy, 275n9
Sutton, Walter, 32
Symbolic interaction, 12, 13, 14, 16, 148, 224, 239
Symbolic systems, 207. *See also* Technosymbolic systems
Syntex, 98

Tacit knowledge, 108, 203, 209, 211, 225, 269n7
Tailoring, 212, 217, 227
Technoscientific objects, 31, 241, 242
Technosymbolic systems, 31, 66–67, 72
Temin, Howard, 48, 49, 51, 136, 210, 264n10, 265n13
Theory as collective work, 12, 135–136, 152
Theory-methods package, 4, 5, 9, 16–18, 137–138, 140–148, 150–153, 155, 159, 170, 181, 200–202, 206–208, 221–222, 224–225; of proto-oncogene-molecular genetics, 227–228, 232
Thomas, W. I., 12
Tinkering, 105, 111, 200, 208, 211, 233, 272n7
Tissue culture techniques, 24, 28, 37–53, 64, 65, 66, 73, 207, 219
Todaro, George, 135, 264n10
Traces of continuity, 16, 136
Training, 15, 111, 113, 143, 156, 173–174, 177, 185
Transduction by retroviruses, 119, 120
Transformation: viral, 45–53; spontaneous, 47; at molecular level, 62–64; of proteins. *See also* Cell: transformation
Translation, 123–127, 156, 160, 175, 200, 208, 227, 239; mutual, 50
Transplantation biology, 25, 27–30, 35–36, 42, 53, 56, 64, 117
Traweek, Sharon, 212, 239, 258n20, 272n2, 277n1, 278n3, 279n13
Treatment/therapy, 2, 4, 23, 37, 54, 74, 129, 143, 178, 218. *See also* Chemotherapy
Triangulation, 151, 212, 213, 219
Tumor(s): biology, 21, 25; definition of, 23; etiology, 24–25; transmission/heritability of, 24–25, 26; immunology, 25, 27, 28; production, 56, 59; naturally occurring, 118, 131
Tumor virus, 16, 25, 34, 46, 51, 52, 64–66, 69, 80, 153, 180, 207. *See also* DNA tumor viruses; Retrovirology/retroviruses; Virology
Turner, Stephen, 97–99

Ultracentrifuges, 186, 187, 201
Uncertainty, 11, 77, 166, 200, 208, 212
Unintended consequences/outcomes, 35, 66, 78, 84, 158, 159, 269n8
United States Congress, 9, 15, 133–134, 151, 160, 256n11, 267n2
Unit of analysis, 3, 4, 5, 135, 150, 151, 216, 218, 264n12
Universal parameters, desire for, 43, 44
University and industry collaborations, 2, 9, 76–77, 113, 172, 173, 257n15, 273n11; DNA studies, 86–88
University of California, 32, 49, 61, 77, 99, 115, 118, 232, 257n15; Berkeley Cancer Research Laboratory, 27, 36, 180
University of Chicago, 12

Varmus, Harold, 1, 6, 17, 50–53, 118–121, 123–125, 127, 130, 135, 136, 142, 160, 179, 180, 206, 208, 210, 222, 224, 225, 229
Venture capital, 15, 75, 76, 87, 88, 100, 113
Viral Cancer Program. *See* National Institutes of Health (NIH)
Viral oncogenes, 20, 34, 46, 47, 48, 50, 54, 64, 69, 122, 161, 162, 167; history of, 116, 117–121, 160, 219
Virogene-oncogene hypothesis, 3, 118–119, 122
Virology, 64, 167, 210; tumor, 117–121, 126–127, 130–135, 153,
Virus(es), 24, 36, 128; infectious virus theory, 25, 36, 119; from genetics to, 34–36, 210; assembly, 53; cultures, 117; transformation studies, 117, 118, 128
Vogt, Marguerite, 46, 48, 49
Vogt, Peter, 49

Warburg Hypothesis, 25
War on cancer, 131
Waterfield, Michael, 121
Watson, James D., 18, 49, 60, 84, 132, 133, 233
Webs of relationships, 12
Weinberg, Robert, 63, 123–128, 157, 160, 225, 228, 229, 276n27
Weismann, August, 214, 215
Whitehead, Alfred North, 33
Wigler, Michael, 63, 124, 160
Willmer, E. N., 38–39, 45–46
Wilson, E. B., 214

Wimsatt, William, 214
Woolgar, Steve, 144, 214, 275n8
Work organization, 9–10, 153, 156, 173, 174, 175, 182, 184–187, 198–200, 206, 208, 229–230
Worlds of practice, 16–17, 155, 204, 221
Wright, Susan, 74, 75, 87
"Xavier" laboratory, 20–21, 161–166, 170–173, 187–200, 228, 230

Yanagisako, Sylvia, 258n20
"Yuzen" laboratory, 20–21, 158, 166–174, 176

Zinder, Norton, 132
Znaniecki, F., 12